£95.

✳

D1486294

Structural Detailing in Timber

Structural Detailing in Timber

A comparative study of international codes and practices

M.Y.H. Bangash

Whittles Publishing

CRC Press
Taylor & Francis Group

Published by
Whittles Publishing Limited,
Dunbeath,
Caithness, KW6 6EY,
Scotland, UK
www.whittlespublishing.com

Distributed in North America by
CRC Press LLC,
Taylor and Francis Group,
6000 Broken Sound Parkway NW, Suite 300,
Boca Raton, FL 33487, USA

ISBN-13 978-1870325-53-0
USA ISBN 978-1-4200-6834-4

The publisher and author have used their best efforts in preparing this book, but assume no responsibility for any injury and/or damage to persons or property from the use or implementation of any methods, instructions, ideas or materials contained within this book. All operations should be undertaken in accordance with existing legislation and recognized trade practice. Whilst the information and advice in this book is believed to be true and accurate at the time of going to press, the author and publisher accept no legal responsibility or liability for errors or omissions that may have been made.

Every effort has been made to correctly acknowledge illustrative material. The author and publishers ask that any omissions be reported to them for incorporation in future printings.

Typeset by Compuscript Ltd., Shannon, Ireland

Printed by Bell & Bain Ltd., Glasgow

Contents

Preface

Timber is probably the oldest of man's building materials. It has an unmatched variety of historic form, reaching in an unbroken line from the split oak logs to modern glulam space frames, lamella roof and arena domes. It still continues to be used both in domestic construction and for commercial use.

The main feature of this book is its handling of the structures and respective codes. A sound background is necessary for understanding structural timber. Readers who wish to carry out an in-depth study in specific "formal method" have been directed to other references. Each chapter is strictly self-contained. A step by step approach is adopted to make the reader understand the detailing of a timber element as an exercise and a global case study.

The book introduces codes and practices in Britain, Europe, America and Russia.

This timber detailing manual has been prepared to provide practical and up-to-date information on various aspects of timber construction for educators, designers, draughtsmen, detailers, fabricators and all others who have an interest in structural timber. This text covers the fundamentals of drawing and drafting practices, connections and fastening and structural detailing of various timber elements. The types of structures covered represent the bulk of the typical fabricator's work in timber structures such as buildings, bridges, grid-works, shells, folded plates etc. Some examples of timber detailing in CAD form are included. A total of seven chapters cover in detail the comparative study of timber detailing of British, European, Russian and American Codes and practices. Some relevant structural timber detailing from Russian practice is included in the text.

Many of the drawings included are typical and, with minimal alteration, can be adopted directly from the book.

This book should serve both as a premier for the trainee detailer and as a reference manual for more experienced personnel. Engineers, architects, surveyors, and contractors will find the book useful for daily use and practice.

M.Y.H. Bangash

Acknowledgements

The author wishes to express his appreciation to friends, colleagues and some students who have assisted in the early developments of this book by suggesting relevant changes. The author has received a great deal of assistance, encouragement and inspiration from practicing engineers and contractors, particularly those for whom he has acted as consultant. The author is indebted to all those people and organizations who are referred to in this book and to the following, in particular, for making this book a reality:

Blondeau Ingenirie, France
Bernard Quirot, Architect, Paris, France
SARI Development, Structural Engineers and Contractors, Paris, France

E. Aasheim, The Norwegian Institute of Wood Techology, P.O. Box 113, n, 0314 Oslo, Norway
M.P. Ansell, University of Bath School of Materials Science, Claverton Down, Bath BA2 7AY, United Kingdom
P. Aune, University of Trondheim. The Norwegian Institute of Technology, Dept. of Structural Engineering, Rich. Birkelands vei la, 7034 Trondheim, Norway
J.P. Biger, Bureau Veritas, Civil and Environmental Engineering, 17 bis Place des Reflets, 92400 Paris la Defense Cedex 44, France
G. Bignotti, Holzbau Ag-S.p.a., P.O. Box 224, 39042 Brixen/Bressanone., Italy
G. Bonamini, Universita degli Studi di Firenze, Inst. Di Assest e Techn. Forestale, Via S. Bonaventura 13, 50145 Firenze, Italy
H. Bruninghoff, Gesamthochschule Wuppertal, Pauluskirchstrasse 7, 42285 Wuppertal, Germany
A. Ceccotti, Universita degli Studi di Firenze, Dipartimento di lngegneria Civile, Via di S. Marta 3, 50139 Firenze, Italy
J. Chilton, University of Nottingham, School of Architectures, University Park, Nottingham NG7 2RD, United Kingdom
J. Fischer, Lignum – Schweizerische Arbeitsgemeinschaft fur das Holz, Falkenstrasse 26, 8008 Zurich, Switzerland
H. Hartl, Zivilingenieur fur Bauwesen, Kenyongasse 9, 1070 Wien, Austria
B. Johannesson, Chalmers University of Technology, Dept. of Structural Engineering, Stecl and timber structures, 41296 Goteborg, Sweden
G. Johansson, Chalmers University of Technology, Dept. of Structural Engineering, Steel and Timber structures, 41296 Gotenborg, Sweden
A. Kevarinmaki, Helsinki University of Technology, Laboratory of Structural Engineering and Building Physics, Rakentajanaukio 4a, 02151 Espoo, Finland
M.H. Kessel, Fachhochschule Hildesheim/Holzminden, Fachbereich Bauingenieur-wasen Labor fur Holztechnik, Hohnsen 1, 31134 Hildesheim, Germany
J-W.G. van de Kuilen, Delft University of Technology, Faculty of Civil Engineering, Timber structures, P.O. Box 5048, 2600 GA Delft, Netherlands
H.J. Larsen, Danish Building Research Institute, Dr. Neergaarsvej 15, 2970 Horsholm, Denmark
A.J.M. Leijten, Delft University of Technology, Faculty of Civil Engineering, Timber Structures, P.O. Box 5048, 2600 GA Delft, Netherlands
C.J. Mettem, formerly Timber Research and Development Ass., Stocking Lane, Hughenden, Valley, High Wycombe, Buckinghamshire HP14 4ND, United Kingdom
L. Mortensen, University of Aalborg, Dept. of Building Technology and Structural Engineering, Sohrlgaardsholnk|vej 57, 9000 Aalborg, Denmark
J. Natterer, Ibois-Civil Engineering department, GCH2 Ecublens, 1015 Lausanne, Switzerland
N. Nebgen, Ingenieurburo fur Holzbau, Dieselstrasse 12, 72770 Reutlingen, Germany
P. Racher, C.U.S.T.-Genie Civil, P.O. Box 206, 63174 Aubiere Cedex, France
B. Roald, The Norwegian Institute of Wood Technology, P.O. Box 113, Blindern, 0314 Oslo, Norway
J.L. Sandoz, Ibois – Civil Engineering department, GCH2 Ecublens, 1015 Lausanne, Switzerland
C. Short, Colin Short Associates Consulting Engineers, Brookfield, Glen Road, Delgany, Greystones, County Wicklow, Ireland
K.H. Solli, The Norwegian Institute of Wood Technology, P.O. Box 113. Blinded, 0314 Oslo, Norway

G. Steck, Fachhochschule Munchen, Fachbereich 02, Karlstrasse 6, 80333 Munchen, Germany

K. Schwaner, Arbeitsgemeinschaft Holz e.V., Postfach 300141, 40401 Dusseldorf, Germany

S. Thelandersson, Lund University, Division of Structural Engineering, P.O. Box 118, 221 00 Lund, Sweden

P. Touliatos, National Technical University of Athens, 15 Dinokratous Str, 10675 Athens, Greece

L. Uzielli, Universita degli Studi di Firenze, Dipartimento di Ingegneria Civile, Via di S. Marta 3, 50139 Firenze, Italy

G. Vidon, Socotec, Dept. Parois-Isolation, Les Quadrants, 3 Avenue du Centre, 78182 Saint Quentin en Yvelines Cedex, France

The author is deeply grateful to those men and women in the following American manufacturing and industrial organizations who cooperated so completely whenever the author requested for help:

Agricultural Experiment Station, Oregon State College
American Lumber and Treating Co.
American Institute of Timber Corp.
ANSYS UK Ltd.
Automatic Building Components Inc.
Ben C. Gerwich, Inc
Tannewitz Works
Timber Structures, Inc
U.S. Corps of Engineers
National Lumber Manufacturers Assc. USA
National Safety Council, USA
Northfield Foundry and Machine Co. USA
Oliver Machinery Co. USA
Porter Cable Machine Co. USA
Rilco Structural Wood Products, USA
Rockwell Mfg. Co. USA
Stanley Tools, USA
U.S. Dept. of Agriculture, USA
U.S. Dept. of the Army, USA
U.S. Forest Service
Upson-Walton Company, USA
STRUCAD London, UK
STRAD–PRO/dse, Birmingham, UK
Canada Mortgage and Housing Corporation

The author is indebted to the following Russian organizations for presenting their drawings and details:

Marka Company, Pionerskaya, Maikop, Adygeya, Russia 385020

Anuradha Timbers International, 244 Chinnatokatta, New Bowenpally Secenderabed – 500 033, Andhra Pradesh, India

DEKO-WOOD, Donker, Harstes, 8600 AC Sneek

EuroForest, Mead House, Bentley, Farnham, Surrey, GU10 5HY

ICTT, International Centres for Telecommunication Technology, Inc, 100 East Campus Drive, Terre Haute, Indiana 47802, USA

International Center for Timber Trade, The International Environment House 2, 7 Che, in de Belexert 1219 Chatelaine, Geneva, Switzerland

National Timber Company, 42 Koziyenskaye Str. Vologda, 160035, Russia

The author is grateful for the enormous support given by the following individuals and organisations, without whom this work could not have been achieved:

Rob Thomas, Librarian of the Institution of Structural Engineers, London, UK

Prof Dr-Ing J. Eibl, Karlsruhe, Germany

Prof Dr-Ing U. Quast, University of Hamburg, Germany

Prof Dr-Ing E, Cusens, University of Leeds, UK

Mr Khalid Chaudry, Director, SC. Consultant, Berkshire, UK

Miss Rose Manroy, Library Manager, Institution of Civil Engineers, London, UK

Mr S. Baig, ASZ Partners Ltd., Ilford, Essex, UK

AceCad Software Ltd., Derby, UK (StruCad)

Staad-Pro Software, Birmingham, UK

Metric conversions

Overall geometry	
Spans	1 ft = 0.3048 m
Displacements	1 in = 25.4 mm
Surface area	1 ft^2 = 0.0929 m^2
Volume	1 ft^3 = 0.0283 m^3
	1 yd^3 = 0.765 m^3

Structural properties	
Cross-sectional dimensions	1 in = 25.4 mm
Area	1 in^2 = 645.2 mm^2
Section modulus	1 in^3 = 16.29 × 10^3 mm^3
Moment of inertia	1 in^4 = 0.4162 × 10^6 mm^4

Material properties	
Density	1 lb/ft^3 = 16.03 kg/m^3
Modulus of elasticity and stress	1 lb/in^2 = 0.006895 MPa
	1 kip/in^2 = 6.895 kN/m^2

Loadings	
Concentrated loads	1 lb = 4.448 N
	1 kip = 1000 lbf = 4.448 kN
Density	1 lb/ft^3 = 0.1571 kN/m^3
Linear loads	1 kip/ft = 14.59 kN/m
Surface loads	1 lb/ft^2 = 0.0479 kN/m^2
	1 kip/ft^3 = 47.9 kN/m^2

Prefixes in SI units
G = giga 10^9
M = mega 10^6
k = kilo 10^3
m = milli 10^{-3}
Pa = Pascal

Chapter 1
Introduction

This chapter is devoted exclusively to timber detailers/ draughtsmen and their responsibilities towards engineers, architects and fabricators. Their major function is to serve as an intermediary between the planners and executors of a project. It is therefore important that they should have a clear understanding of the engineers' intent and the ability to translate it into a graphic representation. The detailers/draughtsmen must have knowledge of the various processes involved, including:

1. types of timber structures to be built and method,
2. creating a permanent record of the designers and engineers intent, including design, calculations and sketches,
3. construction and fabrication of the timber structural elements and components,
4. the conveyance of information on all aspects of detailing on the lines given by the engineer or an architect, using manual and computer aided means,
5. presentation of information with clarity and accuracy,
6. project organisation and the timber detailer's role in it.

1.1 Detail drawings

Timber structures are represented by means of elevations, plans and cross-sections with, where necessary, enlarged sections of particular areas of the timber structure that require more detail for additional information. The detailers must be familiar with the instrumentation, codes and methods employed in the production of drawings. Whether it is a building, a bridge or any other timber structure, the elevation and plan must be to an appropriate scale that shows, by means of suitable annotations, the sizes and shapes of the members. Where support bearings and special end member connections are involved, they must be provided in sufficient detail (ie an enlarged scale) to enable the timber worker, the blacksmith and the carpenter to construct components to a reasonable degree of accuracy.

With the advent of sophisticated timber, structural prefabrication became essential, bringing with it the need to supplement the arrangement drawings with detail drawings of all individual members and components. These are known as shop detail drawings and are usually prepared by the timber manufacturing company in its own drawing office for use in its workshops. They are based on the layout and arrangement drawings supplied by the owner, or by the consulting engineer appointed to carry out the design. The job of the detailer is to check the information and drawings from the workshop on the fabrication of timber components based on the requirements of a typical timber code and the design drawings provided by their office. It is in the checking of these drawings that the structural timber detailers find their role and are able to play a vital part in the sequence of events within timber structural engineering.

The detailer should ensure that:

1. the information is clearly transmitted,
2. shop drawings are correct with regard to stylised representation involving the use of standardised, abbreviated notation and special symbols,
3. any variations have been discussed with the designer,
4. large amounts of technical data have been recorded and presented in a concise manner.

1.2 Function of the timberwork detailer

The role of the timberwork draughtsman or detailer will now be examined more closely. When the contract is placed with a timberwork fabricator, and assuming that the timberwork detailer works for the fabricator, their duties can be categorised as follows:

1. The consulting engineer's drawings and specifications are passed on by the company management to the drawing office, where the drawing office manager assesses the extent, complexity and time content of the job.
2. The section leader confers with a senior detailer who forms a team of experts in specific areas.
3. The first objective is to prepare a list of timber materials from the layout drawings provided by the consulting engineer, enabling the contractor to reserve the items from stock or to place orders with timber merchants or wood mills.
4. The detailers proceed with the preparation of the timberwork detailing drawings, providing an accurate representation of structure components, namely beams, girders, trusses, columns, bracings, stairways, platforms, rails, brackets, girths,

purling etc. The detailing of other structures such as bridges, towers, tanks etc follows a similar pattern.

5. An experienced senior draughtsman or detailer must supervise drawings and carefully scrutinise them. It is essential to correct the errors at this stage, as correcting errors during fabrication in the shop or during erection on site will be infinitely more expensive. Draughtsmen should be critical of their own work, to ensure that their drawings are error-free to the best of their ability.

6. The detail drawings are sent to the fabrication shop. Work then begins on drawing the material from stock, cutting it to exact length, drilling or punching the necessary holes and assembling the various parts by means of bolting or welding to make up the components or subassemblies ready for transport.

7. The drawing office, whether using a manual system or computer-aided facilities, must now proceed with the preparation of erection drawings, showing timber framework in skeleton form (elevations, plums and cross-sections). These drawings should be checked by the senior detailer and endorsed by a qualified structural engineer. The timber erector will refer to these drawings during the assembly of the structure on site. The position of each component is identified by a distinguishing erection mark. In the fabrication shop, such erection marks are hand-marked, painted or tagged onto the timber components.

8. All drawings are updated to incorporate any revisions that have occurred during the progress of the job and a complete set of prints is handed to the engineer for filing. These serve as a record of the work and are useful for future reference.

1.3 Project organisation

It is important to consider the role of the detailer in the overall management and technical organisation involved in the timber construction project.

1. When the owner appoints the architect and engineer for the timber project, they assume the following roles:
   a) The architect prepares the preliminary and detailed planning of drawings and specifications and sends them to the structural engineer involved with the design of the timber structure.
   b) The structural engineer then prepares preliminary design drawings and, with the help of quantity surveyors, indicates costs to the architects, or directly to the owners where architects are not involved.

2. The detailer acts as a liaison between the engineer and the contractor and is responsible, whether working for the engineer or timberwork contractor, for the preparation of the general arrangement (GA) drawings, shop drawings and the erection drawings in some complicated projects.

3. The timber detailer liases with the timberwork contractor during erection of timberwork on site, ensuring that the engineer/designer is fully informed of day-to-day erection problems, particularly non-compliance with detailed drawings.

4. The timber detailer is responsible for keeping the logbook and other recording arrangements, including storing in an electronic or mechanical retrieval.

1.4 Drafting practice based on British codes

Full drawings are prepared by structural engineers acting as consultaonts as part of the tender documentation. The architects are involved in the preparation of the site and other general arrangement plans. The main contractors are involved in the preparation of temporary work drawings, including shoring and firework. During the contract, minor amendments may occasionally be applied to the drawings, or additional details added. The drawings are updated as the project progresses. The drawings, which are distributed to other engineers including service providers and contractors, are prints taken from the original drawings on tracing paper, referred to as negatives. Negatives are provided with thick borders as a precaution against tcaring. Plastic film, on the other hand, provides a smooth hardwearing surface. Almost all drawings are done in ink. A typical drawing sheet contains the following data in the panel on the right-hand side of the drawing:

Starting from the top	*Example*
NOTES	Specification, etc.
REVISION	751/10 Rev D (details of amendments)
NAME OF THE ENGINEER	Bangash Consultants
NAME OF THE CLIENT/ ARCHITECT	Bangash Family Estate
DRAWING TITLE	Bangash Estate Centre Foundation Layout
SCALES	1:20, 1:50, and 1:100
DRAWN BY	Y. Bangash
DATE	13 July 2004

1.4.1 Drawing number

The drawing numbers may run in sequence such as 751 or 1, 2, 3 or 100, 101 etc. The International Standard Organization (ISO) recommends A or B ranges for paper sizes. Most common are A1 (594 × 841 mm) and B1 (707 × 1000 mm). For structural detailing in timber, A2 (420 × 594 mm) size is recommended. For small sketches, detailing and specifications, design teams and contractors use A4-sized (210 × 297 mm) sheets. All major drawings and site plates carry a compass direction.

1.4.2 Drawing instruments

The most general instruments required for good drawings are: a drawing board, bookcase pencils, clutch pencils, automatic pencils, technical drawing pens, erasers, scales, set squares, templates and stencils. A description of these is excluded from this text as they are well known.

1.4.3 Linework and dimensioning

The structure is viewed 'square on' to give a series of plans, elevations and sections. The two basic types are: first-angle projection and third-angle projection. Dimensioning varies from country to country. Some examples are given later on in this section and in other sections of the book.

1.4.3.1 Line thickness

See Table 1.1 for line thickness (based on ISO line thickness) recommended for concrete drawings.

Table 1.1

		Colour Code
General arrangement drawings	0.35 mm	Yellow
Concrete outlines reinforcement drawings	0.35 mm	Yellow
Main reinforcing bars	0.70 mm	Blue
Links/stirrups	0.35–0.70 mm	–
Dimension lines and centre lines	0.25 mm	White

The line thickness increases with the ratio $12:\sqrt{2}$, for example, $0.25 \times \sqrt{2} = 0.35$.

1.4.3.2 Dimensioning

As stated in Section 1.4.3.1, dimension lines of 0.25 mm thickness are shown in several ways (see Fig. 1.1, for example). A gap is necessary between the dimension line and the structural grid. Dimensions are given in different ways. In SI units, dimensions are given in various countries as follows:

Britain (BS 1192)	All dimensions shown, 1700 for 1700 mm
Sweden	1700 mm rather than 1700
Switzerland	1.700 m
Italy	1700
Japan	1.7 m and 1700
Germany	1.7 m
USA	Major dimensions in ft (feet), smaller dimensions in inches
Pakistan/India	Same as USA on some projects, also m and mm.

Figure 1.1

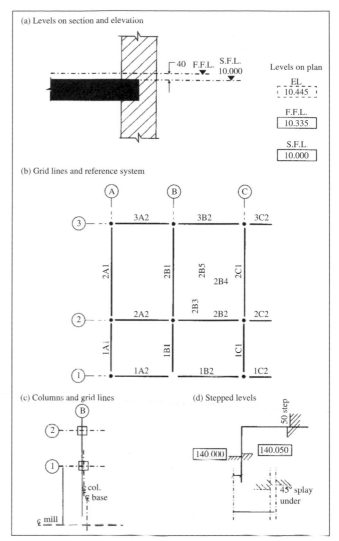

Figure 1.2 *Grids and levels for timber buildings.*

1.4.4 Grids and levels

A point on any drawing can be located by a grid reference. A grid is a series of vertical and horizontal lines on the plan of the structure, sometimes referred to as building grids. They may not have identical spacing but it is preferable that the spacing is constant in the same row between the grid lines, identifiable with letters and numbers. On sections and elevations, various levels are marked. Typical examples are shown in Fig. 1.2 for grids and levels and a proper notation is shown for reference beams and columns.

1.4.5 Sections and elevation markers

An exact style cannot easily be determined as it varies from country to country. It is not important what style is used, but that it is simple and clear. The markers are located on the plane of the section or elevation with indicators pointing in the direction of the view. The sections markers must be shown in the correct direction and the letters must read from the bottom of the drawing. Some of them are shown later on various drawings and details, either with horizontal and vertical

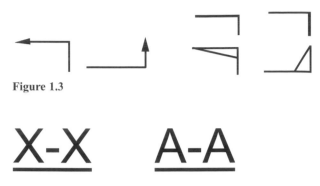

Figure 1.3

X-X   A-A

Figure 1.4

thick lines or arrowheads of the types shown in Fig. 1.3. In some important cases two thick lines are shown. Where sections are indicated they are marked as shown in Fig. 1.4.

Similar markers can be seen on different drawings. The author has deliberately changed these markers on drawings to give the reader a choice of any marker that he or she wishes to adopt.

1.4.6 Grading stamps: American practice

Each piece of graded lumber is stamped with an assigned grade, species (or species group), moisture condition at time of surfacing, identity of the mill that produced the lumber, and possibly other information. By the grade stamp, the builder can be assured of obtaining the specified quality of lumber and the owner or the building inspector can be assured that the material actually being furnished conforms to the construction specifications.

An example of a grade stamp for visually graded lumber is given in Fig. 1.5. In this example stamp, the number 12 at the upper left identifies the mill that produced the lumber. (Sometimes the mill is identified by name rather than number.) The logo in the circle at the lower left certifies that the grading was carried out under Western Wood Products Association (WWPA) rules and supervision. The species group is represented in the square at the lower right. In the bottom centre, the label "KD" reveals that the wood was kiln dried

Figure 1.5

Figure 1.6

(seasoned) when it was surfaced. The "1 & BTR" designation shows that the piece bearing this stamp is of Grade No.1 & Better. It is not necessary for all parts of the grade stamp to be in the same position or form shown by this illustration. Also, the stamp may contain a value for allowable fibre stress. If it does, the value shown is the single-member allowable stress for bending about the major axis.

Figure 1.6 also shows an example grade stamp for machine stress-rated lumber. This stamp, too, indicates which mill produced the lumber and graded it, under which agency rules and supervision it was graded, its moisture condition when it was surfaced, the species or species group, and finally, an allowable bending stress and modulus of elasticity. In this example, E is 1.5 million pounds per square inch (psi), and the allowable stress for major-axis bending is 1650 psi.

1.4.7 Caution

The grade stamp on a piece of lumber applies only to the entire piece that was graded. A piece cut off from the original graded piece can be of either the same grade, a higher grade, or a lower grade than the stamp indicates. The reason for this is that in grading the lumber, the location of knots and other defects is considered.

For example, consider the Beam-and-Stringer rules outlined above. Assume that a 6.1 m (20 ft) length of selected structural timber 1.83 × 3.68 m (6 × 12 ft) has been ordered. It is now decided to cut that length in half. However, it is not necessarily true that one now has 2 × 3.048 m pieces of the original element. If the original member had a 1:12 grain slope in the end third of its 6.1 m length, this grain slope would now appear in the central third of one of the new 3.048 m lengths, at a location where the limit is 1:15. Consequently, the grade of one of the shorter pieces is lower than the grade of the original 6.1 m piece. Further, it may be that only one of the two pieces has the necessary stamp to verify to the building inspector that it is graded lumber. The other 3.048 m piece could be rejected as upgraded.

1.4.8 Size and scaling of drawings

Table 1.2 indicates the size and scaling of drawings.

Drawings must contain the following general information:

1. Descriptive title for drawing
2. Name and location of project
3. Architect's, engineer's and general contractor's names
4. Name, address and phone number of precast concrete manufacturer
5. Initials of drafter
6. Initials of checker
7. Date of issue
8. Job number
9. Number of each sheet
10. Revision block.

Table 1.2 Size and scale of drawing: American practice

Type	Prefix	Size (inches)	Scale
Erection drawing	E	24 × 36	1/8 in. min
Keyplan and general notes	K or E	24 × 36	1/8 in. or proportion
Elevations	E	24 × 36	1/8 in., 3/16 in., 3/4 in.
Erection plans	E	24 × 36	1/8 in., 1/4 in.
Sections	E or S	24 × 36	1/2 in., 3/4 in., 1 in.
	E or S	8 × 11	1 in., 3 in.
Connection details	E or CD	24 × 36	1/2 in., 3/4 in., 1 in.
	E or C D	8 × 11	1 in., 3 in.
Anchor layouts	E or AL	24 × 36	1/8 in., 3/16 in., 1/4 in.
Hardware details	H	8	× 11 3/4 i n., 1in., 11 in., 3 in.
Piece drawings	PD or use	18 × 24	1/2 in., 3/4 in., 1 in.
	piece mark	11 × 17	or proportion itself
Shape drawings	SH	18 × 24	3/4 in., 1 in. or
	SH	11 × 17	Proportion
Reinforcing tickets	R	18 × 24	3/4 in., 1 in. or
	R	11 × 17	Proportion
Handling details	HD	8 × 11	Proportion
	HD	11 × 17	Proportion

1.4.9 Line work

All dimensions and arrowheads should be made using a style that is legible, uniform and capable of rapid execution. Two types of dimensioning methods are used within the timber industry. They are point-to-point and continuous dimensioning. Point-to-point relates to the technique of dimensioning from point a to point b, point b to point c, etc. Continuous dimensioning relates to the technique of referring the location of all points back to the same reference. While this technique minimises the possibility of cumulative errors in locating items, it requires subtraction to find the distance between any two points, which increases the possibility of drafting errors. The following dimensioning practices cover most conditions normally encountered: always give all three primary (overall for height, length and thickness) dimensions; primary dimensions should be placed outside of the views and on the outermost dimension line; secondary dimensions should be placed between the view itself and the primary dimensions.

1.4.10 Lettering

All letters and numbers should be distinct in form to avoid confusion between symbols such as 3 and 5, 3 and 8, 5 and S, 6 and C, 6 and G, 8 and B, 0 and D, U and V etc. The height and boldness of letters and numerals should be in proportion to the importance of the note or dimension. For titles, 3/16 in to 1/4 in is recommended, while 1/8 in should be used for notes and dimensions (see Fig. 1.7). Individual preference should dictate the use of either vertical or slanted lettering, however, one style should be used on a drawing consistently. Often a firm will establish a policy on the lettering type to be used. It may be useful to refer to the project specifications, since occasionally future microfilming requirements may dictate the lettering style to be used.

Use of guide lines is recommended for lettering. Guide lines should be lightweight lines that will not reproduce when the drawing is printed. The use of non-print lead should be considered.

1.5 Drafting practice based on Eurocode 5

Many of the current British Codes have been adapted to European Codes, specifically Eurocode 5 or EC5. Many practical differences exist, emphasised in the Code. The drawing sizes are identical, as are the lifework and the dimensioning. The grids and levels are almost identical to the British Codes. Sections and elevation markers are different, as explained later in the text. At present, there is no convention adopted for the representations of symbols and abbreviations. When comparing noted drawings based on EC5, differences are evident. The grids and levels shown in Fig. 1.2 are basically the same under the timber code EC5. Variations to these are identified on the sample drawings for EC5.

Figure 1.8 shows a typical ground floor section on which familiar grid lines are drawn. All walls and columns are marked with thick black lines and black rectangles respectively. The black circles on the outside of the boundary lines are circular large columns supporting the cantilever zones of the building. The internal columns and their axes are oriented to split the design and architectural appearance. The comma sign reflects the European practice for representing a decimal. The dimensions are marked with •—• rather than an arrow. The walls are shaded generally. All small dimensions on the sections are in centimetres.

Sectional elevations of window jambs and typical wall construction are shown in Fig. 1.8. Black and white

ABCDEFGHIJKLMN
OPQRSTUVWXYZ&
1234567890 ½ ¾ 1⅝6 ⅞8 1⅝9₂
abcdefghijklmnopq
rstuvwxyz ½ 15⁄16 ⅞ 1⁄64

ABCDEFGHIJK
LMNOPQRSTU
VWXYZ123456
7890.,:;=÷+
−±@&*?#×"%'
( )[ ]° ! ¢ $ /

Lettering and symbols and lettering styles (adapted from AISC practice)

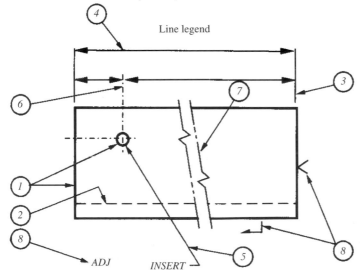

Line legend

	Line type	Pen size
1	object	0.6
2	hidden	0.3
3	extension	0.25
4	dimension	0.25
5	leader	0.25
6	centre	0.25
7	break	0.3
8	symbol	0.35

Figure 1.7 *Lettering and symbols.*

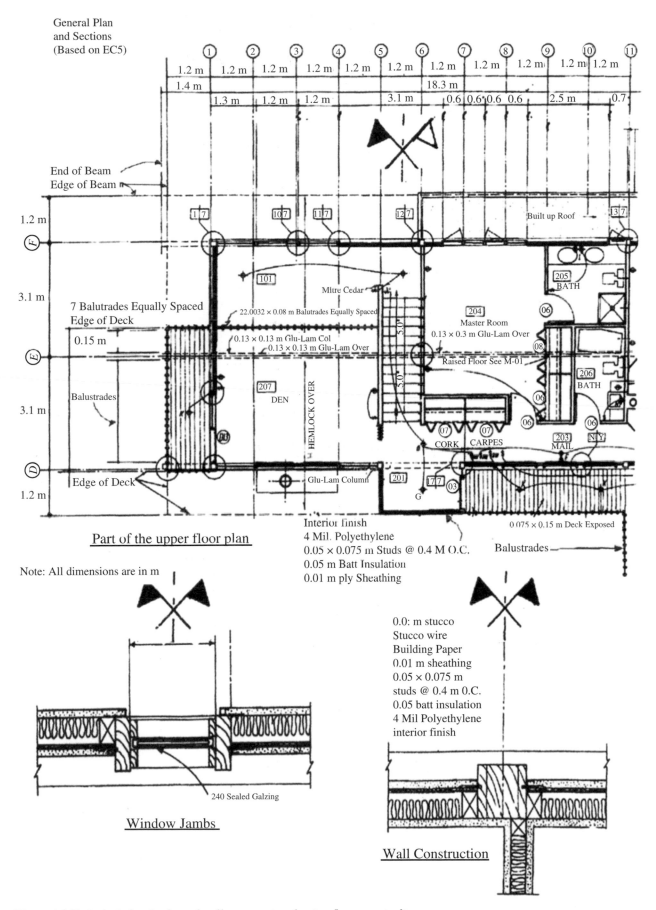

Figure 1.8 *Typical window jambs and wall construction showing flag as centre lines.*

(a) Unsymmetrical   (b) Symmetrical (Mirror Image)

Figure 1.9

cross flags, as in Fig. 1.9a, mark the European practice for unequal elevations. Where the elevation on either side of the centre line is equal, the cross flags are both black as in Fig. 1.9b.

1.6 Definitions based on Eurocode-5 and American codes

For the purposes of this text the following definitions apply:

Char-line Border line between the char-layer and the residual cross-section.

Design fire load density, q_d The fire load density considered for determining thermal action in fire design; the value allows for uncertainties and safety requirements.

Effective cross-section Cross-section of the member in structural fire design used in the ejective cross-section method. It is obtained from the residual cross-section by removing parts of the cross-section with assumed zero strength and stiffness.

Effects of actions, E External or internal forces and moments, stresses, deformations, displacements of the structure (compared to action effects S which comprise only internal forces and moments).

Fire compartment A space in a building, extending over one or several floors, which is enclosed by separating members such that fire spread beyond the compartment is prevented during the relevant fire exposure.

Fire load density The fire load per unit area related to the floor area q_p related to the surface area of the total enclosure, including openings q_1.

Fire protection material A material shown, by fire resistance tests, to be capable of remaining in position and of providing adequate thermal insulation for the fire resistance period under consideration.

Fire resistance The ability of a structure, a part of a structure or a member to fulfil its required functions (load bearing and/or separating function) for a specified fire exposure for a specified period of time.

Global structural analysis (for fire) An analysis of the entire structure, when either the entire structure or only parts of it are exposed to fire. Indirect fire actions are considered throughout the structure.

Indirect fire actions Thermal expansions or thermal deformations causing forces and moments.

Integrity criterion, E A criterion by which the ability of a separating construction to prevent passage of flames and hot gases is assessed.

Load bearing criterion, R A criterion by which the ability of a structure or member to sustain specified actions during a fire is assessed.

Member analysis (for fire) The thermal and mechanical analysis of a structural member exposed to fire in which the member is considered as isolated with boundary conditions. Indirect fire actions are not considered, except those resulting from thermal gradients.

Normal temperature design Ultimate limit state design for ambient temperatures according to ENV 1995-1-1.

Parametric fire exposure Gas temperature in the environment of surfaces of members etc. as a function of time, determined on the basis of fire models and the specific physical parameters describing the conditions in the fire compartment.

Protected members Members for which measures are taken to reduce the temperature rise.

Residual cross-section Cross-section of the original member reduced by the charging depth.

Separating construction Load bearing or non-load bearing construction (eg walls and floors) forming the enclosure of a fire compartment.

Separating function The ability of a separating member floor assembly to prevent fire spread by passage of flames or hot gases (integrity) or ignition beyond the exposed surface (thermal insulation) during the relevant fire exposure.

Standard temperate-time curve The nominal temperature-time curve given in ENV 1991-2-2.

Structural member The load bearing members of a structure, including bracings.

Sub-assembly analysis (for fire) The structural analysis of parts in which the respective pad of the structure is considered as isolated, with appropriate support and boundary conditions. Indirect fire actions within the sub-assembly are considered, but no time-dependent interaction with other parts of the structure. (Notes: 1) Where the effects of indirect fire actions within the sub-assembly are negligible, sub-assembly analysis is equivalent to member analysis. 2) Where the effects of indirect fire actions between the sub-assembly are negligible, sub-assembly analysis is equivalent to global structural analysis.)

Support and boundary conditions Effects of actions and restraints at supports and boundaries when analysing the entire structure or only part of the structure.

Temperature analysis The procedure of determining the temperature development in members on the basis of the thermal actions (net heat flux) and the thermal material properties of the members and of the protective surfaces, where relevant.

Temperature-time cubes Gas temperatures in the environment of member surfaces as a function of time. They may be nominal (conventional curves, adopted for classification or verification of the fire resistance eg the standard time-temperature curve) or parametric (determined on the basis of fire models and the specific physical parameters karat, or defining the conditions in the fire compartment).

Thermal actions Actions on the structure described by the net heat flux to the members.

Thermal insulation criterion, I The insulation needed to protect timber against heat loss.

Notation

A	area
E	modulus of elasticity; effect of actions
$E\,x$	integrity criterion for x minutes in standard fire exposure
F	opening factor
$I\,x$	thermal insulation criterion for x minutes in standard fire exposure
R	resistance
$R\,x$	load bearing criterion for x minutes in standard fire exposure
V	volume
X	material proper
a	distance
b	width
c	specific heat
d	depth; height
f	strength
k	coefficient
l	length
n	number
p	perimeter
q	fire load density related to floor area
r	radius
t	time; thickness; toil
α	angle
β	charring rate
γ	partial coefficients
η	coefficient (always with subscript)
λ	thermal conductivity
ζ	coefficient
ρ	density
ω	moisture content
θ	temperature

Subscripts

char	charring
d	design
ef	effective
f	floor
fi	fire; fire design g up *(what does g stand for???)*
ins	insulation
k	characteristic
m	material proper; bending
max	maximum
min	minimum
mean	mean value
my	minimum *(is this correct?)*
mod	modification
n	normal temperature design
P	panel

pr	protective
par	parametric
r	residual
req	required
st	steel
t	total
w	wood
0	basic value; zero
05	fifth percentile

Units

The following units are recommended in addition to ENV 1995-1-1:

temperature: $^{\circ}$C

specific heat: J/kg/K

coefficient of heat transfer: W/m^2/K

coefficient of thermal conductivity: W/m/K

1.7 Placing drawings based on American codes

Placing drawings, ordinarily prepared by the timber manufacturers and contractors, provide details for fabrication and for the placing of connections. The drawings may comprise bar lists, schedules, bending details, placing devils and placing plans or elevations. They may be prepared entirely manually or may include a computer printout. They are prepared to the same standards as engineering drawings. A general layout is shown on Fig. 1.10. Drawings usually include a plan, elevations, sections and details of a structure, accompanied by schedules for footings, columns, beams and slabs. The plan should be drawn in the upper left corner of the sheet.

They are not for use in building form work (except joist forms when these are supplied by the same fabricator) and consequently the only required dimensions are for the proper location of the connections. Building dimensions are shown on the placing drawings only if it is necessary to locate the connections properly, since the detailer becomes responsible for the accuracy of dimensions when they are given. The placing drawings must be used with the contract (engineering) drawings. Bending details may be shown on a separate list instead of on the drawings.

Upon receipt of the engineering drawings, the timber contractor:

1. Prepares placing drawings
2. Obtains engineer's, architect's or contractor's approval, if required
3. Prepare shop lists and fabricates where necessary
4. Provides coated connectors and connection details if specified.

The detailer is responsible for completing the instructions on the contract documents. When coated connections are

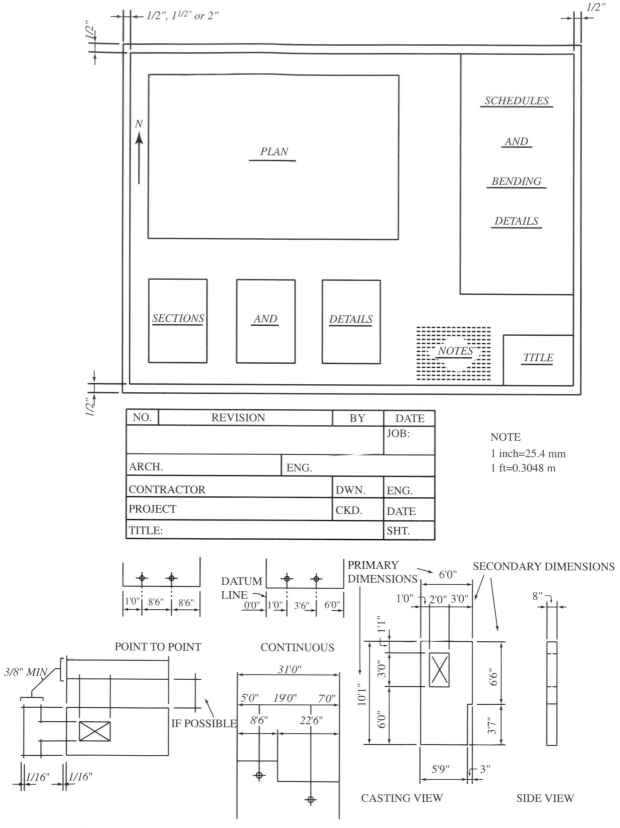

Figure 1.10 *The placing drawing, based on American practice.*

detailed along the uncoated types, the coated should be identified in some way, eg with a suffix E or G or with an asterisk and note stating that all the connections marked as such are to be epoxy coated or galvanized. Epoxy coated connections listed as uncoated connections in schedules or Bills of Materials should also be marked with E or *. The designation G is appropriate for galvanised devils. Placing drawings must show the size, shape, grade and location of coated and uncoated details in the structure.

The connections in the floors and many other parts of the structures can best be shown in tabular form, commonly referred to as a schedule. The schedule is a compact summary of all the connecting devils complete with the number of pieces, shape, size lengths, marts grades and coating information.

To ensure proper interpretation of the engineering drawings and the contractor's requirements, the fabricator's placing drawings are usually submitted for approval to the contractor before shop fabrication has begun.

Joist, beams, girders and sometimes footings that are similar on engineering drawings are given the same designation mark. Where possible, the same designations shall be used on both placing drawings and on engineering drawings. When members that are alike on the engineering drawings are slightly different on the placing drawings, a suffix letter is added to the designation to differentiate the numbers. If part of the beams are marked 2B3, say, on the engineering drawing and actually differ from the others, the placing drawing would show these parts of the beams as 2B3 and the others as 2B3A. In timber joist floors there may be so many variations from the basic joists shown on the engineering drawings that it is necessary to change the basic designations (for example, from prefix J to prefix R, for rib).

Columns and generally footings are numbered consecutively or are designated by a system of coordinates on the engineering drawings. The same designations are used on placing drawings. Connecting

devices must be individually identified on placing drawings. Connectors in elements of a structure may be drawn on placing drawings either on the plan, elevation, or section, or may be listed in a schedule. It is acceptable practice to detail footings, columns, beams and slabs in schedules. There is no standard format for schedules. They take the place of a drawing, such as a beam elevation, and must clearly indicate to the reinforcing bar placer exactly where and how all the material listed must be placed.

1.7.1 Drawing preparation

The effectiveness of a drawing is measured in terms of how well it communicates its intent. To the user, an erection or production drawing is a set of instructions in the form of diagrams and text. With this thought in mind, the drafter can improve the presentation by observing the following:

1. Make all notes on the drawings brief, clear and explicit, leaving no possibility of misunderstanding. Use commands.
2. Make all views and lettering large enough to be clearly legible.
3. Emphasise the specific items for which the drawing is intended. (For instance, when drawing connector tickets, draw the outline of the panel with light lines, but reinforce designations with dark lines.)

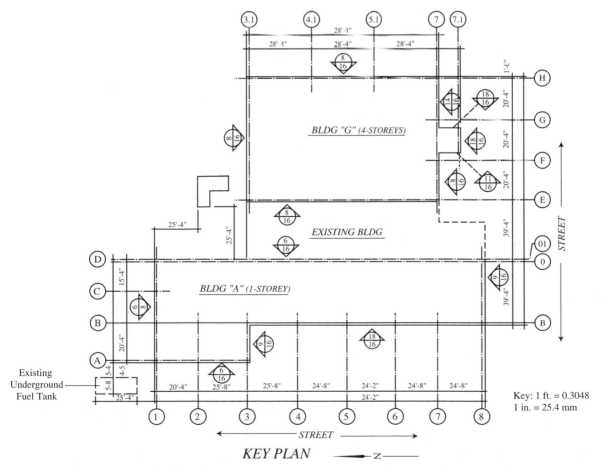

Figure 1.11 *A blown-out key plan, based on American practice.*

4. If drawings are to be reduced photographically, use broader lines and larger lettering.
5. Do not allow the drawings or details to become crowded. Use additional drawings and extra large-scale details when necessary.
6. Highlight special purpose notes so that they are clearly evident, eg ERECTOR NOTE!
7. Use cross-references to other erection drawings as required.

The drafter should become familiar with the following standards in order to use them properly in the preparation of specialised drawings:

1. general information
2. tolerances
3. drawing symbols
4. graphic symbols
5. finish designations
6. welding symbols and charts.

All drawings should have a title block, which is usually pre-printed on the lower right-hand corner (see Fig. 1.10). The information recommended for inclusion in the title block is shown in Fig. 1.10.

1.7.2 Scales and lines

All erection drawings should be drawn to scale. Production drawings generally cannot be drawn to scale since techniques to speed up the process are often used. However, they should be proportionately correct.

All line work falls into one of the following eight extension/dimension categories: object; hidden; extension; dimension (primary and secondary); leader; centre; break and symbols. Varying line weight (density) helps to differentiate between types of lines on a drawing, providing increased clarity and ease of interpretation. Line weight can be varied by making repeated strokes on a line or by using different weight leads. When using ink, line weight is controlled through the use of different pen points. Dimension and extension lines, while being the lightest (thinnest) lines on the drawing, must be dense enough to reproduce clearly when multi-generation copies are made. The drawing must illustrate the appearance of each type of line as they relate to one another on a drawing, and the recommended weight for each line. Table 1.3 lists the symbols used in drawings.

1.7.3 Grids and levels

The American practice is identical to that described in other codes.

1.7.4 Sections and elevation markers

Sections and elevation markers and their contents are identical for the American practices.

Table 1.3 Symbols and abbreviations based on American practice.

Abbreviations and acronyms

AF&PA	American Forest and Paper Association	NDS	National Design Specification
Allow	allowable	oc	on centre
ASD	allowable stress design	OM	overturning moment
B&S	beams and stringers	OSB	oriented strand board
C-TO-c	centre to centre	P&T	posts and timbers
cg	centre of gravity	PDS	Plywood Design Specification
DF-L	Douglas Fir - Larch	PL	plate
Ecc	eccentric	PSL	parallel strand lumber
EMC	equilibrium moisture content	Q/A	quality assurance
FBD	free-body diagram	RM	resisting moment
FS	factor of safety	S4S	dressed lumber (surfaced 4 sides)
FSP	fibre saturation point	SCL	structural composite lumber
glulam	structural glued laminated timber	Sel. St.	select structural
ICBO	International Conference of Building Officials	SJ&P	structural joists and planks
IP	inflection point (point of reverse curvature and point of zero moment)	SLF	structural light framing
J&P	joists and planks	T&G	tongue and groove
Lam	lamination	Tab	tabulated
LF	light framing	TCM	Timber Construction Manual
LRFD	load and resistance factor design	TL	total load
LVL	laminated veneer lumber	trib	tributary
MC	moisture content	TS	top of sheathing
MDO	medium density overlay (plywood)	UBC	Uniform Building Code
MEL	machine evaluated lumber	WSD	working stress design
MSR	machine stress rated lumber	WWPA	Western Wood Products Association
NA	neutral axis	WWUB	Western Wood Use Book

(continued)

Table 1.3 (Continued)

Notation		
a	acceleration	
A	area	$(\text{mm}^2, \text{m}^2)$
A_B	ground floor area of structure	(m^2)
A_g	gross cross-sectional area of a tension or compression member	(mm^2)
A_h	projected area of hole caused by drilling or routing to accommodate bolts or other fasteners at net section	(mm^2)
A_m	gross cross-sectional area of main wood member	(mm^2)
A_n	cross-sectional area of member at a notch; net cross-sectional area of a tension or compression member at a connection	(mm^2)
A_p	in-structure component amplification factor	

Chapter 2

Standard Specifications for Timber and Timber Connections

2.1 General introduction

This chapter is entirely based on standard specifications for timber and timber connections. Three major codes are examined, namely, BS5268, EC-5 and LAFD/ASD. A step-by-step guide through individual codes and practises is provided. In a number of places, a comparative study is given on various issues. Agreements and disagreements where present have been thoroughly explained. There are areas which give additional information which could not be available in a specific code. They are intended to be included in the timber design. Guidance must be sought from the experienced timber designers or researchers. Loadings are clearly explained regarding their use. This chapter gives extensive data on timber species and material properties.

2.2 Loads and material properties based on BS5268 and British design practices

All structural members, assemblies or frameworks in a building, in combination with the floors and walls and other parts of a building, should be capable of sustaining, with due stability and stiffness and without exceeding the relevant limits of stress given in BS 5268 r, the whole dead, imposed and wind loading and all other types of loading referred to in BS 5268. The design requirements of BS 5268 should be satisfied either by calculation, using the laws of structural mechanics, or by load testing.

The design and details of parts and components should be compatible, particularly in view of the increasing use of prefabricated components such as trussed rafters and floors. The designer responsible for the overall stability of the structure should ensure this compatibility even when some or all of the design and details are the work of another designer.

To ensure that a design is robust and stable:

1. the geometry of the structure should be considered;
2. the required interaction and connections between timber load bearing elements and between such elements and other parts of the structure should be assured;
3. suitable bracing or diaphragm effect should be provided in planes parallel to the direction of the lateral forces acting on the whole structure.

2.2.1 Loading

For the purpose of design, loading should be in accordance with BS 6399-1, BS 6399-2 and BS 6399-3 and other relevant standards, including EC-5, American, Canadian and Russian codes where applicable.

2.2.1.1 Accidental damage (including exceptional snowdrift loads)

In addition to designing a structure to support loads arising from normal use, there should be a reasonable probability that the structure will not collapse catastrophically because of misuse or accident. No structure is expected to be resistant to the excessive loads or forces that could arise from an extreme cause, but it should not be damaged to an extent that is disproportionate to the original cause.

Whilst the possibility of accidental damage or misuse of a structure should be considered, specific business design requirements are not usually necessary for buildings not exceeding four storeys, other than those requirements applicable to normal use. However, for buildings exceeding four storeys, the recommendations for reducing the sensitivity of the structure to disproportionate collapse in the event of an accident given in section 5 of Approved Document A to the Building Regulations 2004 applicable to England and Wales), and in the Building Standards Scotland 1990 and the Building Regulations (Northern Ireland) 2004 should be followed.

The effects of exceptional loads caused by 'local drifting of snow' on roofs, are defined in BS 6399-3 and should be based on the number of storeys, and should be checked on the assumption that such loads are accidental.

2.2.1.2 Service classes

Because of the effects of moisture content on material mechanical properties, the permissible property values should be those corresponding to one of the following service classes.

1. Service class 1 is characterised by a moisture content in the materials corresponding to a temperature of 20°C and relative humidity of the surrounding air only exceeding 65% for a few weeks per year. In such moisture conditions, most timber will attain an average moisture content not exceeding 12%.
2. Service class 2 is characterised by a moisture content in the materials corresponding to a

temperature of 20°C and the relative humidity of the surrounding air only exceeding 85% for a few weeks per year. In such moisture conditions, most timber will attain an average moisture content not exceeding 20%.

3. Service class 3, due to climatic conditions, is characterised by higher moisture content than service class 2.

2.2.2 Grade stresses for strength classes and individual species

2.2.2.1 General

Grade stresses for service classes 1 and 2 are given in Table 2.1 for 16 strength classes, and in Tables 2.2, for individual softwood and hardwood species and grades.

Because it is difficult to dry thick timber, service class 3 stresses and moduli should normally be used for solid timber members more than 100 mm thick, unless they are specially dried.

Designs should be based either on the stress for the strength classes, or on those for the individual species and grades. For designs based on strength classes, the material specification should indicate the strength class, whether softwood or hardwood, or, if the choice of material is limited by factors other than strength, the particular species required. The species and visual grades which meet the requirements of the strength class are given in various tables. Machine graded timber, meeting the requirements of BS EN 519, is graded directly to the strength class boundaries and marked accordingly.

A species and grade combination has been assigned to a strength class if the bending stress, mean modulus of elasticity and characteristic density (as defined in BS EN 338) for the combination, and appropriate to a depth (or width for tension) of 300 mm, are not less than the class values.

2.2.2.2 Service class 3 (wet exposure) stresses

Grade stress values for service class 3 should be obtained by multiplying the tabulated stresses and moduli given in Tables 2.1, 2.3 by the modification factor K_2 from Table 2.2.

Table 2.1 Moisture content of timber related to service class

Service Class(as defined in 1.6.4)	Examples of end use of timber in building	Average moisture content likely to attained in service conditions[a] %
3	External uses, fully exposed	20 or more
2	Covered and generally unheated	18
2	Covered and generally heated	15
1	Internal uses, in continuously heated building	12

[a] Moisture content should be measured using a moisture meter with insulated probes inserted 20 mm or one-quarter of the timber thickness, whichever is the lesser.

Table 2.2 Modification factor, K_2, by which stresses and moduli for service classes 1 and 2 should be multiplied to obtain stresses and moduli applicable to service class 3

Property	Value of K_2
Bending parallel to grain	0.8
Tension parallel to grain	0.8
Compression parallel to grain	0.6
Compression perpendicular to grain	0.6
Shear parallel to grain	0.9
Mean and minimum modulus of elasticity	0.8

Table 2.3 Modification factor, K_3, for duration of loading

Duration of loading	Value of K_3
Long-term (eg. dead + permanent imposed[a])	1.00
Medium-Term (eg. dead + snow, dead + temporary imposed)	1.25
Short-term (eg. dead + imposed +wind[b], dead + imposed +snow + wind[b])	1.50
Very short-term (eg. dead + imposed + wind[c])	1.75

For uniformly distributed imposed floor loads K_3 = 1.00 except for type C$_3$ occupancy (see BS 6399-1:1996, Table 1) where for foot traffic on corridors, hallways, landings and stairs, K_3, may be taken as 1.5.

For wind, short-term category applies to class C (15 s gust) as defined in CP 3: Chapter V-2 or, where the largest diagonal dimensions of the loaded area a, as defined in BS 6399-2, exceed 50 m.

For wind, very short-term category applies to classes A and B (3 s or 5 s gust) as defined in CP 3: Chapter V-2 or, where the largest diagonal dimension of the loaded area a, as defined in BS 6399-2, does not exceed 50 m.

Note: Tables for strength class grade stresses are available for various species in the relevant code of practice, to which the reader is referred.

2.2.3 Duration of loading

Table 2.9 of BS 5628 gives the modification factor K_2 by which stresses and moduli for service classes 1 and 2 should be multiplied to obtain stresses for service class 3. Modification factor K_3 is then used for the duration of loading.

2.2.4 Glued laminated timber

2.2.4.1 Introduction

Glued laminated timber (glulam) should be manufactured in accordance with BS EN 386. All timber used for laminated softwood members should be strength graded visually or mechanically in accordance with BS EN 518 or BS 519, respectively, and the additional recommendation for wane, etc given in the National Annex of BS EN 386:1995. All timber used for laminated hardwood members should be strength graded in accordance with BS 5756. (Notes: 1) The stresses given in BS 5268-2 apply only to timber graded in accordance with BS EN 518, BS 519 and BS 5756. Because these standards are primarily for solid structural timber, manufacturers of glued laminated timber may need to specify overriding requirements for wane, fissures and distortion. 2) Glulam strength properties may be determined from the strength class properties and the assignments of many grades and species to that class are given. Glulam designed and manufactured in accordance with the recommendations given in BS 5268-2 but using grades in accordance with BS EN 1912, is acceptable.)

2.2.4.2 Grade stresses for horizontally glued laminated members

The grade stresses for horizontally glued laminated softwood or hardwood members should be taken as the products of the strength class stresses for the relevant grade and species, and the modification factors from Table 2.4 appropriate to the strength class of timber used for the laminations. Appropriate tables for the strength classes can be found in BS EN 386.

For tension perpendicular to grain and torsional shear, permissible stress should be calculated and factors K_{19} and K_{27} should be disregarded. Table 2.4 of BS 5268 applies to members horizontally laminated from one strength class. Members may be horizontally laminated from two strength classes provided that the strength classes are not more than three classes apart and the members are fabricated so that not less than 25% of the depth at both the top and bottom of the member is of the superior strength class. For such members, the grade stresses should be taken as the modification factor from Table 2.4 of BS 5268.

For bending tension and compression parallel to grain grade stresses, these values should then be multiplied by 0.95. The modification factors given in Table 2.4 apply to members having four or more laminations, all of similar thickness.

2.2.4.3 Grade stresses for vertically glued laminated beams

Permissible stresses for vertically glued laminated beams are governed by the particular conditions of service and loading. The modification factors, K_{27}, K_{28} and K_{29}, given in Table 2.5 appropriate to the number of laminations are considered for stresses.

The modification factors given in Table 2.5 apply to the mechanical properties given. For tension perpendicular to grain and torsional shear, permissible stress should be calculated and factor K_{27} should be disregarded.

Table 2.4 Modification factors K_{15}, K_{16}, K_{17}, K_{18}, K_{19} and K_{20} for single grade glued laminated members

Strength classes	Number of laminations[a]	Bending parallel to grain K_{15}	Tension parallel to grain K_{16}	Compression parallel to grain K_{17}	Compression perpendicular to grain K_{18}[b]	Shear parallel to grain K_{19}	Modulus of elasticity K_{20}[c]
C37, C30, D50, D60, D70	4 or more	1.39	1.39	1.11	1.49	1.49	1.03
C22, C24	4	1.26	1.26	1.04	1.55	2.34	1.07
D35, D40	5	1.34	1.34				
	7	1.39	1.39				
	10	1.43	1.43				
	15	1.48	1.48				
	20 or more	1.52	1.52				
C16,	4	1.05	1.05	1.07	1.69	2.73	1.17
C18, D30	5	1.16	1.16				
	7	1.29	1.29				
	10	1.39	1.39				
	15	1.49	1.49				
	20 or more	1.57	1.57				

Table 2.5 Modification factors, K_{27}, K_{28}, K_{29}, for vertically glued laminated members

Number of laminations	Bending, tension and shear parallel to grain K_{27}		Modules of elasticity, compression parallel to grain K_{28}[a]		Compression perpendicular to grain K_{29}[b]
	Softwoods	Hardwoods	Softwoods	Hardwoods	Softwoods and hardwoods
2	1.11	1.06	1.14	1.06	1.10
3	1.16	1.08	1.21	1.08	
4	1.19	1.10	1.24	1.10	
5	1.21	1.11	1.27	1.11	
6	1.23	1.12	1.29	1.12	
7	1.24	1.13	1.30	1.12	
8 or more	1.25	1.13	1.32	1.13	

[a] When applied to the value of the modules of elasticity, E, K_{28} is applicable to the minimum value of E.
[b] If no wane is present, K_{29} should have the value 1.33 and, regardless of the grade of timber used, should be applied to the SS or HS grade stress for the species.

2.2.4.4 Glued end joints in glued laminated timber

Finger joints should have characteristic bending strengths of not less than the characteristic bending strength of the strength class for the lamination when tested in accordance with BS EN 385. Alternatively, end joints in limitations should have bending efficiency ratings (regardless of the type of loading) equal to or greater than the values given. In the case where end joints with a lower efficiency rating than is required for the strength class are used, the maximum bending, tension or compression parallel to the grain stress to which a glued end joint in any individual lamination is subjected, should not exceed the value obtained by multiplying the stress for the strength class given by the efficiency rating for the joint by the appropriate modification factors for size of member, moisture content and duration of loading, and also by the modification factor, K_{30}, K_{31}, or K_{32}, from Table 2.6, as appropriate to the type of loading.

For a vertically glued laminated beam, glued end joints should have an efficiency of not less than that required for the strength class or species and grade of timber unless the permissible stress is reduced accordingly. (Note: It can be assumed that the presence of finger joints or plain scarf joints in a lamination does not affect its modulus of elasticity. The full cross-section of jointed laminations may be used for strength and stiffness calculations.)

Table 2.6 Modification factors, K_{30}, K_{31}, and K_{32}, for individually designed glued end joints in horizontally glued members

Timber	Bending parallel lto grain K_{30}	Tension parallel to grain K_{31}	Compression parallel to grain K_{32}
Softwood	1.63	1.63	1.43
Hardwood	1.32	1.32	1.42

If other types of end joint are used, either the contribution of the laminations in which they occur should be omitted when calculating the strength and stiffness properties of the sections, or their suitability for use should be established by test.

2.2.4.5 Glued laminated flexural members

2.2.4.5.1 General

The permissible stresses for glued laminated timber flexural members are governed by the particular conditions of service and loading as given by the modification factors for flexural timber members and by additional factors. In addition to the deflection due to bending, the shear deflection may be significant and should be taken into account.

2.2.4.5.2 Camber

Subject to due regard being paid to the effect of deformation, members may be pre-cambered to offset the deflection under dead or permanent loads, and in this case, the deflection under live or intermittent imposed load should not exceed 0.003 of the span.

2.2.4.5.3 Curved glued laminated beams

For curved glued laminated beams with constant rectangular cross-section, the ratio of the radius of curvature, r, to the lamination thickness, t, should be greater than the mean modulus of elasticity, in Newtons per square millimetre (N/mm^2), for the strength class divided by 70, for both hardwoods and softwoods.

2.2.5 Plywood

2.2.5.1 General

This section provides recommendations for the use of sanded and unsanded plywoods manufactured in accordance with the standards listed in the Bibliography, and subject to quality control procedures.

The strength of plywood depends mainly on total thickness, and on the species, grade, arrangement and thickness of the individual plies. When specifying plywood for structural work, reference should be made to the type, grade, panel nominal thickness, number of plies, and whether a sanded or unsanded finish is required. Within a specification of panel nominal thickness and number of plies, a range of constructions, combining plies of different thickness, may be available. The tabulated section properties are based on the full cross-section ignoring the effect of orientation of veneers.

The tabulated grade stresses and moduli are based on strength and stiffness tests on plywoods of specified thickness and lay-up, allowing for the effect of orientation of veneers, and are expressed in terms of the tabulated section properties. The strength or stiffness capacity of a particular plywood is obtained by multiplying the appropriate stress or modulus value by the appropriate section property. The species used to manufacture these plywoods are always available.

Alternatively, grade stresses and moduli may be derived by the design equations developed from the characteristic values obtained from the appropriate European Standards for structural plywood, methods of testing and derivation of characteristic values.

2.2.5.2 Durability

Although all of the plywoods covered by BS 5268-2 are bonded with an exterior-type adhesive, this does not mean that they are necessarily suitable for use in damp or wet exposure conditions for long periods. Consideration should be given to the natural durability of the wood species from which the plywoods are made. Generally, plywoods used for permanent structures in damp or wet conditions, unless inherently durable, should be adequately treated against decay (see BS 5268-5 and DD ENV 1099). American plywood designated Exposure 1 is manufactured with exterior-type adhesive, but is not recommended for use in prolonged damp or wet conditions, as the inclusion of D-grade veneer in backs and inner plies of Exposure 1 board could affect localised glue line performance. For applications subject to permanent outdoor exposure, a board of exterior bond durability classification should be used.

2.2.5.3 Dimensions and section properties

The section properties of plywoods are given in various manufacturers' lists and are based on the minimum thickness of plywood permitted by the relevant product standards and apply to all service classes.

2.2.5.4 Grade stresses and moduli

Generally tables are available for the grade stresses and moduli for plywoods. They apply to long-term loading in service classes 1 and 2 and should be used in conjunction with the corresponding section properties. For other durations of load and for service class 3 conditions, the stresses and moduli should be multiplied by the magnification factor, K_{mod} given in Table 2.7 of the code.

2.2.5.5 Design data derived from characteristic values

Characteristic values should be derived in accordance with:

1. BS EN 789 and BS EN 1058 or from tests in accordance with BS EN 12369-2 for plywoods

Table 2.7 Modification factor, K_{mod} for duration of loading and service class for plywood

Duration of loading	Service class		
	1	**2**	**3**
Long-term	0.60	0.60	0.50
Medium-term	0.86	0.86	0.68
Test duration	1.02	1.02	0.82
Short-term	1.06	1.06	0.86
Very short-term	1.08	1.08	0.88

conforming to European Standards listed therein; or

2. BS EN 789 and BS EN 1058 for plywood conforming to those sections of prEN 13986 that relate to plywood for structural components; or

3. BS EN 789 and BS EN 1058 for plywood conforming to those section of BS EN 636-1, -2 or -3 that relate to structural applications.

Specifiers should also take into account the appropriate bonding quality and biological durability for the particular end use application to which the plywood is being put (see the relevant sections of BS EN 636-1, -2 -3 as appropriate). Characteristic strength values are defined as the population 5th percentile value obtained from tests with duration of 300 s using test pieces at an equilibrium moisture content resulting from a temperature of 20°C and a relative humidity of 65%. Characteristic moduli values are defined as either the 5th percentile value or the mean value, under the same test conditions as defined above.

Characteristic strength values should be converted into grade strength values for the permissible stress basis of design by:

$$X_d = k_{mod} X_k/(1.35\gamma_M) \qquad (2.1)$$

where

X_d is the grade strength value;
X_k is the characteristic strength value;
k_{mod} is the modification factor for duration of loading and service class given in Table 2.7;
γ_M is the material partial safety factor (eg 1.2 for plywood).

When a modification factor, k_{mod}, greater than unity is used in accordance with this clause, the design should be checked to ensure that the permissible stresses are not exceeded for any other condition of loading that might be relevant. Characteristic moduli values should be converted into grade moduli values for the permissible stress basis of design by:

$$E_d = E_k/(1+k_{def}) \qquad (2.2)$$

where

E_d is the grade modulus value;
E_k is the characteristic modulus value (mean or minimum value as relevant) obtained from tests;
k_{def} is the modification factor for creep deformation and service class given in Table 2.8.

Table 2.8 Modification factor, K_{def}, for elastic or shear modulus for plywood

Duration of loading	Service class		
	1	2	3
Long-term	0.80	1.00	2.50
Medium-term	0.11	0.14	0.62
pest duration	0.00	0.00	0.00
Short-term	0.00	0.00	0.00
Very short-term	0.00	0.00	0.00

Table 2.9 Modification factor, K_{36}, by which the grade stresses and moduli for long-term duration and service classes 1 and 2 for plywood should be multiplied to obtain values for other durations and/or service class 3

Duration of loading	Value of K_{36}			
	Service classes 1 and 2		Service class 3	
	Stress	Modulus	Stress	Modulus
Long-term	1.00	1.00	0.83	0.57
Medium-term	1.33	1.54	1.08	1.06
Short- and very short-term	1.50	2.00	1.17	1.43

The deflections or deformations of a structural member subject to a combination of loads of different duration should be determined by considering the load in each category as acting separately, and calculating the deflections or deformations induced by each, using the appropriate moduli values.

2.2.5.6 Plywoods for roof and floor decking

The design of plywood for decking on roofs and floors should be determined:

1. by calculation using grade stresses and moduli (Table 2.9); or, where used for buildings,
2. by testing on samples of typical roof or floor assemblies with the product supported on joists, under impact and concentrated static loads (of the appropriate contact area) in accordance with BS EN 1195 to satisfy the performance specification and requirements in BS EN 12871 and 8.9. The application of plywood for these purposes should comply with BS EN 12872.

The requirement in BS EN 12871 to verify the design by calculation for uniformly distributed loading need not be made where all of the following conditions apply.

1. The specified imposed loading is in accordance with BS 6399 but excluding loading for vehicle traffic areas.
2. The ratio of uniformly distributed load (in kN/m^2) to the related concentrated load (in kN) does not exceed 1.5.
3. The maximum span of the panel product in the roof or floor decking is not more than 1.00 m.

This method of design does not apply to grade and vehicle traffic areas and roofs for special services such as helicopter landings.

2.3 Loads and material properties based on Eurocode-5 and European practices

EC5 addresses types of actions (forces) and three types of design situations and these are given in Tables 2.10 and 2.11, respectively. The two design situations for ultimate limit states may each be subdivided into verifications of strength and of stability or static equilibrium, because all these relate to safety rather than serviceability.

2.3.1 Characteristic value of actions

Most actions do not have a single known value, but come in a range of values. In general, the range can be statistically described, and can be defined by a single characteristic value, which is set at a level which will be exceeded only with a certain calculated probability.

Table 2.10 Types of action

Type of Action	Symbol*	Definition	Examples
Permanent	G_k	Invariable during the design life of the structure	Self-weight of the structure and its fittings
			Self-weight of fixed equipment
Variable	Q_k	Fluctuating, moveable or indefinite loads	Imposed ceiling, floor and storage loads
			Snow
			Wind
			Concentrated load from a man
			Traffic loads
Accidental	A_k	Abnormal loads	Explosions
			Vehicles or ship impact
			Local drifted snow
			Earthquake **

* In EC5 the suffix "k" indicates a characteristic value.
** Rules for seismic design are given in EC8. The classification of earthquake as accidental loads may change

Table 2.11 Types of design situation

Design situations	Limit state		Associated types of action
Fundamental design situations	Ultimate	Strength	Permanent + variable
		Stability	
Accidental design situations	Ultimate	Strength	Accidental + permanent + variable
		Stability	Permanent + variable following an accident
Serviceability calculations	Serviceability		Permanent + variable

EC5 designs are based on these characteristic values, and these will now be explained in more detail.

2.3.1.1 Permanent actions

For timber structures, permanent actions are confined to the weights of the building materials. For general materials, it is usually considered adequate to take as characteristic values the weights quoted in BS 648 or in the manufacturers' literature. In the case of timber and timber-based materials, however, weights are normally calculated from their mean densities (Clause 2.2.2.2(3)).

The mean densities of solid timber strength classes and glulam strength classes are given in prEN 338 and prEN 1194, respectively, as are the lower (5th percentile) characteristic density values.

2.3.1.2 Variable actions

Variable actions normally have two possible values: an upper characteristic value, which is the maximum value that a designer needs to consider, and a lower value which is usually zero. For example, the snow loads specified in BS 6399: part 5 are upper characteristic values corresponding to snowfall which is likely to be exceeded only once in 50 years. A designer will use these upper values to verify the strength of a rafter, but in checking wind uplift on a roof, the lower value of zero will be used.

2.3.1.3 Accidental actions

Accidental actions may be specified as upper characteristic values, as in earthquake loading, or else as design values which may be used without further modification, as for local drifted snow or vehicle impact loads.

2.3.2 Partial factors

The characteristic values of actions are converted to design values by means of partial factors. Two principal kinds of factor are used, and these are shown in Table 2.12.

2.3.2.1 Partial safety factors

Partial safety factors, commonly called 'gamma factors' because they are identified by the Greek letter γ, are used

as safety factors in ultimate limit state design situations. The subscript "F" indicates a general action or load. More specific subscripts such as "G" for permanent actions or "Q" for variable actions may be used instead. The value of γ_F depends on the type of action, the design situation under consideration and, in some cases, on the type of structure. For example, in normal structures, permanent loads (G_k) are increased by a factor of $\gamma_G = 1.55$ for strength verifications, whereas variable loads (Q_K) are increased by a factor of $\gamma_G = 1.5$ because they are known with less certainty.

Table 2.13 gives the values of the various gamma factors for use in timber structures. The factors γ_G and γ_{GA} are used to modify permanent loads in fundamental and accidental design situations, respectively, and the factor γ_Q to modify variable loads in fundamental ones. In accidental design situations, the factor γ_A is in theory used to modify accidental loads, A_k, but since these are normally specified by their design value, A_d, γ_A is set at 1.0. The factors for favourable effects are applied only to loads which relieve stress or tend to restore equilibrium. They may be distinguished from the factors for unfavourable effects by writing them as $\gamma_{G, inf}$ and $\gamma_{Q, inf}$ ("inf" = inferior or lower). The factors for un-avoidable effects may be distinguished as $\gamma_{G, sup}$ and $\gamma_{Q, sup}$ ("sup" = superior or higher). Most commonly, the factors for unfavourable effects are used, and such distinctions are unnecessary.

The reduced coefficients shown in Table 2.13 may be used for strength verifications in structures for which reduced safety factors are considered to be appropriate. Eurocode 5 recommends their use for single-storey buildings with moderate spans which are only occasionally used for example storage buildings, sheds, greenhouses, agricultural buildings and silos, ordinary lighting masts, light partition walls and cladding.

2.3.2.2 Coefficients for representative values of actions

In many design situations, more than one variable load occurs simultaneously. When considering the racking resistance of a ground floor timber frame wall panel, for example, a designer will have four simultaneous variable

Table 2.12 Principal factors used to convert characteristic values of actions to design values

Symbol	Type of factor	Purpose
γ_F	Partial safety factor for actions	Increases characteristic loads by a safety factor to allow for uncertainties
ψ	Coefficient for representative values of actions	Reduces variable loads (eg snow), when they are in combination with other variable loads, to appropriate values

Table 2.13 Partial safety factors for loads in ultimate limit state calculations from Eurocode 5 and the UK National Application Document

Verifications		Fundamental Situations				Accidental Situations	
		γ_G		γ_Q		γ_{GA}	γ_Q
		Favourable	Unfavourable	Favourable	Unfavourable		
Strength	Normal Values	1.0	1.35	0.0	1.5	1.0	1.0
	Reduced Values	1.0	1.2	0.0	1.35		
Overall stability		0.9	1.1	0.0	1.5	1.0	1.0

Table 2.14 Coefficients for representative actions

Variable action	Building type	Ψ_0	Ψ_1	Ψ_2
Imposed floor loads	Dwellings	0.5	0.4	0.2
	Other occupancy classes*	0.7	0.6	0.3
	Parking	0.7	0.7	0.6
Imposed ceiling loads	Dwellings	0.5	0.4	0.2
	Other occupancy	0.7	0.2	0.0
Imposed roof loads	All**	0.7	0.2	0.0
Wind loads				

* Institutional and educational, public assembly, offices, retail, industrial and storage as defined in BS 6399: Part 1: 1984, Table 1
** as listed and defined in BS 6399: Part 1: 1984, Table 1

loads to consider: wind, snow, imposed floor load and imposed ceiling load. As variable loads, these are all specified as upper characteristic values, but Eurocode 5 recognizes that it is unlikely, if not impossible, for the maximum value of all four to occur simultaneously. The Code, therefore, provides a set of reduction factors which are used to obtain representative values of variable actions when they occur in combination with other variable actions, or in a rarely occurring accidental design situation (Table 2.14).

2.3.2.2.1 Ψ_0 Combination coefficient

In fundamental design situations (permanent and variable loads only), the designer considers the full characteristic value of one variable action (eg wind) and multiplies all of the others by the factor Ψ_0.

2.3.2.2.2 Ψ_1 Frequent coefficient

This gives a greater reduction than Ψ_0, producing values for variable loads which may be regarded as average values for dominant loads in accidental design situations, or for post secondary loads in serviceability calculations.

2.3.2.2.3 Ψ_2 Quasi-permanent coefficient

This gives the greatest reduction of all, producing values which may be regarded as the minimum values of secondary variable loads when they are in combination with accidental loads. It is used only in accidental design situations, and for wind and snow its value is zero.

2.3.2.3 Expressions for design values of actions

In the light of the information which has been given, it should be possible to make sense of the expressions given in Eurocode 5 and Eurocode 1 for calculating the design values of actions. In these expressions, the letters j and i identify individual permanent and variable actions, and the Greek letter Σ has the conventional meaning of "the sum of". The reference numbers (eg 2.3.2.2a) all refer to EC5.

The design value of a combination of actions is calculated from the appropriate expression given in the

next section. When more than one variable action is involved, the design value is the most critical value of the expression, taking each variable action in turn as the dominant one and the others as secondary ones. This critical value is termed the 'design load' for ultimate limit states, or the 'service load' for serviceability limit states.

2.3.3 Ultimate limit states

2.3.3.1 Fundamental design situations

$$\sum \gamma_{G,J} G_{k,j} \quad + \quad \gamma_{Q,1} Q_{k,1} \quad + \quad \sum \gamma_{Q,I} \Psi_{0,i} Q_{k,i} \quad (2.3)$$

i > 1

Sum of factored permanent loads | Factored dominant variable load | Sum of other factored variable loads

2.3.3.2 Alternative expressions for fundamental design situations[1]

The design value of a combination of loads may be taken as the larger of the two expressions which follow

$$\sum \gamma_{G,J} G_{k,j} \quad + \quad 1.5 Q_{k,1} \quad (2.4)$$

Sum of factored permanent loads | 1.5 x the most unfavourable variable load

$$\sum \gamma_{G,i} G_{k,j} \quad + \quad 1.35 \sum_1^n Q_{k,i} \quad (2.5)$$

Sum of factored permanent loads | 1.35 x all of the variable loads

2.3.3.3 Accidental design situations

$$A_d + \sum \gamma_{GA,j} G_{K,J} \quad + \quad \Psi_{1,1} Q_{K,1} + \sum \Psi_{2,i} Q_{K,i} \quad (2.6)$$

i > 1

Design value of factored accidental load | Sum of permanent loads | Factored dominant variable load | Sum of other factored variable loads

2.3.4 Serviceability limit states

$$\sum G_{K,j} \quad + \quad Q_{k,1} \quad + \quad \sum \psi_{1,i} Q_{k,i} \quad (2.7)$$

i > 1

Sum of deformations produced by unfactored permanent loads | Deformation produced by unfactored dominant variable load | Sum of deformations produced by other factored variable loads

2.3.4.1 Components of variable loading

Wind, snow, the concentrated load on a roof, the concentrated load on a ceiling, the imposed ceiling load and the imposed floor loads should each be treated as

---

[1]This is a simplified but approximate method which assumes a value for Ψ_0. The design value obtained may be from 8% too low to 22% too high.

a separate variable action[2]. The reduction factors for floor loads in multi-storey buildings that are permitted by BS 6399 Part 1 are not used in designs to EC5. See Table 2.15.

2.3.4.2 Calculation for simple and accidental design loads

Table 2.16 gives clear explanations for obtaining simple and accidental design loads.

Table 2.15 EC5 load-duration classes

Load-duration class	Accumulated duration of load during the design life of the structure [(+)]	Examples of loading
Permanent	More than 10 years	Self-weight
Long-term	6 months to 10 years	Uniformly distributed imposed ceiling loads Storage
Medium-term	1 week to 6 months	Imposed floor loads
		Imposed roof loads other than snow, with general access
		Falsework
Short-term	Less than 1 week	Wind
		Snow*
		Imposed concentrated load for maintenance and repair only
		Concrete formwork
		Loads following accidents
Instantaneous	Instantaneous	Explosions
		Vehicle impact
		Earthquake (see Eurocode 8)

[+] The load duration class is not only determined by the estimated duration of the characteristic load, but also to a lesser extent by the duration of loads below their characteristic value.
* Note that snow loads in the United Kingdom should be treated as short-term.

Table 2.16 Simple and accidental design loads

Design Calculations For Simple Design Load	Design Calculations for Accidental Design Load
1. Calculate uniformly distributed load on joist. Conditions: Joist supporting a floor of tongued and grooved bonding with a plaster board ceiling soffit beneath it.	1. A glulam roof beam supports two purlins. Due to local drift of snow of a 2.5 kN/m² (BS 6399), find design load for a strength verification
	Summary of design data
Dead weight of floor boarding plasterboard and joists = 0.34 kN/m²	Permanent points loads from purlins Gk,1 – 4.9 kN each
Imposed floor loadings, From BS 6399 : Part 1 : 1984 = 1.50 kN/m	Permanent duration udl from ceiling dead load + self weight of beam Gk,2 – 4.9 kN/m
For joists at 0.6 m centres:	Long-term uniformly distributed imposed load from ceiling Qk,1 = 0.60 kN/m
Permanent load, G_k (Permanent duration) = 0.34 × 0.6 = 0.900 kN/m	Short-term point load from purlins due to local drifted snow Ak = 12 kN each
Variable Load, Qk (medium-term duration) = 1.5 × 0.6 = 0.900 kN/m	From 2.25, γGA = γa = 1.0
Using the values in Table 4 and expression 2.3.2.2a	From Table 2.26, ψ1 for the ceiling load Qk,1 = 0.4 With one variable load, expression 2.3.2.26 becomes
for permanent duration the design loads	$$A_d + \sum \gamma_{GA,j} G_{K,1} + \psi_{1,1} Q_{k,1}$$
$F_{d,permanent}$ = 1.35 × 0.204 = 0.275 kN/m $\quad\quad\;\; \gamma_G \quad\;\; G_k$ for medium-term duration	where $A_{d = \gamma_A A_K}$
	hence the accidental design load (short-term load case)
$F_{d,medium-term}$ = (1.35 × 0.204) + 1.5 × 0.9 $\quad\quad\quad\quad\;\; \gamma_G \quad\; G_k \quad\quad \gamma_Q \quad Q_k$ = 1.625 kN/m	(1.0 × 12.00) + (1.0 × 4.19) + = each point load (Kn) each point load (Kn)
	(1.0 × 0.62 + (0.4 × 0.60) + udl (kN / m) udl (kN / m)
	= 16.19 kN each point load + 0.86 kN/m udl

[2] The recommended combinations of load which a roof designer should check may be found in BS 5268 Part 3.

2.4 Design values for material properties based on EC5

Guidance Document 2 explains how to calculate the design values of loads using Eurocode 5. This Guidance Document explains how to calculate the design values of material properties for solid timber, glulam, wood-based panel products and metal fasteners.

Two principal kinds of factor are used to calculate design.

2.4.1 Characteristic values, X_k

A timber designer requires data on the density and mechanical properties of the members which have been specified, and some means of adjusting these properties for moisture content, load duration and any other factors which may influence them. For design work to Eurocode 5, the data on properties are provided in the form of characteristic values, which are derived directly from the results of laboratory tests.

To determine the characteristic value of *strength properties*, specimens are loaded to failure, and a 5th percentile characteristic value is calculated from the mean value of the failure loads and their standard deviation. This value, X_k, is set at a level such that only 5% of a set of similar specimens selected at random are likely to have a lower strength. *Stiffness properties*, measured in load/deflection tests, are specified as both 5th percentile ($E_{0, 05}$ or $G_{0, 05}$) and 50th percentile or mean values ($E_{0,mean}$ or $G_{0,mean}$). These correspond to the values E_{min}, G_{min}, E_{mean} and G_{mean} used in BS 5268: Part 2. Densities are also specified as both 5th percentile (ρ_k) and 50th percentile (ρ_{mean}) values, the former being used in joints calculations, and the latter in self-weight calculations.

2.4.1.1 Timber properties

EC 5 shows the timber materials for which characteristic values are currently available, and the standards in which they are published. The National Application Document (NAD) relates grade/species combinations commonly used in the UK to the European strength classes used in prEN 338. For example, glulam and wood-based panel products, the current designations respectively are PrEN1194 and ENTC 112.406.

2.4.2 Design values, X_d

For timber materials, all of the above properties are measured in specimens conditioned at 20°C AND 65% relative humidity, in tests lasting approximately 5 min. The mechanical properties are therefore applicable only to these service loads. The design value, X_d, of a material property is calculated from its characteristic value, X_k, as follows:

$$X_d = \frac{k_{mod} X_k}{\gamma_M} \qquad \text{for strength properties} \qquad (2.8)$$

or

$$X_d = \frac{X_k}{\gamma_M} \qquad \text{for stiffness properties} \qquad (2.9)$$

For certain strength properties, other factors are also applicable, but first an explanation of k_{mod} and γ_M will be given.

2.4.3 Strength modification factor, k_{mod}

As stated in GD 2 (Table 3.1.6 of EC5), five different load duration classes are defined: permanent, long term, medium term, short term, and instantaneous. The Code also defines three different service classes, which are related to the ambient temperature and relative humidity (Clause 3.1.5). These correspond to equilibrium moisture contents in softwoods of approximately:

- 12% or less for Service Class 1 (20°C, 65% r.h. rarely exceeded)
- 20% or less for Service Class 2 (20°C, 85% r.h. rarely exceeded)
- Above 20% for Service Class 3 (conditions leading to a higher moisture content).

For timber materials, the value of k_{mod} depends on the load duration, the service class and the type of material. For metal components, $k_{mod} = 1.0$. Table 2.17 shows the load duration for various grades.

2.4.4 Partial safety factor for material properties, γ_M

The values specified in EC5 for γ_M are shown in Table 2.18.

2.4.5 Other modification factors

Other modification factors are applicable to certain material strength properties. For precise details, reference must be made to Eurocode 5, for a list of the factors.

2.4.6 Timber species

Examples are given in EC 5 for appropriate service class.

2.5 Loads and material properties of timber based on AITC standards and other American practices

2.5.1 Loads

Buildings or other structures and all parts thereof should be designed to support all loads safely including dead loads, that may reasonably be expected to affect the structure during its service life. These loads should be as stipulated by the governing building code or, in the absence of such a code, the loads, forces, and combination of loads should be in accordance with accepted engineering practice for the geographical area under consideration.

Table 2.17 Material/load duration for classes 1, 2 and 3

Material/ load-duration class	Service class		
	1	**2**	**3***
Solid and glued laminated timber			
Plywood to prEN 636-1*, -2**, -3			
Permanent	0.60	0.60	0.50
Long-term	0.70	0.70	0.55
Medium-term	0.80	0.80	0.65
Short-term	0.90	0.90	0.70
Instantaneous	1.10	1.10	0.90
Particleboards to BS EN 312-6* and -7*			
OSB to BS EN 300 Grades 3 and 4*			
Permanent	0.40	0.30	–
Long-term	0.50	0.40	–
Medium-term	0.70	0.55	–
Short-term	0.90	0.70	
Instantaneous	1.10	0.90	–
Particleboards to BS EN 312-4* and -5*			
OSB to BS EN 300 Grades 2*			
Hardboards to BS EN 622-5*			
Permanent	0.30	0.20	–
Long-term	0.45	0.30	
Medium-term	0.65	0.45	–
Short-term	0.55	0.60	–
Instantaneous	1.10	0.80	–
Medium boards and Hardboards to BS EN 622-3*			
Permanent	0.20	–	–
Long-term	0.40	–	–
Medium-term	0.60	–	–
Short-term	0.80	–	–
Instantaneous	1.10	–	–

* not to be used in Service classes 2 and 3
** not to be used in Service class 3
*** Service class 3 should be used for solid (not laminated) timber more than 100 mm thick, unless it is specially dried.

Table 2.18 Values of γ_M

Design situation	Material	γ_M
Strength properties		
Fundamental (permanent + variable loads only)	Timber and wood-based materials	1.3
	Steel used in joints	1.1
Accidental (accidental + permanent + variable loads)	All materials	1.0
Stiffness properties		
All	All materials	1.0

2.5.1.1 Dead loads

This is defined as the vertical load due to all permanent structural and non-structural components of a building such as walls, floors, roofs, partitions, stairways, and fixed service equipment.

2.5.1.2 Live loads

This is defined as the load superimposed by the use and occupancy of the building or other structure, not including the wind load, snow load, earthquake load or dead load.

2.5.1.3 Floor live loads

In the absence of a governing building code, the minimum uniformly distributed live loads or concentrated loads in *Minimum Design Loads for Buildings and Other Structures* (ASCE) are recommended.

2.5.1.4 Roof live loads

The minimum roof live loads should be as stipulated by the governing building code. Roof live loads used in design should represent the designer's determination of the particular service requirements for the structure, but in the absence of a governing code, in no case should they be less than the recommended minimum.

Minimum roof live loads for flat, pitched, or curved roofs are as recommended by ASCE. Roofs should be designed to resist either the tabulated minimum live loads, applied as balanced or full unbalanced, or the snow load, whichever produces the greater stress. Roofs to be used for special purposes should be designed for appropriate anticipated loads.

2.5.1.5 Roof snow loads

Roof snow loads may be based on a determination of the ground snow load as specified by the governing building code or the roof snow loads may be specified in the code. Maximum snow load maps may also be used to determine the actual snow loads to be expected on the roof surface. As previously indicated, the roof snow load is a function of various factors.

These factors can be accounted for in design by applying appropriate snow load coefficients to the basic ground snow loads. Specific snow load coefficients have been developed to relate roof snow load to ground snow load based on comprehensive surveys of actual conditions.

For the design of both ordinary and multiple series roofs, either flat, pitched, or curved, minimum roof snow loads may be determined by multiplying the ground snow load of 300 psf (14.364 N/m^2) by the appropriate coefficients based on the roof geometry and climatic conditions. Specific coefficients for these roof configurations are given by ASCE. Because unbalanced loading can occur as the result of drifting, sliding, melting, and refreezing or physical removal of snow, structural roof members should be designed to resist increased loading in certain areas or other unbalanced loading conditions. For example, the roof should resist the full snow load as defined above distributed over the entire roof area, the full snow load distributed on any one portion of the area, and dead load only on the remainder of the area, depending on which load produces the greatest stress on the member considered.

Duration of load is the cumulative time during which the maximum design load is on the structure over its entire life. A 2-month duration is generally recognised as the proper design level for snow.

2.5.1.6 Wind loads

Much research has been conducted to evaluate wind effects on various structures, and has resulted in the establishment of design coefficients that account for building shape and wind direction. In addition, extensive studies of basic wind velocities related to geographical location have resulted in the development of detailed wind speed maps for the United States. Other studies of surface resistance relative to the degree of land development and gust characteristics at a given location have provided a method for a further refinement of the basic wind velocity and its effects on structures. Additional work relates the dynamic behaviour of structures to wind forces that are attributable to gusting and turbulence.

Sources of wind load analysis information include the ASCE manual on wind loads, which contains a number of other references on this subject.

The duration of maximum wind load is relatively short during the life of a structure. Ten minutes is the generally accepted load duration resulting in a duration-of-load adjustment factor of 1.6. Note that this is the duration-of-load adjustment and should not be confused with probability factors.

2.5.1.7 Basic wind speeds

Basic wind speeds for observed air flows in open, level country at a height of 33 ft above the ground have been developed. For the design of most permanent structures, a basic wind speed with a 50-year mean recurrence interval should be applied. However, if in the judgment of the engineer or authority having jurisdiction, the structure presents an unusually high degree of hazard to life and property in case of failure, the basic wind speed may be modified appropriately. Similarly, for temporary structures or structures having negligible risk to human life in case of failure, the design wind speed may be reduced.

2.5.1.8 Cyclic loading

Tests to date indicate that wood is less sensitive to repeated loads than are crystalline structural materials such as metals. In general, the fatigue strength of wood is higher in proportion to the ultimate strength values than are the endurance limits for most structural metals. In present design practice, no factor is applied for fatigue in deriving working stress values for wood, nor is it considered necessary to do so. For these reasons, normal duration of load is used for cyclic loading design.

Where estimation of repetitions of design or near-design loads is indicated to approximate one million or more cycles, the designer should investigate the visibility of fatigue failures in shear or tension perpendicular to grains. When shear governs design and fatigue is a distinct possibility, shear stresses should be reduced

by 10% or changes should be made on the basis of a detailed analysis as indicated by examination of USDA Forest Products Laboratory Report No. 2236, *Fatigue Resistance of Quarter-Scale Bridge Stringers in Flexure and Shear (7)* and/or ASCE Paper No. 2470, *Design Consideration for Fatigue in Timber Structures (8)*. Analyses made on the research available indicate that bending failure due to fatigue is not usual and need not be considered in ordinary design situations. Tension perpendicular to grain loading of wood should be avoided, especially when cyclic loading is expected.

2.5.1.9 Vibration

Vibration is closely related to impact and cyclic loading. The effects of vibration can usually be neglected in timber structures, except in those portions made of materials for which impact or vibration forces may be critical or where the vibration may be objectionable to the human occupancy. However, vibration may cause loosening of threaded connections used in timber structures. If there is a feasibility of fatigue failure due to vibration, it should be considered, and care should be taken to avoid notches, eccentric connections, and similar design conditions.

2.5.1.10 Minimum roof load combinations

In designing, the most severe realistic distribution, concentration, and combination of roof loads and forces should be taken into consideration. Table 2.19 gives examples of roof load design combinations that should be checked and examples of the application of the duration-of-load factor for different load combinations.

Table 2.19 Examples of load combinations for design[a]

Roof Load Combination

Wind →

Windward Side C_D	Leeward Side	Duration of Load	Duration of Load Factor
DL	DL	permanent	0.90
DL + LL	DL + LL	7 days	1.25
DL[b]	DL + LL[b]	7 days	1.25
DL + SL	DL + SL	2 months	
DL + $\frac{1}{2}$ SL[c]	DL + SL[c]	2 months	1.15
DL + WL	DL + WL	10 minutes	1.6
DL + EL	DL + EL	10 minutes	1.6

[a] Symbols: DL = dead load, LL = live load, SL = snow load, WL = wind load, EL = earths load. Special configurations, locations, and occupancies of structures may require investigation of (a) full unbalanced SL or (b) combinations of WL with LL and DL, or WL with partial SL apt! It may be assumed that wind and earthquake loads will not occur simultaneously.
[b] Full unbalanced loading.
[c] Half unbalanced loading.

In order to determine the critical load combination and associated duration-of-load factor, each load combination may be divided by the applicable duration-of-load factor. The highest value determines the critical combination of loads. The following example illustrates this method.

Example 2.1 A beam is to be designed for the following loads. Determine the critical combination of loads and applicable duration-of-load factor.

$DL = 10$ psf $(0.48$ kN/m$^2) = 0.9$
$LL = 12$ psf $(0.57$ kN/m$^2) = 1.25$
$SL = 20$ psf $(0.96$ kN/m$^2) = 1.15$
$WL = 5$ psf $(0.24$ kN/m$^2) = 1.6$
$DL/0.9 = 11.1$ psf $(0.53$ kN/m$^2)$
$(DL + LL)/C_D = (10 + 12)/1.25 = 17.6$ psf $(0.84$ kN/m$^2)$
$(DL + SL)/C_D = (10 + 12)/1.15 = 26.1$ psf $(1.25$ kN/m$^2)$
$(DL + WL)/C_D = (10 + 5)/1.6 = 9.4$ psf $(0.45$ kN/m$^2)$
$(DL + WL + SL/2)/C_D = (10 + 5 + 20/2)/1.6$
  $= 15.6$ psf $(0.75$ kN/m$^2)$
$(DL + SL + WL/2)/C_D = (10 + 20 + 5/2)/1.6$
  $= 20.3$ psf $(0.97$ kN/m$^2)$

For these loads and the loading combination checked, the combination of dead load plus snow load with duration-of-load factor of 1.15 is critical.

2.5.1.11 Earthquake loads

In areas subject to earthquake shocks, every building or other structure and every portion thereof should be designed and constructed so as to resist stresses produced by lateral forces as provided by the governing building code regulations. Sources of earthquake load design information include the current Uniform Building Code.

2.5.2 Duration of load
2.5.2.1 Wood response to load with time

The time period during which a load acts on wood members is an important characteristic in determining the load a member can carry safely. This effect is termed the duration-of-load or duration-of-stress influence.

There is evidence that an intermittent load produces a cumulative effect on strength. Hence, the total accumulated duration of an intermittent full-load condition should be used to calculate the duration-of-load effect.

The following is the recommended duration-of-load factors and the resulting allowable stresses must be multiplied by them:

  0.90 for permanent load (more than 10 years)
  1.15 for 2 months' duration of load, as for snow
  1.25 for 7 days' duration of load
  1.60 for wind or earthquake
  2.00 for impact

In my design experience, there are several loads on the structure, some acting simultaneously, and each with a different duration. Each increment of time during which

the total load is constant should be treated separately, with the most severe condition governing the design.

2.5.3 Fire-resistive design

There are many approaches for providing fire safety in any structure. This involves a combination of (1) preventing fire occurrence, (2) controlling fire growth, and (3) providing protection to life and property. All need systematic attention in order to provide a high degree of fire safety economically. The building design professional can control fire growth within the structure by generating plans that include such features as protecting occupants either confined to or exiting the structure, confining fire in compartmented areas, and incorporating fire suppression and smoke-heat venting devices at critical locations.

Controlling construction features to facilitate rapid egress, protection of occupants in given areas, and the prevention of fire growth or spread are regulated by codes as a function of building occupancy. If the design professional rationally blends protection solutions for these items with the potential use of a fire-suppression system (sprinklers, for example), economical fire protection can be achieved.

Though attention could be given to all of the protection techniques available to the building design professional, the scope here will be limited to the provisions that prevent fire growth and limit the fire to compartments.

2.5.3.1 Planning

Generating the plans for a building for prescribed occupancy is a challenge due to the varying requirements of the three major regional building codes: Building Officials Conference of America (BOCA), National; International Conference of Building Officials (ICBO), Uniform; and Southern Building Code Congress International, Inc. (SBCCI), Standard. Canada is regulated by a separate code.

The buildings constructed using wood are generally code classified under types as type I, type III, type IV, type V (111), and type V (000).

Type I. This construction is generally required to be of non-combustible materials having fire endurance ratings of up to 4 h, depending on the size and location of the building. Some circumstances provide for the use of wood in the walls in non-combustible wall wood joists and non-combustible types of construction. The Uniform Building Code, for example, allows the use of fire-retardant-treated wood framing for non-bearing walls if such walls are over 5 ft (1.52 m) from the property line.

Type III. This construction has exterior walls of non-combustible material, roofs, floors, and interior walls and partitions of wood frame. As in wood frame, they are also subdivided into two classes that are either protected or unprotected.

Type IV. This construction consists of exterior walls of non-combustible materials and columns, floors, roofs,

Table 2.20 Minimum sizes of wood for type IV construction

Roof Decking	
Lumber (thickness)	2 in nominal
Plywood (thickness)	1$^{1/8}$ in net
Floor decking thickness)	3 in nominal
Roof framing	4 by 6 in nominal
Floor framing	6 by 10 in nominal
Columns	8 by 8 in nominal

Note: 1 in = 25.4 mm.

and interior partitions of wood of a minimum size. These sizes are listed in Table 2.20.

Type V. This construction is defined as having exterior walls, bearing walls, partitions, floors, and roofs of wood stud and joint framing of 2-in (38.1-mm) nominal dimension. These are subdivided into two classes, type V (111) or type V (000) construction. Protected construction calls for having load-bearing assemblies of 1 h fire endurance.

2.5.3.2 Fire-rated assemblies

It was shown in the preceding section that some occupancies require the use of fire-rated assemblies or members to prevent collapse or fire spread from one compartment of a building to another or from one building to another. Members and assemblies are rated for their ability either to continue to carry design loads during fire exposure or to prevent the passage of fire through them.

Such rating is done by either calculation or experiment for both members and assemblies. The exposure is defined as that given in ASTM E119.

A 1 h fire-resistance rating for wall, floor, and floor-ceiling assemblies incorporating nominal 2-in structural lumber can be accomplished through the use of non-combustible surfaces (such as gypsum wallboard). Fastening of these surface materials is critical for ceiling membranes, however, and is carefully specified. For some wood wave assemblies, 2-h ratings have been achieved.

Where fire endurance is expected to occur on only one side of a wall, as on the interior side of an exterior wall, the wall is assigned a rating based on the interior membrane materials. The membrane on the outside or non-fire-exposed side may consist of sheathing,

Table 2.21 Membrane on exterior face of wood stud walls

Sheathing	Paper	Exterior finish
9.5-mm T&G lumber		Lumber siding
8-mm exterior grade plywood	Sheathing paper	Wood shingles and shakes
13-mm gypsum wallboard		7-mm plywood exterior grade
9.5-mm gypsum wallboard		7-mm hardwood
		Metal siding
		Stucco on metal lath
		Masonry veneer
None		9.5-mm exterior grade plywood

sheathing paper, or a combination from Table 2.21, and is assigned an incremental time for fire endurance of 15 min.

2.5.4 Physical material properties

As is common with other materials, wood expands when heated, but not generally to the same degree as materials such as metal. The thermal coefficient of expansion for wood parallel to grain may, for example, be as little as one-three hundredths that of steel. Coefficients perpendicular to grain may be on the order of 5 to 10 times that parallel to grain.

Wood is a good insulator, that is, it has a high resistance to heat flow. In normal wood, thermal conductivity is about the same in either the radial or the tangential direction, but it is 2 to 3 times greater in the longitudinal direction. The average thermal conductivity perpendicular to grain for moisture contents up to about 40% may be expressed by the following empirical equation:

$$k = S\,(1.39 + 0.028\,M) + 0.165 \qquad (2.10)$$

where k = thermal conductivity, Btu.in/h/ft^2.°F
S = specific gravity based on volume at current moisture content and weight when oven dry.
M = moisture content expressed as percentage of oven dry weight.

40% thermal conductivity may be expressed by

$$k = S(1.39 + 0.038\,M) + 0.165 \qquad (2.11)$$

The thermal conductivity of wood is much less than that of many other structural materials notably about one-sixteenth that of sand and gravel concrete and only about one-four hundredth that of steel.

2.5.4.1 Density

The density of wood is determined principally by two factors – the amount of wood substance per unit volume and the moisture content. Other factors, such as the content of extractives and minerals, have minor effects on density. The density of wood, exclusive of water, differs greatly within as well as between species. Typically, density falls in the range of 320 to 720 kg/m^3 (20 to 45 lbm/ft^3). The range actually extends from something like 160 kg/m^3 (10 lbm/ft^3) for balsa to more than 1040 kg/m^3 (65 lbm/ft^3) for some other tropical species. Many characteristics of wood are affected by density. Since it is the wood substance in the fibres that imparts strength and stiffness, wood of high density is stronger and stiffer than wood of low density, other factors such as moisture content being equal. Woods of high density typically shrink and swell more with changes in moisture content than do woods of low density.

Since it is the wood substance that imparts strength to a piece of wood, the greater the proportion of wood substance (ie the higher the specific gravity), the higher the mechanical properties are expected to be.

The relationship between mechanical properties and specific gravity has the form

$$S = KG^n \qquad (2.12)$$

where S = the value of any particular mechanical property
G = specific gravity
K, n = constants depending on the particular property being considered.

For differences between species, the values of n range from 1.00 to 2.25, and for within-species variation, they range from 1.25 to 2.50.

2.5.4.2 Moisture in wood

Wood in the living tree is associated with moisture. The moisture content of the wood varies widely among species, among individual trees of a single species, and between sapwood and heartwood. In the green wood, the moisture content varies from as little as 30% (based on the ovendry weight of the wood) to about 200%. Most uses of wood require that the moisture content be fairly low, with the general rule being that before use, the wood should be dried to a moisture content as close as possible to that which it will be expected to reach in service. Thus flooring for use in homes in the northern part of the United States would be dried to a lower moisture content than would bevel siding for the same home.

Average shrinkage values are normally given for a large number of species. These values may be used to estimate dimensional changes associated with changes in moisture content, assuming the relationship of shrinkage to moisture content change is linear and that shrinkage begins at a moisture content of 30% except for a few species. Dimensional change estimates may be made with the following equation:

$$\Delta D = \frac{D_I(M_F - M_I)}{\frac{30(100)}{S_T} - 30 + M_I} \qquad (2.13)$$

where ΔD = change in dimensions
D_I = dimension at start of change
M_F = moisture content at end of change, percent
M_I = moisture content at start of change, percent
S_T = tangential shrinkage from green to ovendry, percent.

Changes in the moisture content below the fibre saturation point result in changes in the mechanical properties, with higher properties at the lower moisture contents, except that properties related to impact (toughness, work to maximum load in static bending, and height of drop in impact bending) may remain the same or may actually decrease at the lower moisture contents. Estimates of the effect of moisture differences on the properties of clear wood may be obtained by the following equation:

$$P_M = P_a \left(\frac{P_a}{P_g} \right)^{-(M-a)/(M_p-a)} \qquad (2.14)$$

where P_M = property at some moisture content M
P_a = property at some moisture content a
P_g = value of property for all moisture contents greater than moisture content M_p (slightly below the fibre saturation point), at which property changes due to drying are first observed.

2.5.4.3 Grain direction

The fibre direction may not be parallel to the axis of the structural member for a number of reasons. First, the piece may not have been cut parallel to the fibre direction when it was sawed from the log. In some cases, the fibres in a tree form a spiral rather than being parallel to the length of the trunk. Of course, there may also be local irregularities in the grain direction, such as might result if the log were crooked. Whatever the cause, however, cross grain can reduce strength, since a stress parallel to the axis of the piece will have components both parallel and perpendicular to the fibre (grain) direction. Since the stress perpendicular to grain is much smaller than that parallel to grain the perpendicular-to-grain component will limit the strength in a direction parallel to the axes. The steeper the grain slope, the lower the strength parallel to the axis will be, as indicated by the empirical Hankinson equation:

$$N = \frac{PQ}{P\sin^n \theta + Q\cos^n \theta} \qquad (2.15)$$

where N is the strength at an angle θ to the grain, and P and Q are the strengths parallel and perpendicular to the grain, respectively. The exponent n is an empirically derived constant ranging from about 1.5 to 2.5.

The strength of wood members loaded at an angle to the grain is a percentage of the strength of a straight-grained member for several properties.

2.5.5 Sawn lumber

Lumber in North America is subdivided into three types based on thickness:

1. Boards: Less than 2 in (38.1 mm) in nominal thickness
2. Dimension: From 2 to 4½ in (38.1 to 101.6 mm) in nominal thickness
3. Timber: Nominal thickness of 5 in (114.3 mm) or more.

Lumber to be used in structural applications is graded on an estimated strength and stiffness property basis. Though the softwood species most commonly comprise structurally graded lumber, a few hardwoods (aspen, yellow polar, and cottonwood) are structurally graded as well. One can expect to see more hardwoods graded for this use in the future. Dimension lumber and timbers are sorted into structural grades.

Boards, dimension lumber, and timbers are all graded using a "visual" grading technique, that is, the surface of each piece is examined visually by trained and

Table 2.22 Adjustment factors for clear wood stress

| Property content | Average or | Safety Factor Multiplier | | Seasoning factor for 12% moisture |
		5th percentile	Softwood	Hardwood
Bending Strength	5th percentile	0.476	0.435	1.35
Compression				
Parallel to grain	5th percentile	0.526	0.476	1.75
Perpendicular to grain	Average	0.667	0.667	1.50
Horizontal shear	5th percentile	0.244	0.222	1.13
Modulus of elasticity	Average	1.095	1.095	1.20

experienced technicians who assess the influence of natural wood characteristics on the end use of the piece. Dimension lumber and timbers are primarily graded based on the influence that these characteristics will have on their strength and stiffness in a given structural use. In this case, structural lumber is defined as being visually stress-graded.

2.5.6 Glued-laminated timber

Glued-laminated (glulam) timber finds its major application for flexural members, columns, arches, truss members, and decking. Because a degree of homogenisation is achieved when glue laminating structural lumber into larger sections, glulam structural members and products have higher allowable stresses and more reliable stiffness than sawn timber. In addition, they are available in curved or rectilinear forms of larger cross-sections and longer lengths than sawn timber. Glulam is produced from kiln-dried lumber and hence does not have nearly the magnitude of problems of shrinkage and secondary stress generation that often occurs in the joints from sawn timbers.

Glulam timbers utilize graded structural lumber as the laminate materials, which are bonded with wet-use adhesives. Conventional lengths of lumber are finger jointed together to create the length of lamination required for any given member.

i) Grading. In addition to the grading rules for structural lumber previously described, the rules for the visual grading of structural-laminating lumber are more restrictive with respect to skip and the amount of wane permitted.

Mechanical grading (E-rated) combined with visual grading [defined as visual stress rating (VSR)] of laminating lumber is being used increasingly to sort for higher-stiffness lumber. The 302 laminating grades are being increasingly selected using both MSR and E-rating techniques to adapt to the ever changing lumber resource.

ii) Clear wood design strength. The determination of clear wood strength is discussed in the structural lumber section. Clear wood strength for green lumber is used to generate such properties. Based on these properties, allowable clear wood design strength is determined.

Basically, the lower 5th percentile or average green wood level of the strength property desired is determined. This is multiplied by the seasoning factor to correct the stress level from a green to a dry (12% moisture content) condition and divided by an appropriate factor of safety for the strength property. Table 2.23 lists the strength properties and the corresponding multiplier factors. Other comments, adjustments, or procedures applicable for a given property follow.

Bending: The property generated requires multiplication by 0.743 to adjust the stress to that for a uniformly loaded 305-mm (12-in) simple beam with a 21:1 span-to-depth ratio. Clear dry wood design stress values are given in Table 2.23 for three species groups.

Tension: The tensile clear wood design stress is defined as five-eighths of the clear wood design strength in bending of 0.305-m (12-in)-deep members.

Modulus of elasticity: The factor in Table 2.22 adjusts the E value to that for a beam of 100:1 span-to-depth ratio and uniformly loaded. Table 2.23 may be used for E values for the species group given.

Modulus of Rigidity: The modulus of rigidity G can be considered as a constant fraction of the minimum modulus of elasticity E of all of the laminations in a glulam member.

Horizontal Shear: The procedure is as indicated for structural lumber. The 5th percentile horizontal shear clear wood stress is calculated using the procedure given in ASTM D 2555 and multiplied by 0.244. This

Table 2.23 Clear wood design stresses in bending based on large beam

Species	Growth Classification[2]	Clear wood design stress in bending[2] lb/in^2	MPa	Modulus of elasticity 10^6 lb/in^2	MPa
Douglas fir	Medium grain	3000	20.7	1.9	13,100
	Close grain	3250	22.4	2	13,800
	Dense	3500	24.2	2.1	14,500
Southern pine	Coarse grain	2000	13.8	1.5	10,300
	Medium grain	3000	20.7	1.8	12,400
	Dense	3500	24.1	2	13,800
Hem-fir	Medium grain	2560	17.7	1.7	11,700
	Dense	3000	20.7	1.8	12,400

is then adjusted to obtain the allowable horizontal shear stress.

Lamination Effects: For bending members of vertically laminated lumber comprised of two or more laminations, the allowable design bending stress f_b is given by

$$f_b = \bar{E}\left(\frac{f}{E}\right)_{min} \qquad (2.16)$$

where $\bar{E}=$ is the thickness-weighted average of the component laminations E values and $(f/E)_{min}$ is the lowest value of the ratios of the allowable design stress f to the modulus elasticity E for each grade of lumber in the beam.

Average E values applicable to horizontally bending members are 95% of all the values calculated using transformed section analysis. The average E values applicable to vertically laminated beams are 95% of the average of the laminations.

Radial Tension (Perpendicular to Grain): This is defined as one-third of the value for horizontal shear to obtain the clear wood design value. Douglas fir and larch are limited, however, to 0.103 kN/m² (0.103 MPa) for loads other than from wind or earthquake loading.

Compression (Perpendicular to Grain): The allowable stress level is that of the structural lamination, tension or compression face subjected to that load condition.

Strength Ratios: As for width graded structural lumber, knot location and size, slope of grain, and splits determine the reduction to be applied to clear wood stress levels.

(a) Knots: For laminated beams, the influence of knots located throughout the beam is accounted for by considering the location of the knots as measured from the neutral axis. Knots affect strength less if located near the neutral axis. The influence of the knots is best measured by their moment of inertia I_k. Tests of glulam beams have provided an empirical relationship between the moment of inertia I_k when knots are present within 152 mm of a cross-section and the gross cross-sectional moment of inertia I_g. Procedures to determine these are specified in ASTM D 3737.

The strength ratio in bending SR_b for a horizontally laminated beam is given by

$$SR_b = (1+3R)(1-R)^3\left(1-\frac{R}{2}\right) \qquad (2.17)$$

where $R = I_k/I_g$

The knot-influenced bending strength ratio for vertically laminated member is given by

$$SR = C_1(SR_1^{0.81})(N^\alpha)\left(1-\frac{1.645\Omega_1}{N^{1/2}}\right) \qquad (2.18)$$

where C_1 = empirical constant dependent upon SR_1 level
SR_1 = strength ratio for individual piece of lumber loaded on edge (taken from ASTM D 245)
α = 0.329 (1–1.049 SR_1)
N = number of laminations (not to exceed five); use
N = 5 for members with five or more laminations'
Ω_1 = coefficient of variation in bending strength for one lamination: 0.36 for visually graded structural lumber and 0.24 for machine stress-graded lumber.

The compression parallel-to-grain knot-strength ratio is governed by the percentage of the cross-section occupied by the largest knot for individual laminations. The knot-strength ratio is given by

$$SR_k = \left(\frac{1}{4}\right)(Y^3 - 4Y^2 - Y + 4) \qquad (2.19)$$

where Y is the knot size expected at the 95.5th percentile of the distribution of knots for a given grade of structural lumber. Its diameter is expressed as a fraction of the dressed width of lumber used for the lamination.

The strength ratio in tension parallel to the grain as governed by knots is given by

$$SR_k = 1 - Y_2 \qquad (2.20)$$

where Y_2 is the maximum edge knot size permitted in the laminating grade, expressed as a fraction of the dressed width of the wide face of the lumber used.

(b) Splits: Splits, cracks in lumber, or gaps generated by wane at lamination joints, reduce horizontal shear strength. However, shakes, splits, and wane are so restricted through grading that the strength ratio for horizontal shear in horizontally laminated beams can be assumed to be 1.0. In vertically laminated members, one in every four laminations is assumed to have a shake or split that limits the single lamination to a strength ratio in shear of 0.5. The weighted strength ratio for the four laminates combined is 0.875. For two or three laminations, the weighted shear-strength ratios are 0.75 and 0.83, respectively. When species having different shear properties are combined into vertically laminated members, a weighted average of clear wood stress is used to establish a single clear wood shear stress level.

Chapter 3

Timber Design Calculations: A Comparative Study of Codes

3.1 Introduction

The increasing recognition of timber as a structural material demands that timber detailers should look into the basic concepts of a typical design code in timber. It is important not to neglect to a large extent the importance of basic design calculations which can empower the timber detailers prior to embarking on the components and full prototype timber structures.

Three major codes are considered, namely BS5268-2000, EC5 and LRFD. Design examples of timber elements for these codes are given with full explanatory notes.

3.2 Design examples based on BS5268

Example 3.2.1

A beam of 10 m span is loaded in such a way which resists a total actual (F_c) load of 21.30 kN. The EI capacity of this simply supported timber beam is 10,100 kN m^2. Determine the bending deflection of this beam at mid-span.

Solution 3.2.1

$\delta m = \Delta$ = mid-span deflection due to bending

$$= \frac{5F_c L^3}{384 EI} = \frac{5 \times 21.3 \times 10^3}{384(10100)}$$

$$= 0.0275 \text{m} = 27.5 \text{ mm}$$

Example 3.2.2

Determine the permissible bending stress to grain $\sigma_{m, adm, II}$ for a timber beam 50 mm × 200 mm deep Canadian Douglas-fir larch grade 55. Use the following data:

Strength classification   C24
$\sigma_{m, g, II}$ = Grade stress parallel to grain   7.5 N/mm^2
   Class 2 service load, $K_2 = 1$
   short duration $K_3 = 1.5$
   The section is the true rectangle $K_6 = 1$

Depth factor $K_7 = \left(\frac{300}{200}\right)^{0.11} = 1.045$

No load sharing is considered $K_8 = 1$
   Knowing the admissible stress, calculate the allowable bending moment.

Solution 3.2.2

$\sigma_{m, adm, II}$ = permissible bending stress
      $= \sigma_{m, g, II} (K_2 \times K_3 \times K_6 \times K_7 \times K_8)$
      $= 7.5 \times 1.0 \times 1.5 \times 1.0 \times 1.045 \times 1.0$
      $= 11.75$ N/mm^2

M = Bending moment allowed
      $= Z_{xx \, prov} (\sigma_{m, adm, II})$
      $= \frac{bh^2}{6} (\sigma_{m, adm, II})$
      $= \frac{50 \times 200^2}{6} \times 11.75 = 3916666.7$ Nmm

      $= 391.67$ kN m

Example 3.2.3

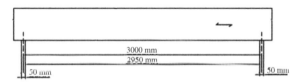

Figure 3.1 *Solid timber details.*

Figure 3.1 shows a solid timber beam with an effective span of 3.0 m, which supports a floor and exerts a long duration loading of 4 kN/m, including its own weight. This beam rests on 50 mm wide walls. The solid timber beam has a section 75 mm × 250 mm and is made of deep sawn section white wood having grade 55 under service class 1. Show that its design is safe using BS5628. Use the following data:

Solution 3.2.3

Grade stress
BS5268: Part 2, Tables 2 and 7
   Whitewood grade SS   Strength class = C24
Bending parallel to grain $\sigma_{m, g, II} = 7.5$ N/mm^2
Compression perpendicular
   to grain $\sigma_{c.g.pp} = 2.4$ N/mm^2
Shear parallel to grain $\tau_{g.II} = 0.71$ N/mm^2
Minimum modulus of elasticity $E_{min} = 7200$ N/mm^2
Total load $W = 4 \times 30 = 12.0$ kN
A = cross-sectional area = $b \times h = 75 \times 250 = 18750$ mm^2
$$I_{xx} = \tfrac{1}{12} bh^3 = 9.766 \times 10^7 \text{ mm}^4$$

K-factors

Service class 1 (K_2, Table 13) $K_2 = 1$
Load duration (K_3, Table 14) $K_3 = 1$ for long term
Bearing: 50 mm, but
    located <75 mm $K_4 = 1$
    From end of member (K_4, Table 15)
Notched end effect (K_5, $K_5 = 1$ for no notch
    Clause 2.10.4)
Form factor (K_6, Clause 2.10.5) $K_6 = 1$

Depth factor (K_7, Clause 2.10.6) $K_7 = \left(\dfrac{300\,\text{mm}}{h}\right)^{0.11}$

                                              $K_7 = 1.03$
No load sharing (K_8, Clause 2.9) $K_8 = 1$

$$M = \left(\frac{WL_e}{8}\right) = 4.5 \text{ kN m}$$

$$Z_{\text{provided}} = \frac{bh^2}{6} = 781250$$

$$\sigma_{m.a.II} = \frac{M}{Z_{\text{provided}}} = 5.76 \text{ N/mm}^2$$

Permissible $\sigma_{m,\,\text{adm, II}} = \sigma_{m.\,g.\,II} \times K_2 \times K_3 \times K_6$
  bending stress $\times K_7 \times K_8$

                 $\sigma_{m,\,\text{adm, II}} = 7.74 \text{ N/mm}^2$
                 Bending stress satisfactory.

Lateral stability

BS 5268: Part 2, Clause 2.10.8 and Table 16

Maximum depth to
breath ratio, h/b $\dfrac{h}{b} = 3.33$

                 Ends should be held in position.

Shear stress

Applied shear force $F_v = \dfrac{W}{2} = 6.0 \text{ kN}$

Applied shear stress $\tau_a = \dfrac{3}{2}\left(\dfrac{F_v}{bh}\right) = \dfrac{3}{2}\left(\dfrac{6 \times 10^3}{75 \times 250}\right)$

                 $\tau_a = 0.48 \text{ N/mm}^2$
Permissible shear stress, no notch
             $\tau_{admII} = \tau_{g.II} K_2 \times K_3 \times K_5 \times K_8$

             $\tau_{adm.II} = 0.71 \text{ N/mm}^2$
             Shear stress satisfactory.

Bearing stress

Applied load $F_V = \dfrac{W}{2} = 6.0 \text{ kN}$

Applied bearing stress

$$\sigma_{c.a.pp} = \left(\frac{F_v}{b \times b_v}\right) = \frac{6 \times 10^3}{75 \times 50}$$

$$\sigma_{c.a.pp} = 1.60 \text{ N/mm}^2$$

Permissible shearing stress

$$\sigma_{c.adm.pp} = \sigma_{c.g.pp} \times K_2 \times K_3$$
$$\times K_4 \times K_8$$

$$\sigma_{c.adm.pp} = 2.4 \text{ N/mm}^2$$
             Bearing stress satisfactory.

Deflection

No load sharing $E = E_{\text{min}}$

Deflection due to bending

$$\Delta_m = \frac{5 \times W \times L_e^3}{384 \times E \times I_{XX}}$$

$$\Delta_m = \frac{5}{384}\frac{12(3000)^3}{7200 \times 9.766 \times 10^7}$$

$$\Delta_m = 0.006 \text{ mm}$$

Deflection due to shear

$$\Delta_s = \frac{19.2 \times (M)}{(b \times h)(E)} = \frac{19.2(4.5 \times 10^6)}{75 \times 250 \times 7200}$$

$$\Delta_s = 0.64 \text{ mm}$$

Total deflection

$$\Delta_{\text{total}} = \Delta_m + \Delta_s$$
$$\Delta_{\text{total}} = 0.646 \text{ mm}$$
$$\Delta_{\text{adm}} = 0.003L_e$$
$$\Delta_{\text{adm}} = 9 \text{ mm}$$
            Deflection satisfactory.

Therefore a 75 mm × 250 mm sawn section whitewood C24 is satisfactory.

Example 3.2.4

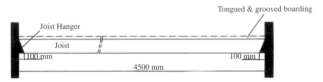

Figure 3.2 *Tongued and grooved boarding with wooden joists on hangers.*

The ground floor is built with a flooring system of tongued and grooved (t & g) boarding and joints (Fig. 3.2). Determine a suitable thickness of the floor boarding and the size of the joists, using the following data:

1. All joints are simply supported at 600 mm centres of timber strength class C22 under service load Class 2. The supports for the joints are 100 mm hangers attached to the concrete wall 4.5 m apart. The effective span from the centroid of the joist-hanger is 4.3 m as such.
2. The floor boarding shall be of timber in strength class C18.
3. Deadweight of t & g boarding = 0.1 kN/m². Imposed loading = 2.5 kN/m².
4. K-factors and grade stresses.

K-factors

Service class 1 (K_2, Table 13) $K_2 = 1$
Load duration (K_3, Table 14) $K_3 = 1$ for long term
Bearing: 50 mm $K_4 = 1$ assumed
  (K_4, Clause 2.10.5)
Notched end effect $K_5 = 1$ for no notch
  (K_5, Clause 2.10.4)
Form factor (K_6, Clause 2.10.5) $K_6 = 1$
Depth factor (K_7, Clause 2.10.6) $K_7 = 1.17$
  for $h \leq 72$ mm $K_7 = 1.03$

Load sharing applies
$(K_8,$ Clause 2.9) $K_8 = 1.1$

Grade Stresses (Floor boarding)
BS 5268: Part 2, Tables 2 and 7
Strength class = C18
Banding parallel to grain $\sigma_{m.g.II} = 5.8$ N/mm^2
Compression parallel to grain $\sigma_{c.g.pp} = 2.2$ N/mm^2,
                                   no vane
Shear parallel to grain $\tau_{g.II} = 0.67$ N/mm^2
Mean modulus of elasticity,
   load sharing $E_{mean} = 9100$ N/mm^2

Solution 3.2.4

Design of floor joists

Grade stresses (Timber joists)
BS 5268: Part 2, Table 7
Strength class – C22
Bending parallel to grain $\sigma_{m.g.II} = 6.8$ N/mm^2
Compression parallel to grain $\sigma_{c.g.pp} = 2.3$ N/mm^2,
                                   no vane
Shear parallel to grain $\tau_{g.II} = 0.71$ N/mm^2
Mean modulus of elasticity,
   load sharing $E_{mean} = 9700$ N/mm^2

Applied bending moment
Load $= (0.1$ kN/m$^2 + 2.5$ kN/m$^2)$ $L_e =$ Joist Spacing
$\qquad \times 0.1 \times (L_e - 0.6)$
$\qquad = 2.6 \times 0.1 \times 0.6$
$\qquad = 0.156$ kN $= W$

$M = \frac{W \times L_e}{8} = \frac{0.156 \times 0.6}{8} = 0.0117$ kN m

$\sigma_{m, adm, II} = \sigma_{m.g.II} \times K_2 \times K_3 \times K_6 \times K_7 \times K_8$
$\sigma_{m, adm, II} = 7.46$ N/mm^2
$Z_{required} = \frac{M}{\sigma_{m.adm.II}}$
$Z_{required} = 1586.36$ mm^3

Required T29 thickness "t"
$t = \sqrt{\frac{6 \times Z_{required}}{b}}, t = 9.70$ mm

Deflection
Load sharing system $E = E_{min}$
                                   $\Delta_{adm} = 0.003 L_e$
Permissible deflection $\Delta_{adm} = 1.8$ mm
                                   $\Delta = \frac{5 \times W \times L_e^3}{384 \times E \times I_{xx}}$
Using $I_{xx} = 2.17 \times 10^4$ mm^4
Therefore from $I_{xx} = \frac{bt^3}{12}$ $t = \sqrt[3]{\frac{12 \times I_{xx}}{b}}$
                                   $t = \sqrt[3]{\frac{12 \times 2.17 \times 10^4}{100}}$
                                   $t = 13.76$ mm

Therefore $t \geq$ (the greater of 13.76 mm and 9.80 mm and allowing for wear), thus $t = 16$ mm.

Adopt 16 mm t & g boarding using class C18 timber.

Floor joist
$\rho = 410$ kg/m^3

$tg = 0.06$ kN/m^2

Self weight $= 0.10$ kN/m^2

$W =$ Total Load (kN)
$\quad = (0.06 + 0.10 + 2.5) \times 0.6 \times 4.3 = 6.86$ kN

$M = \frac{W \times L_e}{8} = \frac{6.86 \times 4.3}{8} = 3.687$ kNm

$\sigma_{m.adm.II} = \sigma_{m.g.II} K_2 \times K_3 \times K_6 \times K_8 = 7.48$ N/mm^2

$Z_{required} = \frac{M}{\sigma_{m.adm.II}} = 4.71 \times 10^5$ mm^3

For lateral stability requirements $h \leq 5b$
Hence $Z_{required}$
$Z_{required} = \frac{b \times (5 \times b)^2}{6}$
$b = \left(\frac{6 \times Z_{required}}{5^2}\right)^{1/3}$
$h = 250$ mm
$b = 50$ mm

$Z_{provided} = \frac{b \times h^2}{6} = 5.2 \times 10^5$ mm$^3 > 4.71 \times 10^5$ mm^3
$Z_{provided} = 5.2 \times 10^5$ mm^3 adopted

$K_7 = \left(\frac{300 \text{ mm}}{h}\right)^{0.11} = 1.03$

Shear stress
$F_V = \frac{W}{2} = 3.43$ kN

$\tau_a =$ applied shear stress $= \frac{3}{2}\left(\frac{F_V}{bh}\right) = 0.49$ N/mm^2

$\sigma_{m.adm.II} = \sigma_{m.g.II} K_2 \times K_3 \times K_6 \times K_7 \times K_8$
$\qquad\qquad = 7.78$ N/mm^2

$\qquad\qquad\qquad\qquad$ Bending stress is satisfactory.

Bearing stresses
$F_V = \frac{W}{2} - 3.43$ kN
$\sigma_{c, adm, pp} = \sigma_{c.g.pp.II} \times K_2 \times K_3 \times K_4 \times K_8 = 3.26$ N/mm^2
$bw_{required} = \frac{F_V}{b\sigma_{adm.pp}} = 28.78$ mm < 100 mm   O.K.

Deflection
$I_{xx} = \frac{1}{12}bh^3 = \frac{1}{12}(50)(250)^3 = 6.51 \times 10^7$ mm^4
$E = E_{mean}$
$\Delta_m = \frac{5}{384} \frac{W}{E} \frac{L_e^3}{I_{xx}} = 12.40$ mm
$\Delta_s =$ Reflection due to shear
$\qquad = \frac{19.2m}{(bh)E} = 0.470$ mm
$\Delta_{Total} = 12.30 + 0.47 = 12.77$ mm
$\Delta_{adm} = 0.003 \times 4.3 \times 100 = 12.9$ mm   O.K.

Section 50 mm $\times$ 250 mm sawn section in C22 timber is adopted.

Example 3.2.5

A truss-roof is to be designed for loads inclusive of wind and snow while using BS6399 Part 3 AND BS 5268: Use the following data:

Span = 5.2 m; Spacing = 0.06 m
Top chord pitch = 11.685°;   Transport height
                                = 1.295 m
                               Transport length
                                = 5.650 m
Overall height = 1.2 m
Effective height of roof = 7.0 m
Top chord left
Overhang = 0.45 m
Wind: h/w = 1.0
       L/w = 1.0
       Internal maximum coefficient = 0.20
       Internal minimum coefficient = –0.30
       Δ_s, altitude above sea = 100 m
       Wind speed = 21 m/s
       Dynamic pressure q_s = 0.754 kN/m²
       Basic snowload = 0.6 kN/m²
       Tank load: 230 litre water tank over three
                  trusses

Design the roof truss for various combination of loads using space frame program or equivalent.

Solution 3.2.5 Wind criteria

Data based on BS 6399
The basic wind speed V_b is given.

The site velocity V_s for any particular direction is given by

$$V_s = V_b \times S_a \times S_d \times S_s \times S_p$$

where

V_b is the direction basic wind speed
S_a is an altitude factor   = 1.10
S_d is a direction factor   = 1.0
S_s is a seasonal factor   = 1.0
S_p is a probability factor   = 1.0
Site factor   = 0.74
S_b = terrain and building factor   = 1.518

Factor S_a
The value of S_a is used to adjust the value of V_b for altitude of the site above sea level.

When topography is not considered significant S_a should be calculated from

$$S_a = 1 + 0.001\Delta_t + 1.2\,\psi_e\,s$$

where

Δ_t is the altitude of the upwind base of significant topography (in metres above mean sea level);

ψ_e is the effective slope of the topographic feature;

s is a topographic location factor.

Calculate member loads, moments, deflections and reactions throughout
A program developed by Mitek Industries Ltd for the proposed Community Hall has been used and Fig. 3.3 covers the complete design.

Geometrical properties
Section assuming 40 mm thick lamination
$h = 40 \times 14 = 560$ mm
$b = 120$ mm width assumed
$b_w = 100$ mm
$L_e = 10$ m
Area $A = bh = 120 \times 560 = 6.44 \times 10^4$ mm²
$I_{XX} = \frac{1}{12}bh^2 = 1.756 \times 10^9$ mm⁴

Lateral stability
$\frac{b}{A} = 4.67$ ends should be held in position and directly connected of t & g decking to the beams.

Shear stress
$$F_V = \frac{W_{med}}{2} = 22.5\,kN$$
$$\tau = \frac{3}{2}\frac{F_v}{b \times h} = 0.51\,N/mm^2 = \tau_a \times //$$
$$\tau_{adm.II} = \tau_{a//} \times K_2 \times K_3 \times K_5 \times K_8 \times K_{19}$$
$$= 2.08\,N/mm^2 > 0.51$$
Therefore O.K.

Bearing stress
$$F_V = \frac{W_{med}}{2} = 22.5\,kN$$
$$\sigma_{c.app} = \text{applied shear stress} = \frac{F}{b \times b_w} = 1.875\,N/mm^2$$
$$\sigma_{c,adm.pp} = \sigma_{c.g.pp} \times K_2 \times K_3 \times K_4 \times K_8 \times K_{18}$$
$$= 3.68\,N/mm^2 > 1.875\,N/mm^2$$
Therefore O.K.

Deflection
$$E = E_{mean} \times K_{20} = 1.16 \times 10^4\,N/mm^2$$
$$\Delta_m = \frac{5W_{med} \times L_e^3}{384 \times E \times I_{xy}} = 34.00\,mm \quad \text{(Due to bending)}$$
$$\Delta_s = 1.48\,mm = \frac{19.2 \times M_{med}}{b \times h \times e}$$
$$\Delta_{Total} = \Delta_m + \Delta_S = 34.00 + 1.48 = 35.48\,mm$$
$$\Delta_{adm} = 0.003L_e = 30\,mm$$

Camber for each beam is required.
Camber required $= \Delta_{long\,term}$ loading
$$= \frac{5W_{long}L_e^3}{389EI_{xx}} = 13.54\,mm$$
Provide 15 mm camber
Δ_{live} = deflection due to live load = $\Delta_{Total} - \Delta_{long\,term}$
$$= 35.48 - 13.54 = 21.94\,mm > 15$$
Therefore O.K.

ADOPT SECTION: 120 mm × 560 mm C24; 14 laminations of single grade glulam beams are satisfactory.

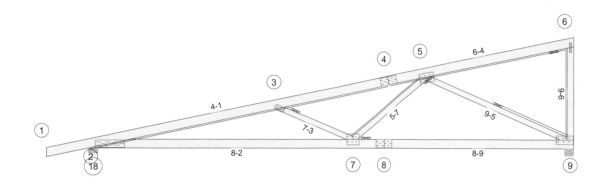

Truss Input Parameters:

Span	5200	Spacing	600
Top Left Pitch	11.9	Top Right Pitch	11.9
Transport Height	1295	Transport Length	5650
Left Cantilever	0	Right Cantilever	0
Left Stub	0	Right Stub	0
Left Overhang	450	Right Overhang	0
Left Heel Height	99	Right Heel Height	1200

General Specification:

British Standard BS 5268 Structural Use of Timber
- Part 2: Code of Practice for Permissible Stress Design, Materials and Workmanship 2002.
- Part 3: Code of Practice for Trussed Rafter Roofs

British Standard BS 6399: Loading of Buildings:
- Part 1: Code of Practice for Dead and Imposed Loads 1996.
- Part 2: Code of Practice for Wind Loads 1997.
- Part 3: Code of Practice for Imposed Roof Loads 1988.

Timber Properties

Grade	Bending N/mm²	Tension N/mm²	Tens Perp. N/mm²	Comp N/mm²	C. Perp N/mm²	Shear N/mm²	Emin kN/mm²	E kN/mm²
C16	5.3	3.2	0.20	6.8	2.2	0.6	5.8	8.8
C24	7.5	4.5	0.23	7.9	2.4	0.7	7.2	10.8
TR26	10.0	6.0	0.36	8.2	2.5	1.1	7.4	11.0

Materials:

Timber:

Top Chord	35x97	R/W TR26
Bottom Chord	35x97	R/W TR26
EV	35x72	R/W TR26
Web	35x60	R/W TR26

Plates:

Connector Plate Technical Specification Sheets are available for download at www.mii.com/europe/

Designed by:

MiTek Industries Limited
MiTek House,, Grazebrook Industrial Park,, Peartree Lane,, Dudley,, West Midlands, DY2 0XW
Phone: 01384 451400   Fax: 01384 451411
Date: 23 Jul. 2008

Standard Loading:

Top Chord Dead	1.175	Slope	kN/m²
Man Load Top	0.900	Vertical	kN
Site Snow (So)	0.600	Plan	kN/m²
Bottom Chord Dead	0.250	Slope	kN/m²
Bottom Chord Imposed	0.250	Slope	kN/m²
Man Load Bottom	0.900	Vertical	kN

Standard Snow Loads:

Snow loads are calculated from the following data using the general method.

Basic snow load (Sb) = 0.600kN/m².

Shape coefficients have been taken from figure 3 of BS6399: Part 3. *

A minimum snow load of 0.600 kN/m² (modified for pitch) has been used.

* Note: Other figures of the Code may also have been used.

Figure 3.3 *Timber truss design using computer aided method (Courtesy MiTek Industries Ltd., West Midlands, UK).*   *(continued)*

Figure 3.3 (Continued)

Standard Wind Loads: Wind loads are calculated from the following data:-

Effective Height of Roof (He)	7.000 m
Terrain and Building Factor (Sb)	1.518
Internal Maximum Coefficient	0.200
Internal Minimum Coefficient	-0.300
Altitude Above Sea Level	100.0 m
Altitude Factor (Sa)	1.100
Seasonal Factor (Ss)	1.000
Probability Factor (Sp)	1.000
Directional Factor (Sd)	1.000
Wind Speed	21.0 m/s
Dynamic Pressure (qs)	0.754

Standard Snow Loads:

Snow loads are calculated from the following data using the general method.

Basic snow load (Sb) = 0.600kN/m².
Shape coefficients have been taken from figure 3 of BS6399: Part 3. *
A minimum snow load of 0.600 kN/m² (modified for pitch) has been used.

* Note: Other figures of the Code may also have been used.

Tank Load:
230 litre water tank over 3 trusses

Load Cases Considered:

1	: Long Term	Dead+L.Term
2	: Medium Term (Snow Uniform)	Dead+L.Term+Snow
3	: Short Term (Man Bottom)	Dead+L.Term+75%ManPL+Snow
4	: Wind Along + Max Internal Pressure	Dead+ExtWind+IntWind
5	: Wind Left + Max Internal Pressure	Dead+ExtWind+IntWind
6	: Wind Right + Max Internal Pressure	Dead+ExtWind+IntWind
7	: Wind Along + Snow + Internal Suction	Dead+L.Term+50%Snow+ExtWind+IntWind
8	: Wind Left + Snow + Internal Suction	Dead+L.Term+50%Snow+ExtWind+IntWind
9	: Wind Right + Snow + Internal Suction	Dead+L.Term+50%Snow+ExtWind+IntWind
10	: Medium Term Snow Left	Dead+L.Term+Snow
11	: Medium Term Snow Right	Dead+L.Term
12	: Short Term Man Top	Dead+L.Term+75%ManPL

Additional Concentrated Loads :

Type	X	Chord	Load (kN)	Direction	From
Dead	2829	Bottom	0.450	Down	
Dead	5164	Bottom	0.450	Down	

Design Results (Member):

Member	Type	Timber	CSI	LC	SSILC		Length & Angle	Buckling length IP	OP	Maximum F(kN)	M(kNm)	V(kN)	Start F (kN)	M (kNm)	End F (kN)	M (kNm)
1-2	TC	35x97 TR26	0.104 IP	2			455 12	909	360	0.09	-0.087	0.470	0.01	-0.001	0.10	-0.107
2-3	TC	35x97 TR26	0.886 IP	2	0.315	2	1733 12	1676	360	-11.40	0.319	1.078	-11.51	0.179	-11.17	-0.243
3-5	TC	35x97 TR26	0.681OP	2	0.306	2	1616 12	735	360	-8.41	-0.311	1.047	-8.73	-0.124	-8.38	-0.466
5-6	TC	35x97 TR26	0.420 IP	12	0.252	2	1589 12	1551	360	-0.08	0.427	0.864	-0.20	-0.105	0.15	0.119
9-6	TC	35x72 TR26	0.049 IP	12			996 90	996		-1.25			-1.25		-1.25	
2-7	BC	35x97 TR26	0.825 IP	3	0.271	3	2779 0	1945		12.53	0.480	1.114	12.53	-0.146	12.53	-0.412
7-9	BC	35x97 TR26	0.645 IP	3	0.241	3	2335 0	1796		6.37	0.455	1.050	6.37	0.180	6.37	-0.207
3-7	WB	35x60 TR26	0.230OP	2			858 -23	772		-2.98			-2.98		-2.98	
7-5	WB	35x60 TR26	0.189 IP	1			1036 40	932		3.07			3.07		3.07	
5-9	WB	35x60 TR26	0.540 IP	2			1650 -24	1485	*1	-6.18			-6.18		-6.18	

LC=Load case, SSI=Shear Stress Index, CSI=Combined Stress Index, IP=In Plane, OP=Out of Plane, F=Force, M=Moment, V=Shear, *=No. of Braces

Figure 3.3 (Continued)

Design Results (Joint):

Joint Member	Timber	LC	Coordinates X	Y	Force kN	Moment kNm	C.S.I.
2	35x97 TR26	2	-5	49	-1.995	-0.107	0.128
3	35x97 TR26	2	2029	479	-11.171	-0.243	0.624
5	35x97 TR26	2	3609	814	-8.377	-0.466	0.804
6	35x97 TR26	2	5164	1093	0.165	-0.001	0.006
7	35x97 TR26	3	2829	49	12.534	-0.412	0.753
9	35x97 TR26	2	5164	49	5.654	-0.184	0.410

Maximum Summations:

		Member			Joint
Top Panel C.S.I.	0.886	19-3	Top Joint C.S.I.	0.804	5
Bottom Panel C.S.I.	0.825	18-7	Bottom Joint C.S.I.	0.753	7

Summary Of Maximum Deflections:

Type	Member/ Joint	Load Case	Deflection Values (mm) Actual	Permissable
Local Deflection of rafters in LC1	2-3	1	2.72	6.93
Local Deflection of rafters overhangs in LC1	1	1.0	3.62	3.60
Local Deflection of ceiling ties in LC1	2-7	1	3.68	8.35
Local Deflection of ceiling ties in LC3	2-7	3	11.37	16.74
Overall Ceiling tie deflection in LC1, 2, 10 or 11	2-9	2	7.94	15.36
Vertical unsupported node deflection in LC1	3	1	6.08	12.00
Horizontal unsupported node deflection in (LC4-LC9)	7	8	0.72	18.00

M

Maximum Reactions (kN) and Minimum Bearings (mm):

Joint	LC	Vertical	LC	Horizontal	LC	Uplift	Restrict	Min. Brg	Input Brg
2	3	-4.967	6	0.570				36	100
9	3	-4.897		0.000				35	100

Reactions / Load Case:

Joint	Load Case	Vertical	Horizontal
9	1	-3.323	
	2	-4.248	
	3	-4.897	
	4	-0.683	
	5	-1.565	
	6	-0.562	
	7	-2.842	
	8	-3.724	
	9	-2.721	
	10	-3.992	
	11	-3.323	
	12	-3.998	
18	1	-3.210	
	2	-4.319	
	3	-4.967	
	4	-0.904	0.516
	5	-1.876	0.123
	6	-0.770	0.570
	7	-2.619	0.267
	8	-3.592	-0.126
	9	-2.486	0.321
	10	-4.011	
	11	-3.210	
	12	-3.847	

Figure 3.3 (Continued)

Plating Calculations:

Joint	Joint Type	Plate Type	Plate Size	Location X	Y	Member Ref.	Area (mm²) Req.	Act.	Angles LTN	LTG	Steel (mm) Req.	Act.
2	TMB1	M20	830	74	10	1-4	5463	8144	17.18	5.23	160	312
						2-8	5744	7853	19.52	19.52		
3	TMW	M20	510	0	0	1-4	1454	1454	0.00	34.90	30	89
						3-7	1306	1453	0.00	0.00		
4	CS	B90	518	90	1	1-4	6276	6320	3.88	3.88	67	90
						4-6	6276	6320	3.88	3.88	67	90
5	CMWW	M20	1015	66	43	4-6	4114	4910	11.19	11.19	121	152
						5-7	1934	2007	28.30	0.00		
						5-9	2989	3023	35.88	0.00		
6	CMV	M20	310	0	0	4-6	970	981	0.00	78.05	17	26
						6-9	737	981	0.00	0.00		
7	CMWW	M20	1113	0	0	2-8	2752	5912	16.47	16.47	77	127
						3-7	1384	1770	22.95	0.00		
						5-7	1977	2206	40.26	0.00		
8	CS	M20	918	89	6	2-8	4327	6232	0.88	0.88	84	90
						8-9	4327	6232	0.88	0.88	84	90
9	CMVW1	M20	918	66	45	5-9	2912	3020	23.92	0.00		
						6-9	876	1862	90.00	0.00		
						8-9	3395	5849	30.14	30.14	80	178

LTN=Load to Nail, LTG=Load to Grain, Req=Required,

Design Notes:

- This design conforms to the input data entered. Please ensure that the dimensions, pitch, loadings and other design considerations have been correctly entered.
- Stiffness matrix analysis has been used.
- Load sharing factor 1.1 used.
- Truss weight = 21 kg.
- Plate location tolerance of 5mm was used.
- Chords must be restrained by battens or bracing at centres not exceeding the out-of-plane buckling length as found in the Design Results Member table.

3.3 Design examples based on Eurocode EC5

The design of a joist is required for a flat roof. Use the following data:

(a) C24 joist: 75 mm × 200 mm; L = span between centres – 4500 mm
(b) Service class 1 with moisture content 12%; ρ_{mean} = 420 kg/m³; k_{mod} = 0.6

Material properties and weights

$f_{m,k}$ = bending strength = 24 N/mm²
$f_{c,90,k}$ = compression strength perpendicular to grain = 5.3 N/mm²
$f_{v,k}$ = shear strength = 2.5 N/mm²; γ_G = 1.35; γ_{Q2} = 1.15
$E_{0,0.5}$ = mean Young's modulus = 11000 N/mm²
$G_{0,0.5}$ = mean shear modulus = 690 N/mm²
l = bearing length = 50 mm; b = 75 mm; d = 200 mm

beam weight = $\dfrac{9.807 \times 420\ (75 \times 200)}{10^9}$
= 0.06174 kN/m

dead weight for decking, insulation, ceilings = 0.6 kN/m² = 0.6 × 0.6 kN/m = 0.36 kN/m
snow load = 0.75 kN/m²

Actions

G_k = 0.4218 kN/m
$Q_{k,1}$ = 0.75 × 0.6 = 0.45 kN/m
$Q_{k,2}$ = 0.9 kN concentrated load
$G_d = \gamma_G\, G_k$ = 0.05694 kN/m (N/mm)
$Q_{d,1} = \gamma_Q\, Q_{k,1}$ = 0.675 kN/m (N/mm)
$Q_{d,2} = \gamma_Q\, Q_{k,2}$ = 1350 N

Ultimate strength

$$\frac{5 F_{udl} L^3}{384 E_{0,0.5} I_y} + \frac{\phi F_{udl} L^2}{8 G_{0,0.5}(A)} = 7.905 F_{udl}\ (\text{mm})$$

$I_y = 97 \times 10^6$ mm⁴

Based on clause 5.4.6 (2) EC5, K_{ls} = 1.1
For 200 mm deep beam, size factor = K_h = 1.0

Lateral stability

L_{ef} = 1.2L = 5400 mm
$\sigma_{m,crit} = \dfrac{0.75 E_{0.05}}{L_{ef}(h)}\,(b^2)$ = 28.90 N/mm²

$\lambda_{rel,m} = \sqrt{\dfrac{f_{m,k}}{\sigma_{crit}}}$ = 0.911

K_{crit} = instability factor = 1.56 − 0.75λ_{xx} = 0.8768
$L_1 = L − l$ = 4450; a = overhang = 10 mm
k_c = bearing factor = $1 + \dfrac{a(150 − l)}{17000}$ = 1.059

M_d = design bending moment = $\dfrac{G_d L^2}{8}$

$\qquad\qquad\qquad\qquad\qquad\quad = 1441.3 \times 10^3$ Nmm

V_d = design shear load = $\dfrac{G_d L}{2} = 1281$ N

$\sigma_{m,d,y}$ = bending strength = $\dfrac{M_d}{W_y} = 0.845$ N/mm^2

Design bending strength

$f_{m,d,y} = \dfrac{k_{ls}k_h k_{crit} k_{mod} k_{m\,k}}{\gamma_M} = 10.68$ N/mm^2

$\sigma_{m,d,y} < f_{m,d,y}$

$\quad || \qquad\quad ||$

$1.845 < 10.68$ bending strength OK

Design shear strength

Design shear stress $\tau_d = \dfrac{1.5 V_d}{A} = 0.128$ N/mm^2

Design shear strength $f_{v,d} = \dfrac{k_{ls}k_{mod}k_{v,k}}{\gamma_M} - 1.27$ N/mm^2

$\tau_d \quad < \quad f_{v,d}$

$|| \qquad\quad ||$

$0.128 < 1.27$ shear strength OK

Permanent and short term loads

Design values of effects of actions

Bending moment due to u.d.l. $Q_{d1} = (Q_{d1}l_2)/8$

$\qquad = 1.7086 \times 10^6$ Nmm

Bending moment due to point load $Q_{dl} = (Q_{d2}L)/4$

$\qquad = 1.52 \times 10^6$ Nmm

M_d produced due to permanent load = $1.708 \times 10^6 +$
$1.52 \times 10^6 = 3.2286 \times 10^6$ Nmm

Design shear load

Due to u.d.l. $Q_{d,1} = (Q_{d,1}L)2 = 1519$ N

Due to u.d.l. $Q_{d,2} = 1350$ N

V_d from $G_d = Q_{d,1} + Q_{d,2} = 2869$

Design bending strength

Design bending stress = $\sigma_{m,d,y} = \dfrac{M_d}{W_Y} = 4.13$ N/mm^2

$f_{m,d,y} = \dfrac{k_{ls}k_h k_{crit} k_{m,d}}{\gamma_M} = 13.15$ N/mm^2

$\sigma_{m,d,y} < f_{m,d,y}$

$\quad || \qquad\quad ||$

$4.33 < 13.15$ bending strength OK

Shear strength

$\tau_d = \dfrac{1.5 V_d}{A} = 0.2869$ N/mm^2

Design shear strength $f_{v,d} = \dfrac{k_{ls}k_{mod}k_{v,k}}{\gamma_M} = 1.90$ N/mm^2

$\tau_d \quad < \quad f_{v,d}$

$|| \qquad\quad ||$

$0.2869 < 1.90$ N/mm^2

Bearing strength

$\sigma_{c,90,d} = \dfrac{V_d}{bl} = 0.769$ N/mm^2

$f_{c,90,d} = \dfrac{k_{ls}k_{c,90}k_{mod}f_{c,90,k}}{\gamma_M} = 4.28$ N/mm^2

$\sigma_{c,90,dy} < f_{c,90,d}$

$\quad || \qquad\qquad ||$

$0.769 < 4.28$ bearing strength OK

Serviceability limit state

Deflection

Deflection due to u.d.l. = $\dfrac{5 F_{udl} L^3}{384 E_{0,0.5} I_y} + \dfrac{\phi F_{udl} L^2}{8 G_{0,0.5}(A)}$

$\qquad\qquad = 7.905\, F_{udl}$ (mm)

Deflection due to point load = $\dfrac{F_{point} L^3}{48 E_{0,0.5} I_y} + \dfrac{12 F_{point} L}{4 G_{0,0.5}(A)}$

$\qquad\qquad = 0.001898\, F_{point}$ (mm)

Instantaneous deflection

$u_{1,inst}$ (dead load) = $7.905\, G_k = 3.334$ mm

$u_{2,1,inst}$ (imposed load) = $7.905\, Q_k = 3.557$ mm

$u_{2,2,inst}$ (point load) = $0.001898\, Q_{k,2} = 1.708$ mm

Recommended deflection limitation $u_{2,\,inst,\,max} = L/300 =$
$\quad 15$ mm   OK

Final deflection

(Solid timber table 10 in EC5)

$k_{def} = 0.8$ permanent duration – Action

$k_{def} = 0$ short duration

$E_{0,0.5} = 9000$ N/mm^2, C18 timber

Final deflection due to permanent loads

$u_{1,fin} = (1 + k_{def})\, u_{1,inst}$ (Clause 4.1 (4))

$\qquad = (1.8)(3.334) = 6$ mm

$u_{2,fin} = 6.0 + 15 = 21$ mm

Allowable deflection = $L/200 = 27.5$ mm

$U_{net,fin} = (21.0 \times 11000)/9000 = 25.67 < 27.5$ OK

C18 joist could be used.

3.3.2 Example on glued laminated column under axial load based on EC5

A column 4.0 m long is a glulam column of strength class GL24 to support an axial load of 75 kN. The column is restrained in position but not in direction at both ends. Service class 2 is adopted. Using the following data, design this column:

Compression strength parallel to grain

$\qquad f_{c,0,g,k} \qquad\qquad = 24.0$ N/mm^2

Minimum modulus of elasticity parallel to grain

$\qquad E_{0.05,g} \qquad\qquad = 8800$ N/mm^2

Actions

Total dead load (permanent duration) $G_k = 30.5$ kN

Imposed snow load (short term) $Q_k = 46$ kN

Show that this column fails under the axial load.

Partial safety factors for loads and materials

$\qquad \gamma G \quad = 1.2 \qquad\qquad$ Table 2.3.3.1

$\qquad \gamma Q \quad = 1.35$

$\qquad \gamma M \quad = 1.3 \qquad\qquad$ Table 2.3.3.2

Modification factor for short-term loads in Service class 2 for glulam

$$k_{mod} = 0.9 \qquad \text{Table 3.1.7}$$

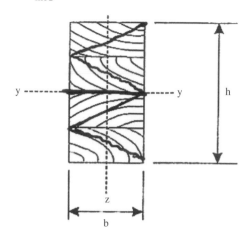

Figure 3.5 *Timber column.*

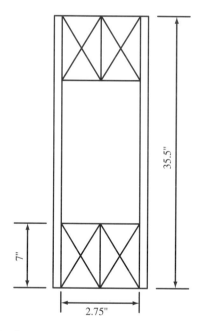

Figure 3.6 *Box beam.*

Estimation of Trial Section

N_d (permanent) $= \gamma_G G_k = 1.2 \times 30.5 \times 1000 = 36600$
N_d (short term) $= \gamma_G G_k + \gamma_Q Q_k = 1.2 \times 30.5 + 1.35 \times 46$
$\qquad\qquad\qquad = 36.6 + 62.1 = 97.7$ kN

Design compression stress

$$\sigma_{c,0,d} = \frac{N_d}{A} = \frac{98.7 \times 10^3}{A} \text{ N/mm}^2$$

Design compression strength

$$f_{c,0,g,d} = \frac{k_{mod} f_{c,0g,k}}{\gamma_M}$$

$$f_{c,0,g,d} = \frac{0.9 \times 24}{1.3} = 16.62 \text{ N/mm}^2$$

$$\sigma_{c,0,d} = 0.5 f_{c,0,g,d}$$

when $\qquad A = \dfrac{98.7 \times 10^3}{0.5 \times 16.62} = 11877 \text{ mm}^2$

Based on Table 5, standard section 65×180 mm
$$A = 11700 \text{ mm}^2$$

Ultimate Limit State

Design value of actions
Design compression force applied to grain
Permanent duration load case
$\qquad N_d$, short term $= \gamma_G G_K = 1.2 \times 30.5 + 1000 = 36480$ N
$\qquad$ Short term load case
$\qquad N_d$, short term $= \gamma_G G_K + \gamma_Q Q_K = (1.2 \times 30.5) +$
$\qquad\qquad\qquad\qquad (1.35 \times 46) \times 1000 = 98.7 \times 10^3$
L_{ef} = effective length columns $= L = 4000$ mm

$$\lambda_y = \frac{L_{ef}}{i} = \frac{L_{ef}\sqrt{12}}{h} = \frac{4000 \times \sqrt{12}}{180} = 76.98$$

Critical compression stress

$$\sigma_{c,crit,y} = \frac{\pi^2 E_{0.005,g}}{76.98^2} = \frac{\pi^2 8800}{76.98^3} = 19.05 \text{ N/mm}^2$$

Relative slenderness ratio

$$\lambda_{rel,y} = \sqrt{\frac{f_{c,0,g,k}}{\sigma_{c,crit,y}}} = \sqrt{\frac{24}{19.05}} = 1.122$$

Since $\sigma_{rel,y} > 0.5$, the combined compression and bending ratio condition must be satisfied.

$\qquad\qquad\qquad\qquad\qquad\qquad$ EC5 Clause 5.2.1(3)

It must be shown that

$$\frac{\sigma_{c,0,d}}{k_{c,y} f_{c,0,g,d}} \leq 1$$

(This is a simplified form of (5.2.1*f*) for columns which are not subject to applied bending stresses.)

Therefore the compression strength modification, k_{cy}, must be calculated.

For glulam $\qquad \beta_c = 0.1 \qquad\qquad$ Clause 5.2.1(4)

$$k_y = 0.5(1 + \beta_c(\lambda_{rel,y} - 0.5) + \lambda_{rel,y}^2$$

$$k_y = 0.5(1 + 0.1(1.122 - 0.5) + 1.122^4) = 1.323$$

$$k_{c,y} = \frac{1}{k_y + \sqrt{k_y^2 + \lambda_{rel,y}^2}} \qquad \text{(EC5 5.2.1g)}$$

$$k_{c,y} = \frac{1}{1.323 + \sqrt{1.323^2 + 1.122^2}} = 0.3270$$

Design value of effect of actions:
Design compression stress

$$\sigma_{c,0,d} = \frac{N_d}{bh} = \frac{98.7 \times 10^3}{65 \times 180} = 8.436 \text{ N/mm}^2$$

Compression stress ratio

$$\frac{\sigma_{c,0,d}}{k_{c,y} f_{c,0,g,d}} = \frac{8.436}{0.3270 \times 16.62} = 1.552 > 1.0 \quad \text{FAILS}$$

since the compression stress ratio is greater than 1.0.

3.4 Design examples based on ASD/LFRD codes

3.4.1 Design examples on built up and composite members

Design a plywood lumber box. Beams span 30 ft (9.2 m) and are at 8 ft (2.5 m) centres. Use hem-fire select structural flanges and structural interior plywood webs. The beams support a roof system with a slope. Use the following data:

Figure 3.7 shows the cross-section (note 1 psf = 47.9 pa).

Suspended ceiling	8 psf
Roofing	7 psf
Insulation	2 psf
Decking and beams	9 psf
Snow	30 psf
	56 psf

Total w = 56 (8 ft) = 448 lb/ft; maximum M = $(448)(30^2)/8$ = 50400 ft-lb. Choose a section:

Use depth approx = span/10 = 30(12)/10 = 36 in. (915 mm)

Base F_t for flanges is 900 psi. (6210 kN/m^2)

Assuming 2 × 8 flanges, the allowable bending stress is

$$F_b' = C_d C_f F_t = (1.15)(1.2)(900)$$
$$= 1240 \text{ psi } (8556 \text{ kN/m}^2)$$

For 2 × 8 and total depth on 36 in., the moment arm between the centroid of the flanges is 36–7.25 = 28.75 in. (712 mm) Equating internal and external moments,

$$1240 A_f (28.75) = 50400(12)$$
$$\text{Req } A_f = 17.0 \text{ in.}^2 \text{ } (10968 \text{ mm}^2)$$

Area provided by two 2 × 8s = 2(10.875) = 21.8 in.^{2}
(14065 mm^2)

Try two 2 × 8s for each flange.

Choose plywood based on the thickness effective for shear, t_s. The following empirical equation can be used to estimate the required thickness for shear.

$$\text{Req } t_s = 5V/(4h F_v')$$

where h is the overall depth. For species Group 1 and Grade Stress Level S-2,

$$F_v' = 1.15(190) = 218 \text{ psi}$$
$$V = 448(30)/2 = 6720 \text{ lb}$$
$$\text{Req} - t_s = 5(6720)/[4(36)(218)] = 1.07 \text{ in}$$

t_s provided by two 3/4-in. panels = 2(0.739) = 1.5.

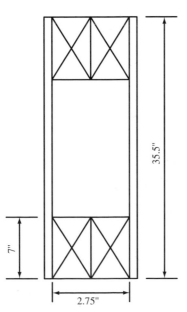

Figure 3.7 *Built up column.*

Try 3/4-in. Structural I plywood for each outside web. The flanges have been surfaced so that a pair of 2 × 8s is 7 in. deep by 2.75 in. wide. Surfacing has decreased the net overall depth to 35.5 in.

Bending check

$I_{\text{flanges}} = 2[(1/12)(2.75)(7)^3 + 2.75(7)(14.25^2)] = 7975 \text{ in.}^4$

Specifying that the plywood face grain is horizontal, the thickness of plywood effective in resisting bending is $A_{//}/12 = 4.219/12 = 0.352$ in.

$$I_{\text{one web}} = (1/12)(0.352)(35.5)^3 = 1312 \text{ in.}^4$$
$$I_{\text{total}} = 7975 + 2(1312) = 10,600 \text{ in.}^4$$
$$I_{\text{net}} = 7975 + 1312 = 9290 \text{ in.}^4$$

For the plywood, allowable bending stress, F_c' is 1.15(1540) = 1770 psi. (The tabular value F_b is not used, because it is for flatwise bending only.)

The allowable bending stress for the flange lumber is 1240 psi. The 1240-psi value controls.

$$I_{\text{req}} = M_c/F_c' = 50400(12)(17.75)/1240 = 8657 \text{ in.}^4$$
$$I_{\text{provided}} = I_{\text{net}} = 9290 \text{ in.}^4 \qquad \text{O.K.}$$

Shear check

$$Q_{\text{flanges}} = bd(h/2 - d/2) = (2.75)(7)(35.5/2 - 7/2)$$
$$= 274.3 \text{ in.}^3$$
$$Q_{\text{webs}} = 2t(h/2)(h/4) = 2(4.219/12)(17.75)(8.875)$$
$$= 110.8 \text{ in.}^3$$
$$Q_{\text{total}} = 274.3 + 110.8 = 385.1 \text{ in.}^3$$

Horizontal shear

For continuous glue along framing parallel to face grain, allowable horizontal shear stress may be increased by 19%. The allowable shear force is

$$V_h = F_v' I_t t_s / Q$$
$$V_h = (1.19 \times 1.15 \times 190)(10600) \times (2 \times 0.739)/385.1$$
$$= 10580 \text{ lb}$$

Actual V = 448(15) = 6720 lb (27.694 kN)   O.K.

Flange-web (rolling) shear

The basic rolling-shear stress is reduced by 50% for flange-web shear (Sect. 3.8.2, ref. 1). The allowable shear force is

$$V_s = 2F_s' dl_t / Q_{\text{flanges}}$$
$$V_s = 2(0.50 \times 1.15 \times 75)(7)(10600)/274.3$$
$$= 23300 \text{ lb} > 6720 \text{ lb}$$

Deflection

Span/depth ratio = 10:1, so factor = 1.5

Total-load Δ = 1.5(5)wl^4/(384 EI$_t$)
= 1.5(5)(448)(30^4)(1728)/384 (1,600,000)(10600) = 0.72 in.

Total-load Δ limit = l/180 = (30/12)/180=2.0 in.O.K.

Live-load Δ = (30/56)(0.72) = 0.39 in.<l/240
= 1.5 in. (38 mm)   O.K.

Bearing stiffeners

Stiffener width required for bearing of stiffener perpendicular to grain of flange:

End reaction = R = 448(15) = 6720 lb (27.694 kN)
$F_{c,\perp}$ (of flange lumber) = 405 psi

Req bearing area = $R/F'_{c,\perp}$ = 6720 lb/405
      = 16.6 in.^{2}

Using a pair of 2 × stiffeners (surfaced to $1^3/_8$ in. each), the required stiffener width is

 16.6/(2 × 1-375) = 6.04 in.^{2}

Double 2 × 8s are OK for bearing perpendicular to grain.

Stiffener width required for rolling shear at each of the two stiffener-web interfaces:

 Req surface area = R/F'_c = 6720(75 × 1.15 × 0.50)
      = 156 in.^{2}

Req stiffener width = 156/(2 × 35.5) = 2.20 in.

Use double 2 × 8s at each end for bearing stiffeners.

3.4.2 Design example of the glue-laminated beams using LRFD approach

The design shall be for solid sawn members applied equally to glulams with a difference between the adjustment factors C_v and C_c. Design the glulam beam using the following data:

The beam is a curved (arched) glulam beam
Simple span = 36 ft (11 m)
Load = 20 kips concentrated at the centre (1 kip = 1000 lb = 4.448 kN)
The radius of curvature of the beam
Centre-line = 44 ft (13.4 m) with moisture content ≥ 16%
The rise of the beam at mid-span = 3.85 ft (1.17 m)
The length of the arch to the centre of support = 37.09 ft (11.3 m)
Weight/ft = 70 lb (21.3) or = 70(37.09/36) = 72 lb/ft of span of beam length.

C_c, assume 0.95
C_v, assume 0.95
Bending moments are:
Due to own weight, M = 0.072 $(36^2)/8$ = 11.7 ft-k (15.6 kNm)
By concentrated load, M = 20 (36/4) = 180 ft-k (244 kNm)
Approximate total M = 191.7 ft-k (259.86 kNm)
CM for bending is 0.8 for moisture content exceeding 16%.
The adjusted allowable bending stress, using the above factors, is
F'_b = 2.2(0.95)(0.8) = 159 ksi (1096.305 MN/m^2)
Approximate S_{req} = 12 (191.7)/1.59 = 1447 in.^{3}
Try a glulam measuring 8.75 in. by 33 in.
Corrected values for this size glulam are determined next.
Use specific gravity of 0.50 and assuming a moisture content of 16%, unit wt = (62.4)(0.50)(1.16) = 36 lb/ft^3.
Corrected wt per ft of beam length = 36(8.75 × 33)/144 = 72 lb/ft, and w = 72(37.09/36) = 74 lb/ft of span due to

weight of the beam itself. Preliminary design of the top bracing system shows that it will weigh about 22 lb per ft, so the total dead load is 74 + 22 = 96 lb/ft.

Total M = 0.096(36^2)/8 + 20(36)/4 = 196 ft-k (265.737 kNm).

Inside radius of the bottom lamination = 12 (44) − 33/2 = 511.5 in.

t/R_i = 1.5/511.5 = 1/341 < 1/125   O.K.

(Note that R_i also complies with the AITC recommendation that it should exceed 27.5 ft.)

 C_c = 1−2000 $(1.5/511.5)^2$ = 0.983
 C_v = 1.09 $[1292/(36 × 8.75 × 33)]^{0.1}$ = 0.885

Adjusted allowable bending stress

F'_b = 2.2(0.983)(0.885)(0.8) = 1.531 ksi

S_{req} = 12(196)/1.531 = 1536 in.^{3} < actual S=1588 in.^{3}
              O.K.

Next, check the section for shear and for radial stress.

The end reaction is approximately 18 (0.096) + 20/2 = 11.73 k. (To be ridiculously precise, the reaction could be divided into an axial compression component. The effect would be trivial in this case.)

Shear, computed at beam depth from the support (assuming bearing length of 4 in.), is

V = 0.096(18−35/12)+10 = 11.45 kip
f_v = 1.5[11450/(8.75×33)] = 59 psi (597 kN/m^2)
< 190(0.875) = 166 psi (1679 kN/m^2) O.K.

Based on the LFRD method, the preliminary unfactored moment for dead load is M = 11.7 ft kips and for live load, it is 180 ft kips.

Load combinations to be considered for the LFRD solution are:

M_U = 1.4M_D = 1.4(11.7) = 16.4 ft-kips (22.235 kNm)
M_U = 1.2M_D + 1.6M_L = 1.2(11.7)+1.6(180)
  = 302.0 ft-kips (409.45 kNm)

The second of these combinations controls, and λ = 0.8.

Tabular reference strength, F_b, is 5.59 ksi

As for the previous example, try an 8.75 × 33-in. glulam.

F'_b = $(F_b)(C_u)(C_c)(C_v)$ = 5.59 × 0.8 × 0.983 × 0.885
  = 33-in.glulam.

M' = $C_L S_X F'_b$ = 1.0 × 1588 × 3.89/12 = 515 ft-k
  (698 kNm)

The design resistance is

λφ $_bM'$ = (0.8)(0.85)(515) = 350 ft-k > M_u = 302 ft-k
  (409.45 kNm)

required for the controlling load combination.

Continue to determine whether the section selected is satisfactory for shear.

For shear or the two load combinations to be considered, shear V_u = (due to loads) is computed at 33 inches (member depth) from the face of the support. Assuming that the required bearing length for the beam will be about 4 inches at each end, the values of factored shear are computed at 33 + 2 = 35 inches from the face of the support. Assuming that the required bearing length for the beam will be about 4 inches at each end, the values of factored shear are computed at 33 + 2 = 35 inches from the centre of the end bearing.

$V_u = 1.4 (0.096) (18–35/12) = 2.03$ kips (D only)
$V_u = 1.2 (0.096) (18–35/12) + 1.6 (10) = 17.7$ kips
  Controls

Next, compute the flexural shear resistance of the section. The adjusted flexural shear resistance, V', is
$$V' = (2/3)F'_v bd$$
In this equation, term F'_v is the adjusted horizontal shear strength (shear resistance strength multiplied by each applicable adjustment factor). Recalling that the MC in use will exceed 16%, Factor $C_M = 0.875$ will be unity.

The reference strength for shear is $F_v = 0.545$ ksi.
$$V' = (2/3)(0.545 \times 0.875)(8.75 \times 33)$$
$$= 91.8 \text{ k}$$

The time effect factor for the controlling load case is $\lambda = 0.8$, and factor Φ_v for shear is 0.75. Thus, the design shear resistance for the section selected in the previous example is
$$\lambda \phi_v V' = (0.8)(0.75)(91.8) = 55.1 \text{kips}$$

3.4.3 Design example: glulam columns based on LFRD/ASD codes

A glulam of western species is selected for use as a column. The section size is 8.75 × 10.5 in. The member is 25 ft long, with weak-way lateral support at mid-height and lateral support both ways at each end. Determine the allowable axial load on the column.

Take: $F_c = 1100$ psi (7579 kN/m²)
     $E = 1300000$ psi $= E_{yy}$

Adjustment factors are CD = 1.15 for load duration and wet service factors, CM = 0.73 for compression and 0.833 for modulus of elasticity, E. The column stability factor must be computed.

The two l_e/d values are:
For the X-axis (strong way) 12(25)/10.5 = 28.6
For the Y-axis (weak way) 12 (12.5)/8.75 = 17.1

The strong way controls. Terms F_{cE} and F_c^*, needed to compute the column stability factor, are

$F_{cE} = 0.418(1300000 \times 0.833)/28.6^2 = 533$ psi (3678 kN/m²)

$F_c^* = 1100 (1.15) (0.73) = 923$ psi (6369 kN/m²)

Substituting in the equation for column stability factor and using $c = 0.9$,

$C_p = 0.537$

The allowable axial load is

$= F_c(C_D)(C_M)(C_P)(8.75 \times 10.5)$
$= 1100(1.15)(0.73)(0.537)(8.75 \times 10.5)$
$= 45560$ lb (202.65 kN)

3.5 Design example for flitched beam based on British practice

(a) Theoretical Equations (see Fig. 3.8)
  Symbols: t = timber
        s = steel

$$M_t = \frac{E_t I_t}{E_t I_t + E_s I_s} M$$

$$\sigma_t = \frac{M_y}{1 + \dfrac{E_s I_s}{E_t I_t}}$$

$I_t = \dfrac{E_s}{E_t} I_s$ is the second moment of area of the equivalent timber beam

$\sigma_s = $ direct stress distribution in steel

$$= -\frac{M_y}{I_t + \dfrac{E_t}{E_s} I_t}$$

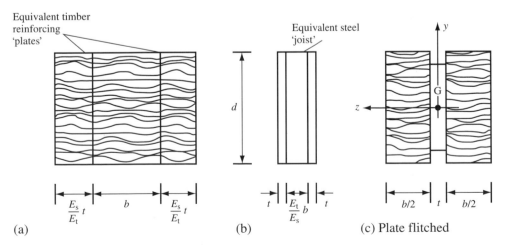

Figure 3.8 *Flitched beam.*

Equivalent timber reinforcing 'plates'

$\dfrac{E_s}{E_t} t$ b $\dfrac{E_s}{E_t} t$

(a)

Equivalent steel 'joist'

t $\dfrac{E_t}{E_s} b$ t

(b)

y

G

z

b/2   t   b/2

(c) Plate flitched

(b) Design example

FLITCH BEAM 1

Span length & partial factors for loading

Span (mm)	Factors for moments & forces			Factors for deflection		
	γ^{fd}	γ^{fl}	γ^{tw}	γ^{dd}	γ^{dl}	γ^{dw}
3500	1.00	1.00	1.00	1.00	1.00	1.00

Loading data (unfactored)

Ref.	Category	Type	Load kN/m	Position mm	Load kN/m	Position mm
1	"Dead"	UDL	1.5	0	–	3500
2	"Imposed"	UDL	0.8	0	–	3500
3	"Imposed"	point load	2.5 kN	1100	–	–

Analysis results – entire span

R_a kN (fac)	R_b kN (fac)	V kN (fac)	M		Deflection: δEI	
			kNm (fac)	Sense	kNm³	Direction
5.6	4.7	5.6	4.9	"Sagging"	6.17	"Down"
$W_{tot} =$	10.3 kN					

Unfactored support reactions

Support A   Dead load −**2.6** kN   Live load −**3.0** kN   Wind load **0.0** kN
Support B   Dead load −**2.6** kN   Live load −**2.1** kN   Wind load **0.0** kN

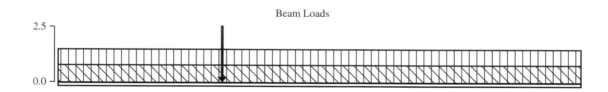

Beam Loads

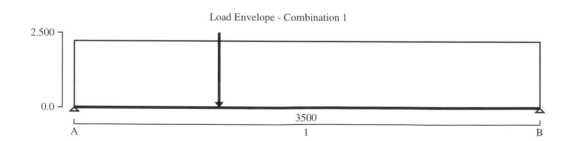

Load Envelope - Combination 1

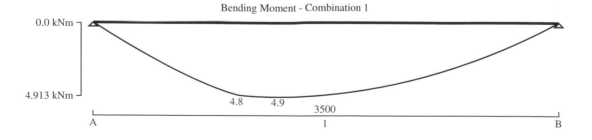

Bending Moment - Combination 1

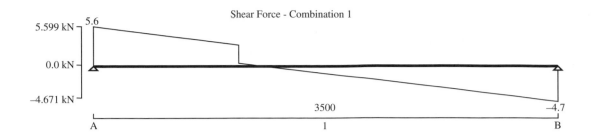

Shear Force - Combination 1

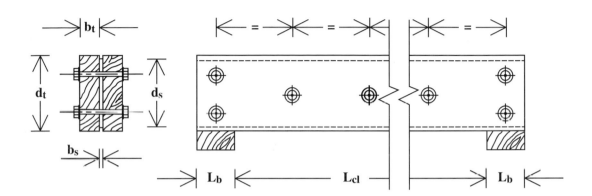

Typical flitch beam details

FLITCH BEAM CALCULATIONS TO **BS5268-2:2002**

Timber section properties

Breadth of timber	$b_t = \mathbf{50}$ mm	Depth of timber	$d_t = \mathbf{175}$ mm
Number of timbers	$N_t = \mathbf{2}$	Strength class	**C16**

Steel section properties

Breadth of steel	$b_s = \mathbf{12}$ mm	Depth of steel	$d_s = \mathbf{175}$ mm
Number of plates	$N_s = \mathbf{1}$	Bending stress	$p_y = \mathbf{165}$ N/mm^2

Check bending stress in timber

Permissible bending stress $\sigma_{m\_adm} = \mathbf{6.186}$ N/mm^2 Applied bending stress $\sigma_{m\_a} = \mathbf{2.536}$ N/mm^2

PASS – Timber stress not exceeded in bending

Check bending stress in steel

Permissible bending stress $p_y = \mathbf{165}$ N/mm^2 Applied bending stress $\sigma_{m\_a\_s} = \mathbf{59}$ N/mm^2

PASS – Steel stress not exceeded in bending

Check shear stress in timber

Permissible shear stress $\tau_{adm} = \mathbf{0.737}$ N/mm^2 Applied shear stress $\tau_a = \mathbf{0.126}$ N/mm^2

PASS – Timber stress not exceeded in shear

Cheak beam deflection

Permissible deflection $\delta_{adm} = \mathbf{10.5}$ mm   Calculated deflection $\delta_{actual} = \mathbf{5.5}$ mm

PASS – Deflection is acceptable

Check bearing stress in timber

Permissible bearing stress $\sigma_{cp1} = \mathbf{2.200}$ N/mm^2 Applied bearing stress $\sigma_{c\_a} = \mathbf{0.560}$ N/mm^2

PASS – Timber compressive stress not exceeded at bearing

Bolting requirements

Provide a minimum of 2 No. 12 mm diameter bolts at each end of the beam

Provide 12 mm diameter bolts at a maximum of 500 mm centres along the length of the beam

3.6 Single lap tension joint using Eurocode-5

Design a single lap tension joint between two timber members of strength class C24 in service class 2. Assume the plain wire nails are drilled directly. Using the following data finally arrive at a suitable nailing pattern:

Data:

Tension between two members $\to 40 \times 125$ mm: ss grade C24

Plain wire nails diameter $\to 3.35$ mm unit 75 mm length

Load (short term load) $\to 2.00$ kN

$f_{i,o,k}$ = Tension strength $\parallel$ to grain = 14 N/mm^2

p_s = Characteristic wood density = 350 kg/m^3 (20% moisture content)

$b_1, b_2 = 40$ mm thick

$h = 125$ mm width of member each

Action (load) = Q_k = 2KN; Area = 40×125 = 5000 mm^2

Partial factors $\begin{cases} \gamma_m = 1.3 - \text{timber} \ (\text{EC5} - \text{Table 2.3.3.2}) \\ \gamma_M = 1.1 - \text{steel} \end{cases}$

Case (A) *Ultimate Limit State*

N_d = design tensile load = $\gamma_Q \, Q_k = 1.5 \times 2000 = 3000$ N

strength modification factors k_h, k_{mod}:

k_h = min. of 1.3 and $\left(\frac{150}{h}\right)^{0.2} = 1.037$

k_{mod} based on Table 3.1.7 = 0.90 (Class 2 short term duration)

$\sigma_{t,o,d}$ design tensile stress = $\frac{Nd}{A} = 0.6$ N/mm^2

$f_{t,o,d}$ = design tensile strength = $\frac{k_{mod} \, k_h \, f_{t,o,k}}{\gamma_m}$
= 10.05 N/mm^2

$\sigma_{t,o,d} = 0.6 < f_{t,o,d} = 10.05$ Tension strength O.K.

Fasteners:

Use clauses 6.2 and 6.3 of EC-5

Adopted initially a nail size 3.35 mm $\times$ 75 mm length

Using EC5 (clause 6.3.1.2e and 6.3.1.2f)

The joint nails and member thickness check

$t_{min} = 7 \times 3.35 = 23.5$ mN < 40 mm *Member* O.K.

or $\frac{(13 \times 3.35 - 30) \times 350}{400} = 17.3$ mm

Check on nail length based on EC5 (Clause 6.3.1.2 (4))

Minimum penetration = 8d = $8 \times 3.35 = 27$ mm

Minimum nail length = 40 + 8d = 67 mm < 75 mm O.K.

$t_1 = b_1 = 40$ mm; t_2 = nail length $- b_1 = 75 - 40$ = 35 mm

Embedded strength EC-5 (Clauses 6.3.1.2(1), 6.3.1.2a, 2.2.3.2a, 6.3.1.2c and 6.2.1n)

$f_{h,1,k} = f_{h,2,k} = \frac{0.082 f_k}{d^{0.3}} = 19.97$ N/mm^2

$f_{h,1,d} = f_{h,2,d} = \frac{k_{mod} \, f_{h,1,k}}{\gamma_m} = 13.83$ N/mm^2

$M_{y,k}$ = moment at yield = 180d$^{2.6}$
= 180 (3.35)$^{2.6}$ = 4172 mm

$M_{y,d}$ = moment at yield = $\frac{M_{yk}}{\gamma_m} = 3793$ Nmm

$\beta = \frac{f_{h,2,d}}{f_{h,1,d}} = 1.0$

$\frac{t_2}{t_1} = \frac{35}{40} = 0.875$

Load carrying capacity of a single nail "R_d"

The lowest value to be adopted

A reference is made to EC-5 clause/equations 6.2.1"a" to 6.2.1"f".

$R_{d,a}$ (Equation 6.2.1a)
$= f_{h,1,d} \, t_1 \, (d) = 1853$ N/nail

$R_{d,b}$ (Equation 6.2.1b)
$= f_{h,1,d} \, t_2 \, \beta = 1621$ N/nail

$R_{d,c}$ (Equation 6.2.1.c) substituting above values in this equation

$$= \frac{f_{h,1,d}}{1+\beta} \left\{ \sqrt{\beta + 2\beta^2 \left\{ 1 + \frac{t_2}{t_1} + \left(\frac{t_2}{t_1}\right)^2 \right\} + \beta^3 \left(\frac{t_2}{t_1}\right)^2} - \beta\left(1 + \frac{t_2}{t_1}\right) \right\}$$

= 723 N/nail

$R_{d,d}$ (Equation 6.2.1.d)

$$= \frac{1.1 f_{h,1,d} \, t_1 d}{2+\beta} \left\{ \sqrt{2\beta \, (1+\beta) + \frac{4\beta(2+\beta) M_{y,d}}{f_{h,1,d} t_1^2}} - \beta \right\}$$

= 917 N/nail

$R_{d,e}$ (Equation 6.2.1.e)

$$= \frac{1.1 f_{h,1,d} \, t_2 d}{1+2\beta} \left\{ \sqrt{2\beta^2 \, (1+\beta) + \frac{4\beta(1+2\beta) M_{y,d}}{f_{h,1,d} \, dt_2^2}} - \beta \right\}$$

= 1639 N/nail

$$R_{d,f} = 1.1 \sqrt{\frac{2\beta}{1+\beta}} \, \sqrt{2M_{y,d} \, f_{h,1,d} \, d}$$

= 917 N/nail

Lowest value of $R_{d,c}$ = 723 N/nail

n = Number of nails = $\frac{N_d}{R_d} = \frac{2000 \times 1.5}{723} = 4.15$ nails

A pattern 5 would look awkward. Make 6 nails and work out the pattern

α = angle of load to grain = $0°$

$a_1 = (5 + |5 \cos \alpha|)d = 10d = 33.5$ mm adopt 35 mm
Minimum spacing along the grain is 35 mm

$a_2 = 5d = 5 \times 3.35 = 17$ mm adopt 20 mm
Minimum spacing across the grain is 20 mm

$a_{31} = (10 + |5 \cos \alpha|)d$
= 15d = 50.3 mm adopt 55 mm
This is the minimum loaded end distance which is taken as 55 mm

$a_{4c} = 5d = 17$ mm (adjust to suite the overall dimension)
Minimum unloaded edge distance

Pattern: a_1 and a_{31} when added as 55 + 35 + 55 = 145 overlap

$a_2 \ (2 \times 20 = 40$ mm$) + 2(a_{4c}) = 125$ mm
$40 + 2 \, a_{4c} = 125$
$a_{4c} = 42.5$ mm
a_{4c} minimum = 20 mm calculated – adjust this dimension

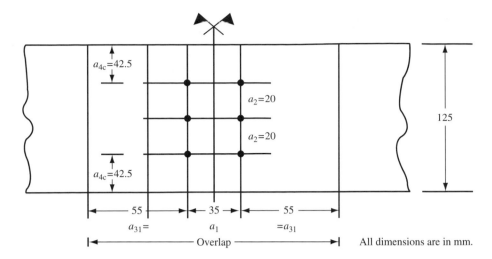

Figure 3.6.1 *Single lap tension joint.*

3.7 Double-bolted lap joint in tension based on EC-5

Design a bolted tension joint in which a timber member 47×125 is sandwiched between two timber members of 30×125 mm each of grade SS. The fastening shall be of M 16 bolts made in grade 4.6. Using EC-5, draw the final layout of this lap joint while using the following data:

Data:

Strength class C18
$f_{t,o,k} = 11$ N/mm^2
$f_k = 320$ kg/m^3
$b_1 = b_2 = 30$ mm
$h = h_1 = h_2 = 125$ mm
See Figure 3.7.1
Q_k = Load for permanent duration = 5000 N
$Q_{k,1}$ = Long term duration = 5000 N
$Q_{k,2}$ = Short term duration = 1875 N

Partial load factors
$\gamma_G = 1.35; \gamma_Q = 1.1$

Material factors
γ_m (timber) = 1.3; γ_m (steel) = 1.1

Combination factors
ψ_o for long term load = 0.7 (for Q_1)
ψ_o for short term load = 0.7 (for Q_2)

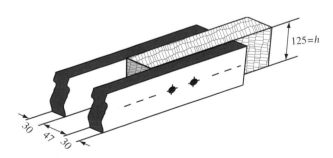

Figure 3.7.1 *Double-bolted lap joint.*

Ultimate Limit State

A reference is made to EC-5 (2.3.2.2a)

Design loads:

(a) Long term $N_{d(long)} = \gamma_G G_k + \gamma_Q Q_{k,1} = 14250$ N
(b) Short term $N_{d(short)} = \gamma_G G_k + \gamma_q [Q_{k,1} + \psi_o Q_{k,2}]$
$\qquad = 16218.75$ N
$\qquad$ or $= \gamma_G G_k + \gamma_Q [Q_{k,2} + \psi_o Q_{k,1}]$
$\qquad\qquad\qquad\qquad$ less than the above.

Parameters for joint
Embedded strength =
$f_{h,1,k} = f_{h,2,k} = f_{h,o,k} = 0.082(1.003d) f_k$
$\qquad\qquad\qquad\qquad\qquad = 22.04$ N/mm^2

(based on EC-5 Clause 6.5.1.2b)
$f_{h,1,d} = f_{h,2,d} = \dfrac{k_{mod} f_{h,o,k}}{\gamma_m} = 11.87$ N/mm^2

(based on Table 3.1.7 of EC-5)
k_{mod} (long term) = 0.7
$f_{h,1,d} = f_{h,2,d} = 15.26$ N/mm^2

(based on Table 3.17 of EC-5)
k_{mod} (short term) = 0.9

Moment at yield $M_{y,k}$ Based on EC-5 Clause 6.5.1.2c

$$M_y = \frac{0.8 f_{u,k} d^3}{6} \quad f_{u,k} = 400 \text{ N/mm}^2 \text{ (EN 20898–1)}$$

$$M_{y,d} = \frac{M_{y,k}}{\gamma_m} = 198600 \text{ Nmm}$$

$$\beta = \frac{f_{h,2,d}}{f_{2,1,d}} = 1.0; \quad t_1 = b_1 = 30 \text{ mm}; \quad \frac{t_2}{t_1} = 1.567$$

$$t_2 = b_2 = 47 \text{ mm}$$

$$d = 16 \text{ mm bolt}$$

Long term (Based on EC-5 Clause 6.2.1)
$$R_{d\text{ⓖ}} = f_{h,1,d} (t_1)(d) = 5697 \text{ N}$$

$$R_{d\text{ⓗ}} = 0.5 f_{h,1,d} (t_2) d\beta = 4463 \text{ N}$$

$$R_{d\text{①}} = \frac{1.1 f_{h,1,d} t_1 d}{2 + \beta} \left\{ \sqrt{2\beta (1 + \beta) + \frac{4\beta (2 + \beta) M_{y,d}}{f_{h,1,d} d t_1^2}} - \beta \right\}$$

$$= 6760 \text{ N}$$

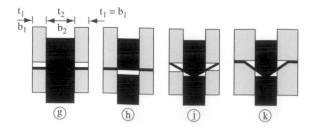

Figure 3.7.1.1 *Types of bolted joints.*

$$R_{d_{\text{(k)}}} = 1.1 \sqrt{\frac{2\beta}{1+\beta}} \ \sqrt{2 M_{\text{y,d}} \ f_{\text{h,1,d}} d} = 9554 \text{ N}$$

$R_{d_{\text{(h)}}} = 4178$ N per shear plane $= R_{\text{long}}$

Short term (same formulae to be used)

$R_{d_{\text{(g)}}} = f_{\text{h,1,d}}(t_1)(d) = 7325$ N

$R_{d_{\text{(h)}}} = 5738$ N; $R_{d_{\text{(j)}}} = 9873$ N

$R_{d_{\text{(k)}}} = 10830$ N

$R_{\text{d,short}} = R_{\text{d,h}} = 5738$ N per shear plane

Number of Bolts (n)

Long term:

$$n = \frac{N_{\text{d(long)}}}{2 R_{\text{d(long)}}} = \frac{14250}{2(4463)} = 1.6$$

Short term:

$$n = \frac{N_{\text{d(short)}}}{2 R_{\text{d(short)}}} = \frac{16218.75}{2(5738)} = 1.4$$

TAKE 2 NO M16.

Bolt spacing

This is based on EC-5 Table 6.5.1.2 $\alpha = 0$
                                                  $\cos\alpha = 1$

Minimum spacing $(a_1) = (4 + 3 \mid \cos\alpha \mid) \alpha$
                                $= 112$ adopt 115 mm
Minimum spacing $a_2 = 4d = 64$ mm   adopt 65 mm
Minimum load distance $a_{3,1} = 7d = 112$
                         or $= 80$ adopt 115 mm
Minimum unloaded edge $a_{4,c} = 3d = 48$ mm
    Adjustment shown in detailing Figure 3.7.1.2
as 57.5

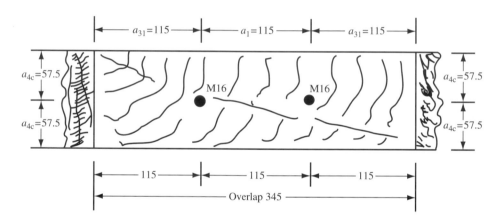

Figure 3.7.1.2 *Bolted lap joint.*

Chapter 4
Data on Timber Fastenings and Joints

4.1 Introduction

For the purpose of raising the load-carrying capacity of wooden structures, various methods of jointing are used: building up (in a lateral direction); scarping or splicing (lengthways); and multiple joints of logs, balks, boards and pieces.

In modern timber structures, the jointing of members is mainly carried out by means of special wooden, metal, plastic and other working connectors. Exceptions to this are compression splices and notches, affected, as a rule, by the direct abutment of the respective members sawn to fit.

Depending on the type of stress in the connectors, the principal varieties of joint are distinguished according to their structural features.

It must be borne in mind that stress features such as compression, bending, tension and shear only refer to the connectors themselves, and not to the joint as a whole.

For example, a joint under tension can be made using connectors under bending stress, ie dowels. The work of the connectors under bearing stress cannot serve as a distinguishing feature, as it characterizes the surface action and reaction of the jointed parts, common to all types of notching, keying, dowelling and connectors under tension; for example tie-bolts usually transfer the forces through a washer, interacting with the surface of the timber with a bearing stress in the same manner. It is only in glued joints that the principal forces are transferred by shear, and not by the bearing stress on the glue line.

The load-carrying capacity of built-up wooden structures depends to a large extent on the method of jointing the members.

The joints connecting members in *tension* present the greatest difficulty, due mainly to local strength-reduction. Dangerous local stresses *between the force and the grain* are taken account by the formulas or graphs. Under the term force is understood here the elementary forces (stresses) of bearing applied to unit bearing area; these forces are considered as applied perpendicularly to the area since the relieving action of friction is not taken into account. The angle α_{bear} is taken as equal to the largest angle (up to 90°) between the perpendicular to the bearing area and the direction of the grain of the members in contact on that area.

In this chapter, design data are examined on connections and joints based on ANSI/LRED, BS 5628,

Eurocode-5, Russian and various respective practices. They include specifications for nails, screws, bolts, connectors and a mixture of them. Joints notched, keyed, dowelled and glued are also examined in detail. Wherever possible, explanatory notes, tables, equations and structural details are provided so that an objective assessment of specific design and detailing can be made.

4.2 Fastener types and definitions

4.2.1 Based on ANSI and American practices

Wood screws, shown in Fig. 4.1 have a continuous, helical short-lead thread around a slightly tapered shank from which the thread projects. They are typically manufactured in lengths of 6 to 100 mm with diameters ranging from 0.060 to 0.372 inch. The head may take several different shapes but a slot or recess is always provided for screw-driver or machine-tool insertion for the fastener. The length of the threaded portion is about two-thirds of the overall length. Wood screws are manufactured from steel, brass, or other metals and alloys and with specific finishes such as nickel, blued, chromium, or cadmium.

Wood screws are generally used in light duty applications where they serve the same purpose as nails. They can be removed and retightened and generally yield a more rigid joint than nails or staples. Drive screws, similar to wood screws but with a shallower thread, are used to fasten dry wall and other sheathing materials and to improve racking resistance in light industrial and commercial structures.

Tapping screws (sometimes called sheet metal screw) are similar to wood screws but have threads the full length of the shank and usually a pan or hex head. They are commonly used to fasten metal or plastic components to wood members in manufactured housing and agricultural structures. A typical application is the connection between steel cladding and wood rafters in post and frame structures. While this chapter will focus on wood screws, some information on the resistance of other screw types is presented.

Lag bolts or screws are stouter than wood screws, have a square or hex head with no slot, a plain cylindrical shank section below the head, and a helically

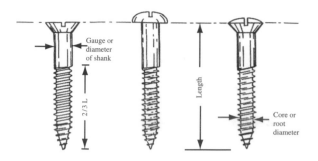

Figure 4.1 *Common wood screws.*

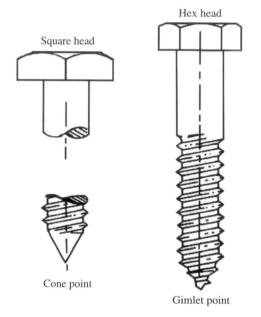

Figure 4.2 *Lag bolts.*

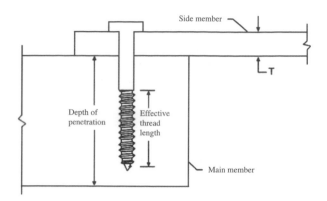

Figure 4.3 *Lag screws connecting two members.*

threaded cylindrical section extending to a pointed tip (Fig. 4.2).

Lag screws are generally made of hot-rolled steel, may be unfinished, electroplated, or galvanised, and are available in diameters 6 mm to 32 mm and lengths of 25 to 406 mm. The length of the threaded portion varies with the length of the screw and ranges from three-quarter inch for the 1- and 1-1/4 inch screws to over half the length for all lengths greater than 10 inches. Lag screws are designed to be inserted by wrench into a pilot hole.

Lag screws are most commonly used to fasten metal components to heavy timber members. Whereas, bolts are generally used for lateral resistance only, lag screws can be used to transmit both lateral and withdrawal loads. Lag screws have found considerable application in recent years due to the increase in glue laminated timber construction. Through-bolts in glulam are often impractical because of the potential length of the hole which may also weaken the wood cross-section. Lag screws are also commonly employed in situations where it is not acceptable to have a nut on the surface or where difficulty in fastening a bolt or drilling a through hole is encountered.

The vast majority of joints using wood or lag screws consist of two members as shown in Fig. 4.3. The side member or cleat may be metal, wood, plastic or other

Table 4.1 Gauge, nominal shank diameter, and lead hole diameter for wood screws. Based on ANSI (1981) and American practices

Gauge of Screw	Shank Diameter of Screw (In.)	Diameter of Lead Hole (In.)					
		Withdrawal Loads		Lateral Loads			
		Group I Species[1]	Groups II, III and IV Species[1,2]	Group I Species		Groups II, III, and IV Species	
				Shank Portion	Threaded Portion	Shank Portion	Threaded Portion
6	0.138	5/64	1/16	9/64	3/32	1/8	5/64
7	0.151	3/32	5/64	5/32	7/64	1/8	3/32
8	0.164	7/64	5/64	5/32	7/64	9/64	3/32
9	0.177	7/64	5/64	11/64	1/8	5/32	7/64
10	0.190	7/64	3/32	3/16	1/8	11/64	7/64
12	0.216	9/64	7/64	7/32	9/64	3/16	1/8
14	0.242	5/32	7/64	1/4	5/32	7/32	9/64
16	0.268	11/64	1/8	17/64	3/16	15/64	5/32
18	0.294	3/16	9/64	19/64	13/64	1/4	11/64
20	0.320	13/64	5/32	5/16	7/32	9/32	3/16
24	0.372	15/64	3/16	3/8	1/4	21/64	15/64

[1] For species in each group see table 1.3.1 for sawn lumber and table 1.3.2 for glued laminated timber.

[2] For group II and IV species, the screw may be inserted without a lead hole.

material with a thickness, *t*. The block or main member is solid wood and we will assume that the screw does not entirely penetrate its thickness.

Lag screws (also called lag bolts) are specified by nominal diameter and length (fractional or decimal equivalent), product name (namely, square or hex head lag screw with cone or gimlet point), and material and finish, if any. Nominal diameter is expressed in inches except for the smallest size which carries a gauge number (#10). The clearest difference between wood and lag screws is that the latter have no slot and must be inserted using a wrench. While there is some overlap in sizes between the two screw types, their thread geometries are slightly dissimilar.

In the United States, the dimensions and manufacturing tolerances of lag screws are standardised by ANSI/ASME B18.2.1. These specifications fix the relationship between thread geometry and lag screw size. They do not address minimum material properties. The dimensions standardised in the ANSI specifications are essentially reproduced in Table 4.1, taken from NDS-91. Minimum thread length is one-half nominal screw length, plus 12 mm or 150 mm, whichever is shorter. The effective, minimum thread length is given in Table 4.2.

4.2.2 Screws and screw joints based on BS 5268 (2002)

Steel screws must comply with BS 1210: Wood screws. Screws must be turned, using a non-corrosive lubricant if necessary, into pre-drilled holes. The top of the countersunk screws must be no more than 1 mm below the timber surface.

The hole for the shank shall have the same diameter as the shank diameter and be no deeper than the shank length. For the threaded portion of the screw, the hole diameter should be 40 or 60% of the shank diameter for softwoods and hardwoods, respectively.

The basic loads may not be increased for additional thicknesses or penetration, but the minimum point side penetration must be at least 60% of the tabulated values.

The basic withdrawal loads for single screws inserted at right-angles to the side grain of dry timber are to be evaluated. The values apply to each 1 mm of penetration and should be multiplied by the actual point side penetration of the particular screw's threaded part. The minimum screw penetration is 15 mm.

No screw turned into end grain may carry a withdrawal load.

4.2.2.1 Steel plate-to-timber joints

The steel plate should have adequate strength and be designed in accordance with typical codes. The hole diameter should be no greater than the screw shank diameter.

Basic loads

For a screw going through a pre-drilled hole in a steel plate into a timber member, the basic lateral load given in the code is multiplied by 1.25, the steel to timber modification factor K_{46}.

4.2.2.2 Plywood-to-timber joints

The plywood must be one of those stated in Chapter 7 (BS 5268: Part 2, clauses 24 to 30), with pre-drilled holes no greater than the screw shank diameter.

Basic loads

The basic single shear lateral loads for single screws in a 6-mm thick plywood-to-timber joint, where the screws are inserted through the plywood at right-angles on to the side grain of dry timber are given in Table 67 (BS 5268: Part 2).

For plywoods over 6 mm thick and up to 20 mm thick, the basic loads are calculated from the expression:

$$F + K_{51}(t-6) \tag{4.1}$$

where F is the lateral load in newtons as given in Table 67, t is the nominal plywood thickness in mm, and K_{51} is a modification factor obtained from Table 66.

For the basic loads to apply, the minimum total screw length should be the greater of the value given in Table 4.2.

Permissible load for a joint

The permissible load for a screwed joint is the sum of the permissible loads for each screw in a joint which equals:

Basic load $\times K_{52} \times K_{53} \times K_{54}$

The modification factor for moisture content K_{52} equals

Long-term loads	1.1
Medium-term loads	1.12
Short- and very short-term loads	1.25

The modification factor for moisture content K_{53} equals

Joints in dry timber	1.0
Joints in green timber or timber which will be under wet exposure	see code

The modification factor for the number of screws in each line K_{54} equals

1.0 for *n* less than 10, or
0.9 for *n* equal to or greater than 10

where *n* is the number of screws in each line. The number of screws of the same diameter acting in single

Table 4.2 Minimum total screw length

Screw diameter (mm)	mm	or	Nominal plywood Thickness plus
4.57	38		29
4.88	44		34
5.59	51		39

and multiple shear must be symmetrically arranged in one or more lines parallel to the line of action of the load in a primarily axially loaded member in a structural framework (clause 5.11). In all other loading cases, where more than one screw is used in a joint, $K_{54} = 1.0$.

The effective cross-section is determined by deducting the net projected area of the pre-drilled holes from the gross area of the cross-section being considered. All screws that lie within a distance of five screw diameters measured parallel to the grain from the cross-section being considered, are assumed to be at that cross-section.

Screw spacing

To avoid undue timber splitting, Table 4.3 gives the minimum screw spacing, using pre-drilled holes.

Table 4.3 Minimum screw spacing

Spacing	Parallel to grain	Perpendicular to grain
End distance	10d	–
Edge distance	–	5d
Distance between lines of screws	–	3d
Distance between adjacent screws in any one line	10d	–

Source: reproduced with permission from BS 5268: Part 2.
Note: d is the screw shank diameter.

Permissible load

The permissible load for a screwed joint is determined from the basic loads for each screw modified in accordance with Section 9.5 (BS 5268: Part 2, clause 42.7).

4.2.2.3 Timber-to-timber joints

The basic single shear lateral loads for single screws inserted at right angles to the side grain of dry timber are given in BS 5268: Part, and are shown in Table 4.4. The screws must fully penetrate the tabulated head side member standard thickness and at least the tabulated point side penetration distance.

4.2.3 Based on Eurocode-5 and European practices

The tables presented are based on design loads (kN) for one steel wood screw in single shear: timber-to-timber point. These tables are applicable to wood screws made from steel with a minimum tensile strength of 550 N/mm^2. The values in Tables 4.5 and 4.6 are the design values (R_d) for service classes 1 and 2. For service class 3, the values are reduced by a factor of 0.8. Screws with diameter greater than 5 mm should be turned into holes which are pre-drilled.

Table 4.4 Basic single shear lateral loads for screws inserted into pre-drilled holes in a timber-to-timber joint

Screw shank diameter mm	Standard penetration mm		Basic single shear lateral load N				
			Strength class				
	Head side mm	Point side mm	C14	C 16/18/20 TR20/C22	C24	TR26/C27 C30/40/50	D35/40/45 50/60/70
3	11	21	192	205	205	232	304
3.5	12	25	260	278	310	321	405
4	14	28	338	361	395	409	518
4.5	16	32	425	454	490	507	643
5	18	35	511	550	593	615	781
5.5	19	39	628	654	705	731	931
6	21	42	734	765	826	856	1092
7	25	49	970	1011	1093	1133	1449
8	28	56	1233	1286	1391	1443	1849
10	35	70	1504	1608	1741	1803	2285

Table 4.5 Lateral design loads for one steel woodscrew in single shear

Timber-to-timber Service class 1 & 2

Permanent, Long-term and Medium-term load duration

Number	Screw shank diameter (mm)	Standard headside thickness (mm)	Timber strength class groups					
			Softwoods			Hardwoods		
			C14 (kN)	C16,18,22 (kN)	C24 (kN)	C27,30,35,40 (kN)	D30,35,40 (kN)	D50,60,70 (kN)
6	3.45	12	0.30	0.32	0.36	0.38	0.51	0.59
8	4.17	16	0.48	0.51	0.58	0.60	0.75	0.86
10	4.88	22	0.71	0.74	0.80	0.83	1.07	1.24
12	5.59	35	1.03	1.08	1.19	1.24	1.60	1.77
14	6.30	44	1.36	1.44	1.58	1.65	1.98	2.19
16	7.01	47	1.62	1.71	1.88	1.97	2.39	2.65

Table 4.6 Lateral design loads for one steel woodscrew in single shear

Timber-to-timber: Service class 1 & 2

Short-term and Instantaneous load duration

Number	Screw shank diameter (mm)	Standard headside thickness (mm)	Timber strength class groups					
			Softwoods				Hardwoods	
			C14 (kN)	Cl6,18,22 (kN)	C24 (kN)	C27.30,35,40 (kN)	D30,35,40 (kN)	D50,60,70 (kN)
6	3.45	12	0.45	0.47	0.51	0.53	0.67	0.78
8	4.17	16	0.66	0.69	0.75	0.77	1.00	1.16
10	4.88	22	0.93	0.97	1.06	1.11	1.45	*1.71*
12	5.59	35	1.41	1.49	*1.59*	*1.64*	*1.96*	*2.17*
14	6.30	44	*1.79*	*1.85*	*1.97*	*2.02*	*2.42*	*2.68*
16	7.01	47	*2.16*	*2.24*	*2.38*	*2.44*	*2.93*	*3.24*

4.2.4 Withdrawal resistance of wood screws

4.2.4.1 Based on American practices

(A) Wood Screws

The tabulated design values are based on the insertion of the screw into a lead hole of proper size. Lubricants such as soap or wax may be used for insertion. No spacing, edge or end distances are specified but splitting must be prevented. All design values assume that failure is screw shank withdrawal from wood. Figure 4.4 shows some design failure mechanisms.

Withdrawal resistance

The allowable design load for a single wood screw in axial withdrawal from side grain of seasoned wood that will remain dry is:

$$W = 2844G2D, \tag{4.2}$$

where:

W = allowable withdrawal load per inch of penetration of the threaded portion (lb/in.),

G = specific gravity based on weight and volume of oven-dry wood,

D = nominal or basic screw diameter (in.)

Wood screws should not be used where withdrawal from end grain is required. The NDS does not have explicit provisions for tapping screw withdrawal design values. Individual building codes or regulatory agencies may prescribe levels based on test data or allow their use in certain applications.

Lateral resistance

The most efficient use of wood screws is for lateral loading. They have a greater lateral capacity than nails of the same diameter but are more costly to install. Wood screws subject to lateral loading must be inserted into a pilot hole to avoid splitting which will reduce lateral resistance.

NDS-86

In the 1986 and earlier editions of the NDS, allowable design loads for a single screw in a two-member joint subject to lateral loads were determined by:

$$P_L = KD^2, \tag{4.3}$$

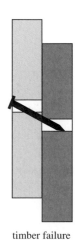
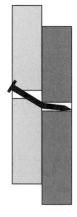
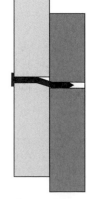

timber failure   combined fastener and timber failure   fastener failure

Figure 4.4 *Design failures.*

where: P_L = allowable lateral load (lb),
          K = species factor;   Group 1 480
                                   Group 2 3960
                                   Group 3 3240
                                   Group 4 2520
          D = nominal or basic screw
                diameter (in.)

Use of these values assumed that the screw is inserted in side grain and is embedded at least seven diameters ($7D$) into the main member with lead holes. For wood screws penetrating less than $7D$, P_L was reduced by multiplication with $C_d = 0.143p$, where p = penetration in shank diameters. Note the screw must have been embedded at least four diameters ($C_d = 0.572$) for any allowable load to have been assigned. P_L should be further modified for changing environments, duration of load, and other factors.

NDS-91

The 1991 version of the NDS contains tables of allowable lateral loads for screws that appear similar to their NDS-86 counterparts. The principal difference is how the numbers were obtained. The designer may use the tabulated values or calculate a value using the equations shown in Table 4.7 (given in NDS-91).

The designer should check fastener strength using the appropriate specifications. Withdrawal from end

grain should be avoided but, if it cannot, then reduce the allowable load by 25% for this situation ($C_{eg} = 0.75$). Be extremely cautious if the joint will be subject to wetting and drying.

4.3 Lag screws

4.3.1 General provisions

Design of lag screw connections is generally specified by the National Design Specification (AFPA 1991) or Timber Construction Manual (AITC 1985). Tables for allowable lateral loads are organised by side member thickness and screw dimensions. It is assumed that the lag screw is inserted into a lead hole of proper size and that both the side member and main member are of similar species.

If more than two screws are in a row, then connection capacity is reduced using the same rationale and method as presented for bolts. Spacing, end, and edge distances are stipulated to be exactly the same as for bolts; reductions for not meeting these requirements are similar.

NDS-91 design values are tabulated in a manner similar to NDS-86. However, the designer may calculate design strength for conditions not found in the tables. This is particularly useful when dissimilar materials are connected.

Table 4.7 Yield equations for NDS-91 allowable lateral load for a single shear wood screw connection with penetration of the screw into the main member, p, equal to or greater than seven diameters ($7D$)

Allowable load shall be the minimum of:

(a) $Z = \dfrac{Dt_s F_{es}}{K_D}$,

(b) $Z = \dfrac{kDt_S F_{em}}{K_D(2+R_e)}$,   where $k = -1 + \sqrt{\dfrac{2(1+R_e)}{R_e} + \dfrac{F_y(2+R_e)D^2}{2F_{em}t_s^2}}$

(c) $Z = \dfrac{D^2}{K_D}\sqrt{\dfrac{1.75F_{em}F_y}{3(1+R_y)}}$,

D = unthreaded shank diameter, in,
t_s = cleat or side member thickness, in,
F_{em}, F_{es} = dowel bearing strength of main and side member, psi
F_y = fastener yield strength, psi
R_e = F_{em}/F_{es}
K_D = 2.2 for $D \le 0.17$,
        = $10D + 0.5$ for 0.17 in. $< D < 0.25$ in.,
        = 3.0 for $D \ge 0.25$ in.

Notes
[1] For metal to wood connections, F_{es} = bearing strength of the metal and allowable load shall be the lesser of equations (b) and (c)
[2] when $D > 0.25$ in. the allowable load shall be determined using the methods for lag screws.
[3] for $p < 4D$ $C_d = 0$
           $4D \le p \le 7D$ $C_d = p/7D$
           $p > 7D$ $C_d = 1$.

4.3.1.1 Withdrawal resistance

Allowable design loads for a single lag screw in withdrawal from seasoned side grain of seasoned wood are given by:

$$W = 1800 \ G^{1.5} \ D^{0.75}, \qquad (4.4)$$

where:

W = allowable withdrawal load per linear inch of penetration of the effective thread (T-E in Table 3.4),

G = specific gravity of species at oven-dry weight and volume,

D = nominal diameter of lag screw shank (in.)

Equation (4.4) is keyed to shank withdrawal failure of a lag screw. AITC recommends the use of 20 ksi as an allowable tension ultimate strength of a lag screw based on a net (root) section analysis.

4.3.1.2 Lateral resistance

Lag screw lateral design values have been widely tabulated in various texts. The basis for these values is the derivation method shown below. This method did not substantially change between the 1944 and 1986 NDS and is also found in the Wood Handbook (USDA 1987). It is a set of procedures based on results from a testing program. The 1991 NDS is EYM-based but was confirmed, in part, using the historical data base for NDS-86.

For a typical lag bolt joint shown in Fig. 4.5 the allowable lateral bolt load for one bolt properly installed in seasoned dry wood that will remain dry is found by

$$P_\mathrm{L} = (K) \ (D^2) \ (\mathrm{CTF}) \ (\mathrm{PF}) \ (\mathrm{GDF}), \qquad (4.5)$$

where: P_L = safe allowable load, in lb, for a single lag bolt in lateral shear,

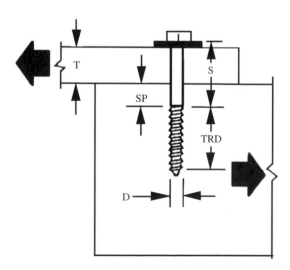

Figure 4.5 *Lag screw joint geometry.*

K = constant depending on wood species,
  Group 1: 2640
  Group 2: 2280
  Group 3: 2040
  Group 4: 1800

D = shank diameter in inches,

CTF = side member thickness factor,

PF = penetration factor,

GDF = grain direction factor.

Use of this formula could be cumbersome and tables were available in NDS-86 and TCM for ready reference. However, it was not uncommon to encounter a situation not covered by the table and some judgment would be applied.

Tables 4.8 to 4.10 provide data on lag screws with shanks while Tables 4.11 and 4.12 present basic withdrawal loads and acceptable penetrations based on BS 5268.

Table 4.8 Lag screw side member thickness factor for various ratios of side member thickness to shank diameter

Shank Diameter	Side Member Thickness Factor
2	0.62
2.5	0.77
3	0.93
3.5	1.00
4	1.07
4.5	1.13
5	1.18
5.5	1.21
6	1.22
6.5	1.22

Table 4.9 Shank penetration coefficients for various ratios of the depth of shank penetration in holding member to shank diameter

Shank Diameter	Side Penetration Coefficients
0	1.00
1	1.08
2	1.17
3	1.26
4	1.33
5	1.36
6	1.38
7	1.39

Table 4.10 Grain direction for perpendicular-to-grain lag screw loading

Shank Diameter	Side Member Thickness Factor
1/4	0.97
3/8	0.75
1/2	0.65
5/8	0.60
3/4	0.55
7/8	0.52

Table 4.11 Basic withdrawal loads per millimetre of pointside of penetration for smooth round wire nails driven at right angles to the grain

Nail diameter mm	Basic withdrawal load N/mm Strength class					
	CI4	C16/18/20/22/TR20	C24	TR26/C27	D35/40/45	D50/60/70
2.7	1.30	1.53	2.08	2.39	5.87	9.77
3	1.44	1.71	2.31	2.65	6.52	10.86
3.4	1.64	1.93	2.62	3.01	7.39	12.31
3.8	1.83	2.16	2.93	3.36	8.26	13.75
4.2	2.02	2.39	3.23	3.72	9.13	15.20
4.6	2.21	2.62	3.54	4.07	10.00	16.65
5	2.41	2.84	3.85	4.42	10.86	18.10
5.5	2.65	3.13	4.24	4.87	11.95	19.91
6	2.89	3.41	4.62	5.31	13.04	21.72
7	3.37	3.98	5.39	6.19	15.21	25.33
8	3.85	4.55	6.16	7.08	17.38	28.95

Table 4.12 Basic withdrawal loads per millimetre of pointside for penetration for screws at right angles to the grain

Screw diameter	Basic withdrawal load N/mm Strength class				
	CI4	C16/18/20 TR20/C22	C24	TR26/C27 C30/35/40	D35/40/45 50/60/70
3	7.57	8.65	11.02	12.32	25.28
3.5	8.83	10.09	12.86	14.37	29.49
4	10.09	11.53	14.70	16.43	33.11
4.5	11.35	12.97	16.54	18.48	31.92
5	12.61	14.41	18.37	20.53	42.13
5.5	13.88	15.86	20.21	22.59	46.35
6	15.14	17.30	22.05	24.64	50.56
7	11.66	20.18	25.72	28.75	58.99
8	20.18	23.06	29.40	32.86	67.42
10	25.23	28.83	36.75	41.07	84.27

where

T	=	Thickness
S	=	as in Fig. 4.5
SP	=	Position of the unthreaded shank that penetrates the main member.
TRD	=	Sum of the threaded length of screw
D	=	Unthreaded shank diameter
SPC	=	Shank penetration diameter
	=	$1-0.1\ (T{-}S)/T$

4.4 Nailed joints

Based on BS 5268

For the loads given here for nails to be valid, the steel wire from which the nails are produced should have a minimum ultimate tensile strength of 600 N/mm². A nailed joint should normally contain at least two nails.

Hardwoods in strength classes D30 to D70 will usually require pre-drilling. The diameter of pre-drilled holes should be not greater than 0.8 times the nail diameter. Skew nailing slightly increases the resistance to withdrawal. Nails loaded laterally should not be skew-driven except at joints where no reversal of stress can occur in service and where the direction of the skew is such that the joint will tend to tighten under load. Opposed double skew nailing is preferable to parallel skew nailing.

4.4.1 Effective cross-section

When assessing the effective cross-section of multiple nail joints, all nails that lie within a distance of five nail diameters, measured parallel to the grain, from a given cross-section should be considered as occurring at that cross-section. Then the effective cross-section should be determined by deducting the net projected area of the nails from the gross area of the cross-section being considered.

No reduction in cross-section need be made for nails less than 5 mm diameter, driven without pre-drilling.

4.4.2 Nail spacing

The end distances, edge distances and spacing of nails should be such as to avoid undue splitting and, unless shown by test to be satisfactory, should not be less than the values given in Table 4.13.

For all softwoods except Douglas fir, the spacing given in Table 4.13 for timber-to-timber joints should be multiplied by 0.8. For nails driven at right angles to the glued surface of pre-glued laminated members, the spacing should be further multiplied by 0.9. In no case, however, should the edge distance in the timber be less than 5d.

4.4.3 Timber-to-timber joints

4.4.3.1 Basic single shear lateral loads

The basic single shear lateral loads for single round wire nails with a minimum tensile strength of 600 N/mm², driven at right angles to the side grain of timber in service classes 1 and 2, are given in Table 4.14. For nails driven into pre-drilled holes in softwood strength classes C14 to C40 and TR26, the values given in Table 4.14 may be multiplied by 1.15. For nails driven into the end grain of timber, the values given in Table 4.14 should be multiplied by the end grain modification factor, K_{43}, which has a value of 0.7. For softwoods where the thicknesses of members or nail penetrations are less than the standard values given in Table 4.14, the basic load should be multiplied by the smaller of the two ratios:

a) Actual to standard thickness of head side member; or

b) Actual penetration to standard point side thickness.

No load-carrying capacity should be assumed where the ratios described in a) or b) are less than 0.66 for softwoods and 1.0 for hardwoods. Where improved nails are used, the ratios may be reduced to 0.50 for softwoods and 0.75 for hardwoods. No increase in basic load is permitted for thicknesses or penetrations greater than the standard values.

Table 4.13 Minimum nail spacings

Spacing	Timber-to-timber joints		Steel plate-to-timber joints	Joints between timber and plywood or particleboard
	Without pre-drilled holes	With pre-drilled holes	Without pre-drilled holes	Without pre-drilled holes
End distance parallel to grain	20d	14d	14d	14d
Edge distance perpendicular to grain	5d	5d	5d	—a
Distance between lines of nails, perpendicular to grain	10d	3d	7d	7d
Distance between adjacent nails in any one line, parallel to grain	20d	10d	14d	14d

NOTE d is the nail diameter.

a The loaded edge distance in the timber should be not less than 5d. The loaded edge distance in the plywood should be not less than 3d. The loaded edge distance in the particleboard should be not less than 6d. In all other cases the edge distance should be not less than 3d.

Table 4.14 Basic single shear lateral load for round wire nails in a timber-to-timber joint

Nail diameter thickness mm	Standard penetration[a] mm	Softwoods (not pre-drilled)				Hardwoods (pre-drilled)		
		Basic single shear lateral load N				Minimum penetration[a] mm	Basic single shear lateral load N	
		Strength class					Strength class	
		C14	CI6/18/20 TR20/C22	C24	TR26/C27 C30/35/40		D35/40/45	D50/60/70
2.7	32	249	258	274	281	22	386	427
3	36	296	306	326	335	24	465	515
3.4	41	364	377	400	412	27	682	644
3.8	46	438	453	481	495	30	709	785
4.2	50	516	534	567	583	34	897	939
4.6	55	600	620	659	678	37	996	1103
5	60	689	712	756	778	40	1155	1279
5.5	66	806	833	885	910	44	1368	1515
6	72	930	962	1022	1051[b]	48	1595	1767
7	84	1200	1240	1318	1365[b]	56	2094	2319
8	96	1495	1546	1643	16891[b]	64	2649	2933

a These values apply to both the headside and pointside penetration.

b Holes should be pre-drilled.

Table 4.15 Nail lengths

Nail diameter (mm)	Length (mm) or	Nominal plywood thickness plus additional length (mm)
2.65	40	25
3.00	45	29
3.35	45	32
3.75	55	38
4.00	60	44

4.4.3.2 Basic multiple shear lateral loads

The basic multiple shear lateral load for each nail should be obtained by multiplying the value given in Table 4.14 by the number of shear planes, provided that the thickness of the inner member is not less than 0.85 times the standard thickness given in Table 4.14. Table 4.15 provides the ratio of nail length to nail diameter. Table 4.16 gives the basic single shear lateral load for round wire nails in a plywood-to-timber joint.

4.4.4 Permissible load for a joint

The permissible load for a screwed joint should be determined as the sum of the permissible loads for each screw in the joint where each permissible screw load, F_{adm}, should be calculated from the equation:

$$F_{adm} = F \times K_{52} \times K_{53} \times K_{54}$$

where the basic load for a screw, F, is taken from 6.5.4, 6.5.5.2 or 6.5.6.2 as appropriate, and where:

K_{52} is the modification factor for duration of loading;

K_{53} is the modification factor for moisture content;

K_{54} is the modification factor for the number of screws in each line.

For duration of loading:

K_{52} = 1.00 for long-term loads;

K_{52} = 1.12 for medium-term loads;

K_{52} = 1.25 for short- and very short-term loads.

For moisture content:

K_{53} = 1.0 for a joint in service classes 1 and 2;

K_{53} = 0.7 for a joint in service class 3;

For the number of screws in each line:

Where a number of screws of the same diameter, acting in single or multiple shear, are symmetrically arranged in one or more lines parallel to the line of action of the load in a primarily axially loaded member in a structural framework then:

K_{54} = 1.0 for $n < 10$;

K_{54} = 0.9 for $n \geq 10$;

where n is the number of screws in each line. In all other loading cases where more than one screw is used in a joint:

K_{54} = 1.0

4.5 Bolts, drift bolts, and pins/dowels

Based on BS 5628

The recommendations contained in 6.6 are applicable to joints utilising black bolts which conform to BS EN 20898-1 and have a minimum tensile strength of 400 N/mm^2 and washers which conform to BS 4320. Advantage can be taken of the end fixity provided by the headed nut and washers to improve the strength of a two-member bolted joint by including the factor K_{2b} in equation G.3 (see Annex G). Values for bolts calculated on this basis are given in Tables 69–74 of the code. For bolts and dowels made of higher strength steel, the equations given in Annex G may give higher load carrying capacities than given in Tables 69–80 of the Code.

Bolt holes should be drilled to diameters as close as practicable to the nominal bolt diameter but in no case should they be more than 2 mm larger than the bolt diameter. Washers with a nominal diameter and thickness of at least 3.0 times and 0.25 times the bolt diameter, respectively, should be fitted under the head of each bolt and under each nut unless an equivalent bearing area is provided, for example, by a steel plate. If square washers are used, their side length and thickness should be not less than the diameter and thickness of the appropriate round washer.

4.5.1 Effective cross-section

When assessing the effective cross-section of multiple bolt joints, all bolts that lie within a distance of two bolt diameters, measured parallel to the grain, from a given cross-section should be considered as occurring at that cross-section. Then the effective cross-section should be determined by deducting the net projected area of the bolt holes from the gross area of the cross-section.

4.5.2 Bolt spacing up

Unless other values are shown to be satisfactory by test, the end distances, edge distances and spacings given in Table 81 should be observed for bolted joints.

4.5.3 Timber-to-timber joints

4.5.3.1 Basic single shear loads

The basic loads for single bolts in a two-member timber joint, in which the load acts perpendicular to the axis of the bolt, and parallel or perpendicular to the grain of the timber, are given in various tables of the code. The basic loads appropriate to each shear plane in a three-member joint under the same conditions, are given in Tables 75–80 of the code. Separate loads are tabulated for long-, medium-, and short-duration loads so no further duration of loading modification factors

Table 4.16 Basic single shear lateral load for round wire nails in a plywood-to-timber joint

Plywood nominal thickness	Nail diameter	Nail length	Basic single shear lateral load N											
			Softwoods (not pre-drilled)						Hardwoods (pre-drilled)					
			Strength class						Strength class					
			C14		C16/18/20/TR20/TR22		C24		TR26/C27/30/35/40		D35/40/50		D50/60/70	
			Plywood group[a]						Plywood group[a]					
mm	mm	mm	I	II	I	II	I	II	I	II	I	II	I	II
6	2.7	40	213	122	218	227	226	236	230	241	275	290	275	305
	3.0	50	252	261	257	267	268	279	273	284	296	346	296	360
	3.4	50	308	318	315	326	323	340	323	346	323	393	323	393
	3.8	75	349	381	349	390	349	407	349	415	349	425	349	425
	4.2	75	374	448	374	455	374	455	374	455	374	455	374	455
9	2.7	45	220	231	229	244	237	257	241	261	282	309	295	324
	3.0	50	261	280	266	286	276	297	281	302	333	362	347	379
	3.4	50	315	335	322	342	334	356	339	362	407	440	425	461
	3.8	75	374	394	382	403	397	420	403	428	490	525	502	551
	4.2	75	437	459	447	470	465	489	473	498	538	619	538	650
12	2.7	45	244	258	250	271	259	284	263	289	305	339	317	353
	3.0	50	281	306	287	312	297	324	302	329	353	390	368	407
	3.4	50	321	337	339	355	353	381	358	387	424	464	442	485
	3.8	75	391	418	399	427	413	443	420	451	503	546	524	571
	4.2	75	452	481	462	491	479	510	487	519	589	635	614	665
15	2.7	50	270	287	277	300	286	316	291	321	335	374	347	389
	3.0	65	308	337	314	343	324	356	329	361	383	424	397	442
	3.4	65	360	390	367	398	379	412	385	419	452	497	469	518
	3.8	75	415	448	424	457	439	474	445	481	528	577	549	602
	4.2	75	476	509	485	520	503	540	511	549	611	664	635	692

[a] Plywood group I comprises American construction and industrial plywood, Canadian Douglas fir plywood, Canadian softwood plywood, Finnish conifer plywood and Swedish softwood plywood. Plywood group II comprises Finnish birch-faced plywood and Finnish birch plywood.

need be applied. Where the load is inclined at an angle α to the grain of the timber, the basic load, F, should be determined from the equation:

$$F = F_{II}F_\perp /(F_{II} \sin^2 \alpha + F_\perp \cos^2 \alpha) \qquad (4.6)$$

where F_{II} is the basic load parallel to the grain, obtained from Tables 69–80 of the code.
$F_\perp$ is the basic load perpendicular to the grain, obtained from Tables 69–80 of the code.

If the load is at an angle to the axis of the bolt, the component of the load perpendicular to the axis of the bolt should be not greater than the basic load given in Tables 69–80, modified where appropriate by the above equation.

For two-member joints where parallel members are of unequal thickness, the load for the thinner member should be used. Where members of unequal thickness are joined at an angle, the basic load for each member should be determined and the smaller load used. For three-member joints, the basic loads given in Tables 75–80 of the code should be used for joints where the outer members have the tabulated thickness and the inner member is twice as thick. For other thicknesses of inner or outer members, the load may be obtained by linear interpolation between the two adjacent tabulated thicknesses. If both inner and outer members have other thicknesses, the load should be taken as the lesser of the loads obtained by linear interpolation of the inner and outer member thicknesses separately.

4.5.3.2 Basic multiple shear loads

The basic load for a joint of more than three members should be taken as the sum of the basic loads for each shear plane, assuming that the joint consisted of a series of three-membered joints.

4.5.4 Steel plate-to-timber joints

4.5.4.1 Steel plate

Steel plates should have a minimum thickness of 2.5 mm or 0.3 times the bolt diameter, whichever is the greater, for the modification factors in 6.6.5.2 of the code to apply. The diameter of the holes in the timber should be as close as practicable to the nominal diameter of the bolt, and in no case more than 2 mm larger.

4.5.5 Permissible load for a joint

The permissible load for a bolted joint should be determined as the sum of the permissible loads for each bolt in the joint, where each permissible bolt load, F_{adm}, should be calculated from the equation:

$$F_{adm} = F \times K_{56} \times K_{57} \qquad (4.7)$$

where the basic load for a bolt, F, is taken from 6.6.4 or 6.6.5.2 as appropriate:
  K_{56} is the modification factor for moisture content;
  K_{57} is the modification factor for the number of bolts in each line.

For moisture content:
  $K_{56} = 1.0$ for a joint in service classes 1 and 2;
  $K_{56} = 0.7$ for a joint made in timber of service class 3 and used in that service class;
  $K_{56} = 0.4$ for a joint made in timber of service class 3 and used in service classes 1 and 2.
For the number of bolts in each line:
  Where a number of bolts the same diameter, acting in single or multiple shear, are symmetrically arranged in one or more lines parallel to the line of action of the load in a primarily axially loaded member in a structural framework (see 1.6.11) then:

$$K_{57} = 1 - \frac{3(n-1)}{100} \quad \text{for } n < 10 \qquad (4.8)$$

$$K_{57} = 0.7 \text{ for } n \geq 10 \qquad (4.9)$$

where n is the number of bolts in each line.
  In all other loading cases where more than one bolt is used in a joint:

$$K_{57} = 1.0$$

4.5.6 Steel dowel joints

4.5.6.1 General

The recommendations contained in 6.6.7 of the code are applicable to plain steel dowels with a minimum tensile strength of 400 N/mm² and a minimum dowel diameter of 8 mm. For the recommendations of 6.6.7 of the code to apply, the tolerance on the specified dowel diameter should be ±0.1 mm and the dowels should be inserted in pre-bored holes in the timber members having a diameter not greater than that of the dowel itself. Where a dowel does not extend to the surface of the outer member, for example, for reasons of appearance, the thickness of the member for calculation purposes should be taken as the actual embedment length of the dowel.

Care should be taken to avoid axial forces being set up in dowelled joints as a result of asymmetrical, eccentric or oscillating loads. Dowels may be used to form steel plate-to-timber joints in accordance with 6.6.5 of the code. Where the steel plates form the outer members of the joint, then the steel plates should be secured in position (eg by threading the ends of the dowel and applying nuts) and it is essential that the dowels have a full bearing on the steel plate.

4.5.6.2 Effective cross-section

When assessing the effective cross-section of multiple dowel joints, all dowels that lie within a distance of two dowel diameters measured parallel to the grain from a given cross-section should be considered as occurring at that cross-section. Then the effective cross-section should be determined by deducting the net projected area of the dowel holes from the gross area of the cross-section.

4.5.6.3 Dowel spacing

The values relating to bolts in 6.6.3 of the code apply.

4.5.6.4 Steel plate-to-timber

The steel plate should have adequate strength and be designed in accordance with BS 5950. The hole diameter in the timber and steel must be no more than the bolt diameter plus 2 mm, and preferably as near the bolt diameter as possible.

4.5.6.5 Basic load

When the timber member is loaded parallel to the grain, the basic load given in Table 69 of the code is multiplied by 1.25, the steel-to-timber modification factor K_{46}. K_{46} is not applied to timber loaded perpendicular to the grain.

4.5.6.6 Permissible joint load

The permissible load for a bolted joint is the sum of the permissible loads for each bolt in the joint which is:

$$\text{Basic load} \times K_{55} \times K_{56} \times K_{57} \qquad (4.10)$$

The modification factor for load duration K_{55} is

Long-term loads	1.0
Medium-term loads	1.25
Short- and very short-term loads	1.50

The modification factor for moisture content K_{56} is

Dry timber	1.0
Green timber or timber which will be under wet exposure conditions	0.7
Jointed green timber which will be under dry exposure conditions	0.4

The modification factor K_{57} for n the number of bolts in each line is (compared with equations 4.8 and 4.9):

0.7 for $n \geq 10$, or

$$1 - \frac{3(n-1)}{100} \quad \text{for } n < 10$$

where n is the number of same diameter bolts acting in single or multiple shear symmetrically placed in one or more lines parallel to the line of action of the load in a primarily axially loaded member in a structural framework (BS 5268: Part 2, clause 5.11).

In all other cases where more than one bolt is used, $K_{57} = 1.0$.

4.6 Based on Eurocode 5 and European practices

(A) Screws as fasteners

Based on EC5, clause 8.7.1 (3) and (4), the characteristic embedded strength f_{hgK} (N/mm^2) shall be treated the same as for nails in the same code. The same rules are applied to panel-to-timber connection within the same code.

4.7 Based on British code on basic withdrawal of loads

The basic withdrawal loads for single nails at right angles to the side of timber in service classes 1 and 2 are given in Table 62 of the code. These apply to each 1 mm depth of penetration, and for a particular nail, should be multiplied by the actual point side penetration achieved. The penetration of the nail should be not less than 155 mm. No withdrawal load should be carried by a nail driven into the end grain of timber. The basic loads given in this clause for each nail should be modified in accordance with 6.4.9 of the code to determine the permissible load for a joint.

A common way of representing bolted connection test results is to plot the normalised bolt bearing stress versus the t_m/D ratio. The normalised bolt-bearing strength is

$$P_n = \frac{F_p}{t_m D f_c}, \qquad (4.11a)$$

where

P_n	= normalised bolt-bearing strength
F_p	= proportional limit strength, lb
t_m	= main member thickness, in.,
D	= bolt diameter, in.,
f_c	= main member compressive strength, lb/in^2.

Several tables, graphs and design data exist to design timber joints under loads for laterally loaded single bolts. A number of possible modes can occur in a dowel-type connection (Fig. 4.6). The failure mode that results in the lowest yield load for a given geometry is the theoretical connection yield load.

Equations corresponding to the failure modes for a three-member joint are generally tabulated. The nominal single bolt value is dependent on the joint geometry (thickness of main and side members), bolt diameter and bending yield strength, dowel bearing strength, and direction of load to the grain. The equations are equally valid for wood or steel side members which is taken into account by thickness and dowel bearing strength parameters. The equations are also valid for various load to grain directions which are taken into account by the K parameter.

The dowel bearing strength is a material property not generally familiar to structural designers. It is determined from tests that relate species specific gravity and dowel

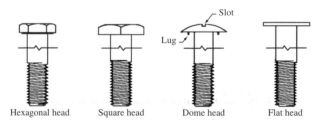

Figure 4.6 *Typical bolts based on U.S. practices.*

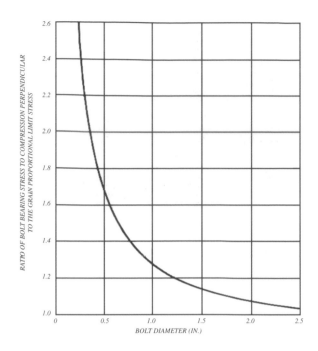

Figure 4.7 *Bearing stress perpendicular-to-grain as affected by bolt diameter.*

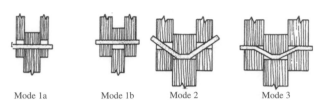

Figure 4.8 *Failure modes in the European yield model.*

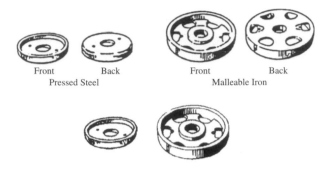

Figure 4.9 *Pressed and malleable iron shear plates (courtesy Cleveland Steel Specialty Company, USA).*

diameter to bearing strength. Empirical equations for these relationships are:

$$F_e = 11{,}200G \text{ (parallel-to-grain)}, \quad (4.12)$$
$$F_e = 6.1G^{1.45} D^{-0.5} \text{ (perpendicular-to-grain)}, \quad (4.13)$$

where F_e is the dowel bearing strength, G the specific gravity based on oven-dry volume, and D is the bolt diameter (in.).

The strength of a multiple-bolt connection is calculated from the single-bolt strength values, multiplied by an adjustment factor that depends on how the load is distributed to each bolt. The adjustment factor reduces connection strength when there are three or more bolts in a row. Figure 4.7 shows the relationship between bearing stress and bolt diameter.

4.7.1 Timber connectors based on US practices

Timber connectors have been in use for over a century and more than 60 different types were patented in Europe and the United States prior to 1930. The first connector patented in the United States was in 1889 and it was a toothed metal plate. The types of timber connectors currently in use are the shear plate, splits ring, spike grid, and circular grid. In the past in the United States, the clamping plate, toothed ring and claw plate were also used. Timber connectors are capable of transferring the largest shear force per unit of any of the mechanical fasteners available.

4.8 Timber connections

The six types of timber connectors discussed in this section are the shear plate, split ring, spike grid, circular grid, clamping plate, and toothed ring. Clamping plates and toothed rings are no longer manufactured in the United

States. Shear plates are intended for use for wood-to-steel connections or for wood-to-wood connections when used in pairs back-to-back. They are placed in pre-cut daps and are completely embedded in the wood.

They are used to attach beams to columns through steel straps, for fabrication of heavy timber trusses using steel splice plates and for connections between glulam members. The force is transmitted from the wood in bearing on the shear plate to the bolt, which in turn transmits the force to the steel side plate. In a wood-to-wood joint, two shear plates are used back-to-back and the force is transferred in the manner described above from one shear plate to the second shear plate.

Shear plates (Fig. 4.9) are manufactured from pressed-steel (a low carbon steel), malleable iron, 316 stainless steel, and fibreglass. Stainless steel and fibreglass shear plates are used in corrosive environments.

Split rings are the most efficient mechanical devices used for joints in timber construction. They are placed into pre-cut grooves that are slightly larger in diameter than the ring in wood-to-wood joints. They are used in trusted rafter construction, heavy beam and girder construction, towers, bridges, and long span roof trusses. Split rings currently available have a double-bevelled cross-section as shown in Fig. 4.10. The original split rings had a so-called "flat" cross-section and were the type of split ring used in the construction of the roof trusses. The single bevelled cross-section followed the "flat" cross-section and the bevelled surface was placed on the inside face of the ring.

The double-bevelled or wedge fit split-ring provides a tighter fit and makes it somewhat easier to install the ring into the pre-cut groove. The split in the ring allows the ring to expand to aid in inserting the ring into the pre-cut groove. A bolt is provided through the centre of the ring and serves to hold the members being joined in place. The bolt does not assist in the transfer of load in the range of design loads.

Figure 4.10 *Single bevelled "flat" cross-section split ring (Courtesy Cleveland Steel Specialty Company, USA).*

Flat   Single Curve

Figure 4.11 *Flat and single curve type spike grid connector (Courtesy Cleveland Steel Specialty Company, USA).*

Split rings are manufactured from carbon steel, stainless steel, silicon bronze, and naval brass. The stainless steel, silicon bronze, and naval brass would be used in corrosive environments. Care is urged in the use of stainless steel in environments that contain a high level of chlorine since the chlorine will react with the stainless steel causing it to corrode.

Spike grids come in two shapes, flat and single curved. These are used with piles and poles in trestle construction, piers and wharves, and transmission lines. The single-curved shape is used to join a flat surface to a pole and the flat shape is used between flat solid sawn timbers (Fig. 4.11).

The circular grid in Fig. 4.12 is used between the top of a pile and the cap to prevent lateral movement. It is also used in trestle and other heavy timber construction similar to the use of a spike grid and it has been used as a tie spacer.

The circular grid is installed by the application of pressure on the pieces to be joined. When it is used between the top of a pile and the cap, several light blows from the pile driver hammer suffice to install the circular grid.

4.8.1 Connectors

4.8.1.1 Material properties and dimensional data

(A) Dimensional data for shear plates

Shear plates are available in two sizes, 2-5/8 inch and 4-inch. The 2-5/8-inch shear plate is used with a 3/4-inch bolt and comes in two gage thicknesses, regular and light. The light gage plate has a hub at the bolt hole to provide for greater bolt-bearing strength. The 4-inch shear plate is used with a 5/8-inch or 7/8-inch bolt.

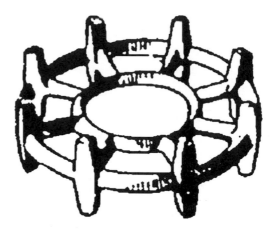

Figure 4.12 *Cicular grid connector (Courtesy Cleveland Steel Specialty Company, USA).*

Shear plates of the 2-7/8-inch size are manufactured from hot-rolled SAE-1010 carbon steel. In addition, for corrosive environment, this size shear plate is also manufactured using 316 stainless steel. The 4-inch shear plate is manufactured using only malleable iron. ASTM standard specification A47, Grade 32510 is used for malleable iron casting.

Row of timber connectors: If several connectors are placed in a row, and the joint is loaded parallel to the row of connectors, the failure mode is by splitting the wood along its grain and parallel to the direction of load. The failure load is typically less than the ultimate load for a single connector times the number of connectors in the row. A group reduction factor needs to be applied when there are more than two connectors in a row.

Displacement of connectors perpendicular-to-grain: When timber connector units are offset so that the centres of the units do not fall along the same wood grain, the allowable value is equal to the single unit value times the number of units. Failure will occur by splitting the lumber along several grains.

(B) Dimensional data for split rings, spike grids and circular grids

Split rings are available in two sizes, 2-1/2 inch and 4 inch. A 1/2-inch bolt is used with the 2-1/2-inch size and a 3/4-inch bolt is used with the 4-inch size.

Spike-grids are available in two styles, flat and single curved and are used with a 3/4- or 1-inch bolt. Circular grids are available in a single size and are used with a 3/4- or 1-inch bolt.

4.8.2 General data and specifications for connector design

4.8.2.1 Coatings

Shear plates, split rings, spike-grids and circular grids are available with a hot-dip galvanised coating. The 4-inch shear plate, spike-grid, and circular grid which are all manufactured from malleable iron are galvanised using ASTM A-153 specifications while the 2-5/8-inch steel shear plate and the split rings are galvanised using ASTM A-123 specifications.

4.8.2.2 Factors that affect allowable connector design values

The factors that affect the design values of a timber connector joint in wood are:

1. size and type of timber connector,
2. species of wood used,
3. moisture content of wood,
4. duration of load,
5. temperature,
6. fire retardant treatment,
7. preservative,
8. angle of load to grain,
9. width of member, treatment,
10. thickness of member,
11. end and edge distances,
12. spacing between connectors parallel and perpendicular to grain,
13. net section of member,
14. number of timber connectors in a row,
15. type of side plates, and
16. type of fastener used with timber connector (bolt or lag bolt).

4.8.2.3 Allowable design values

Allowable loads for split rings and shear plates were obtained by applying a reduction factor to the ultimate test load. It was found that a factor of four gave loads that would not exceed 5/8 of the proportional limit test load. This does not, however, imply a factor of safety of four.

Due to reduction factors needed from a short-duration load (test) to long-duration, and a factor of 3/4, to account for variability of the wood the true factor of safety is more likely to be about 1.75 for a long-duration load.

Figures 4.13 to 4.15 show some test results for various design parameters.

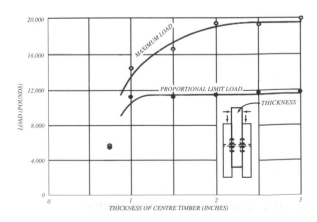

Figure 4.13 *Effect of centre member thickness on proportional limit maximum loads for a 4 inch split ring.*

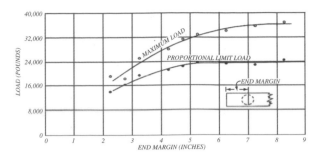

Figure 4.14 *Effect of end distance on load parallel-to-grain for 4-in. split ring.*

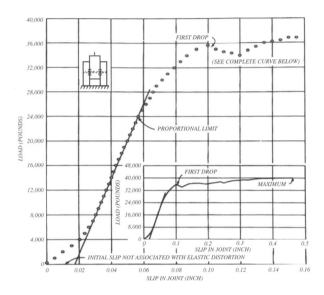

Figure 4.15 *Load-slip for a 4-inch split ring connector joint with 3/4-inch bolt load parallel to grain.*

4.8.2.4 Net section: number of timber connectors in a row and type of sideplates

These are similar for all types of mechanical fasteners and connectors. Table 4.17 provides useful data when calculating the net section for shear plates and split ring.

4.9 Connectors and joints based on BS 5268

4.9.1 Introduction

Associated with each size of split-ring connector is a standard distance, end distance, edge distance and spacing between connectors which permit the basic load to apply. These standard distances are given in Table 4.18–4.21. If the end distance, edge distance or spacing is less than the standard distance, the basic load should be modified (see 6.8.5).

The basic loads parallel and perpendicular to the grain for split-ring connector units are given in Tables 4.18 and 4.19 for timber members satisfying the requirement for strength classes C14 to D70.

Where the load is inclined at an angle α to the grain of the timber, the basic load F should be determined from the equation given in 6.6.4.1 of the code.

Table 4.17 Projected area of connectors and bolts (sq. in.)

Connectors		Bolt Diam.	Placement of Connectors	Total Projected Area in Square Inches of Connectors & Bolts in Lumber Thickness of				
No.	Size			1-1/2	1-1/2	3-1/2	5-1/2	7-1/2
SPLIT	RINGS	1/2	One Face	1.71	2.27	2.84	3.89	5.02
1	2-1/2	1/2	Two Faces	2.60	3.16	3.73	4.78	5.91
2	4	3/4	One Face	3.01	3.86	4.64	6.16	7.79
		3/4	Two Faces	4.85	5.66	6.47	8.00	9.62
SHEAR	PLATES	3/4	One Face	2.00	2.81	3.62	4.14	6.77
1	2-5/8	3/4	Two Faces	2.81	2.68	4.43	5.96	7.58
1	2-5/8	3/4	One Face	1.87	2.68	3.50	5.02	6.65
	LG	3/4	Two Faces	2.56	3.38	4.19	5.71	7.34
2	4	3/4	One Face	3.24	4.05	4.87	6.39	8.01
		3/4	Two Faces	–	6.11	6.93	8.45	10.07

Table 4.18 End distance for split-ring and shear-plate connectors

Type of end distance	Angle[a] of load to grain a degrees	End distance mm			
		Connector size			
		64 mm split-ring or 67 mm shear-plate		102 mm split-ring or 102 mm shear-plate	
		Minimum	Standard	Minimum	Standard
Unloaded	0	64	102	83	140
	45	67	121	86	159
	90	70	140	89	178
Loaded	0 to 90	70	140	89	178

[a] For intermediate angles, values should be obtained by linear interpolation.

Table 4.19 Edge distance for split-ring and shear plate connectors

Type of edge distance	Angle[a] of load to grain a degrees	Edge distance mm			
		Connector size			
		64 mm split-ring or 67 mm shear-plate		102 mm split-ring or 102 mm shear-plate	
		Minimum	Standard	Minimum	Standard
Unloaded	0 to 90	44	44	70	70
Loaded	0	44	44	70	70
	15	44	54	70	79
	30	44	64	70	87
	45 to 90	44	70	70	95

[a] For intermediate angles, values should be obtained by linear interpolation.

The basic loads given in this clause for each split-ring connector unit should be modified in accordance with 6.8.5 to determine the permissible load for a joint.

4.9.2 Bolts and washers

The nominal diameters of the bolts to be used with shear-plate connectors are given in Table 4.29. Bolt holes should be as close as practicable to the nominal diameter of the bolt, and in no case more than 2.0 mm larger than the bolt diameter. Round or square washers should be fitted between the timber and the head of the bolt. The minimum size of washer to be used with each connector is given in Table 4.21. Connectors should not bear on the threads of bolts.

4.9.3 Joint preparation

To prepare a connected joint, the positions of the bolt holes should be set out accurately with reference to the point of intersection of the centre-lines of the members.

One of the following two procedures should be used when drilling the bolt holes:

a) fit the members together in their correct positions and clamp while drilling the bolt holes through all of the members;

b) drill the bolt holes in the individual members using jigs or templates to locate the bolt holes accurately.

Bolt holes should be within 2 mm of their specified position. The contact surfaces of the timber members should be recessed to the dimensions. The recesses for shear-plates may be cut simultaneously with the drilling of the bolt holes.

4.10 Glued joints

4.10.1 Laterally loaded joints

Joints in structural components made from separate pieces of timber, plywood and other panel products

Table 4.20 Basic loads for one split-ring connector unit

Split-ring diameter mm	Nominal bolt size and thread diameter mm	Actual thickness of members[a] mm		Basic load kN — Angle of load to grain											
		Connector on one side only	Connectors on both sides and on same bolt	C14		C16/18/22		C24		C27/30/35/400°		D30/35/40		D50/60/70	
				0°	90°	0°	90°	0°	90°	0°	90°	0°	90°	0°	90°
64	M12	22	32	5.23	3.66	5.38	3.77	5.85	4.09	6.28	4.39	8.92	6.25	10.99	7.70
		25	40	6.32	4.42	6.51	4.55	7.06	4.95	7.58	5.31	10.80	7.55	13.31	9.31
		29	50	7.68	5.38	7.91	5.54	8.58	6.01	9.21	6.45	13.10	9.18	16.15	11.31
102	M20	29	41	10.10	7.04	10.38	7.25	11.22	7.87	12.04	8.45	17.20	12.00	21.20	14.79
		32	50	11.50	8.01	11.82	8.25	12.78	8.95	13.72	9.61	19.50	13.70	24.03	16.89
		36	63	13.50	9.42	13.88	9.70	15.02	10.52	16.12	11.30	23.00	16.10	28.35	19.84
		40	72	14.30	9.98	14.72	10.26	15.98	11.12	17.14	11.94	24.30	17.00	29.95	20.95
		41	75	14.40	10.10	14.82	10.38	16.08	11.22	17.26	12.04	24.50	17.20	30.20	21.20

[a] For intermediate thicknesses, values may be obtained by linear interpolation.

Table 4.21 Sizes of shear-plate connectors and minimum sizes of washers

Nominal size of connector mm	Nominal size and thread diameter of bolt	Minimum size of round or square washers	
		Diameter or length of side mm	Thickness mm
67	M20	75	5
102	M20	75	5

NOTE The sizes given in this table are metric conversions of the imperial sizes given in BS 1579.

Table 4.22 Limiting values for permissible loads on one shear plate connectors unit

Shear-plate diameter mm	Nominal bolt size	All loading except short- and very short-term loading kN	All loading including short- and very short-term loading kN
67	M20	12.9	17.2
102	M20	22.1	29.5

that are fastencd by glue (eg box beams, single web beams, stressed skinned panels, glued gussets) should be manufactured in accordance with BS 6446.

The provision of this clause for glued structural joints applies only to the softwood species listed in Table A.1 (with the exception of excessively resinous pieces), to the plywoods listed in Section 4, and to the other panel products described in Section 5. For gluing of hardwoods and resinous softwoods, advice should be sought from the glue manufacturer.

The provisions of these clauses are limited to the following structural joints:

1. between solid or laminated timber members of which the dimension at right angles to the plan of the glue line is not greater than 50 mm;
2. between solid or laminated timber members of any dimension and plywood or wood particleboard no thicker than 29 mm;
3. between solid or laminated timber members of any dimension or tempered hardboard no thicker than 8 mm;
4. between plywood members of any thickness;
5. between tempered hardboard members of any thickness;
6. between wood particle members of any thickness.

Considcration should be given to the possibility of differential shrinkage, distortion and stress concentration at glued joints. When mechanical fasteners are present in a glued joint thcy should not be considered as contributing to the strength of the joint.

4.10.2 Adhesives

The adhesive used should be appropriate to the environment in which the joint will be used.

Designers should ensure that, in the case of an MR type of adhesive conforming to BS 1204, a particular formulation is suitable for the service conditions and for the intended life of the structure.

4.10.3 Permissible load for a joint

Where more than one connector is used with the same bolt in a multi-member joint, the appropriate permissible load should be calculated for each connector unit, and these loads should be added together.

Under no circumstances should the permissible load for a shear-plate connector unit exceed the limiting values given in Table 4.30. The permissible load for a shear-plate connectored joint, F_{adm} should be determined as the sum of the permissible loads for each connector unit in the joint, where each permissible connector unit load should be calculated as:

$$F_{adm} = F \times K_{66} \times K_{67} \times K_{68} \times K_{69} \quad (4.14)$$

where the basic load for a shear-plate connector unit, F, is taken from 6.9.4.3 or 6.9.5.3, as appropriate, and

K_{66} is the modification factor for duration of loading;
K_{67} is the modification factor for moisture content;
K_{68} is the modification factor for end distance, edge distance and centre spacing.
K_{68} is the modification factor for the number of connectors in each line

For duration of loading:

K_{66} = 1.00 for long-term loads;
K_{66} = 1.25 for medium-term loads;
K_{66} = 1.50 for short- and very short-term loads.

For moisture content:

K_{67} = 1.0 for a joint in service classes 1 and 2:
K_{67} = 0.7 for a joint in service class 3

For end distance, edge distance and spacing:

If the end distance, edge distance and/or centre spacing of the connectors are less than the relevant standard values given in Tables 91–93, respectively, the modification factor, K_{68}, should have the lowest of the values for K_s, K_c and K_d given in Tables 93, 95 and 96, respectively. In no case should the distances and spacing be less than the minimum values given in Tables 91 93. No increase in load is pcrmitted if end distance, edge distance or centre spacing exceeds the standard values.

If shear-plate connectors are used in green hardwood, the standard end distance must be multiplied by 1.5. One-half of this increascd end distance should be taken as the minimum end distance, with a permissible load of one-half of that permitted for the standard end distancc.

For the number of connectors in each line:

where a number of connector units of the same size are symmetrically arranged in one or more lines parallel to the line of action of the load in a primarily axially loaded member of a structural framework then:

$$K_{69} = 1 - \frac{(3n-1)}{100} \quad (4.15)$$
$$\text{for } n < 10$$
$$K_{69} = 0.7 \text{ for } n \geq 10 \quad (4.16)$$

where n is the number of connector units in each line.

In all other loading cases where more than one shear-plate connector unit is used in a joint:
$K_{69} = 1.0$

4.11 Detailing data to Eurocode-5 and European practices

4.11.1 Introduction

EC5 is based on studies undertaken by Working Commission W18 of the International Council for Building Research Studies and Documentation (CIB), in particular the CIB Structural Timber Design Code, produced in 1983. At all stages of its development there has been extensive input from the national standards bodies through their representatives in CEN. The seven-man team which drafted Eurocode 5 included two British members.

In principle, DD-ENV-1995 Part 1.1 contains all of the rules necessary for the design of timber structures. However, it does not contain the material properties, table of fastener loads and other design information found in BS 5268: Part 2. Further to this, the section now contains specific areas on joints in timber structures. Data on fastenings are given for the first time which are in line with the Eurocode5 and European practices.

In EC5, the characteristic values of timber strength properties are converted to design values by dividing them by a partial factor of 1.3 and multiplying them by a factor k_{mod}, which is generally less than unity and which takes account of both load duration and moisture content. There are five load duration classes and three moisture content or *service* classes. For solid timber, glulam and plywood, k_{mod} ranges from 1.1 for *instantaneous* loading in service class 1, down to 0.50 for *permanent* loading in service class 3. k_{mod} is not applied to steel used in joints, for which a partial factor of 1.1 rather than 1.3 is used.

For the calculation of deflection and joint slip in serviceability limit states, EC5 gives rules for the calculation of creep (time-dependent deformation) as well as instantaneous deformation.

4.11.2 Advantages and disadvantages of EC5

4.11.2.1 Advantages

1. The Standard represents the best-informed consensus on timber design in Europe.
2. Its widespread use will facilitate intimation: trade and consultancy.
3. The use of a design method common to all materials will simplify the work of engineers.
4. EC5 facilitates a wider selection of materials and components than BS 5268: Part 2.
5. It gives more guidance than BS 5268: Part 2 on the design of built-up components.
6. There is a unified design and safety basis for laterally-loaded dowel-type joints.
7. In general, designs carried out to the Eurocode standard should have more uniform reliability.
8. Provided that the appropriate NAD is used, British designs will be acceptable in other European countries.
9. Checking designs from other European countries for structures in the UK will no longer require knowledge of foreign codes.

4.11.2.2 Disadvantages

1. Design time will generally be greater, at least until designers are familiar with the format. However, EC5's preference for formulae

makes programming easy, so the provision on computerised design programs should overcome this disadvantage.
2. It will be necessary to have a number of additional reference documents to hand.

4.11.3 Nails and screws

In European practices, various nails and screws are used ranging from ordinary round wire nails to self-tapping connectors. These are shown in Fig. 4.16.

The ongoing development of nails and connectors has also resulted in load-bearing systems made from nailed boards. Factory-produced nailed lattice girders are now highly economic propositions for short spans. Boarded beams or frames are commonly used for agricultural buildings. Larger spans are possible thanks to the use of new nailing techniques and pre-drilling in conjunction with thicker planks.

The shape of the nail head and the layout of the nails should not be neglected for situations with high aesthetic demands. In shear wood, screws behave just like nails, but their pull-out loads are higher and the appearance of the head is an improvement over nails. The development of self-tapping connectors that can be installed with hand-held plant without needing to drill a pilot hole has made the use of these wood screws very economical and hence successful.

In many instances, components can be fixed in position with a tension-resistant connector by choosing a suitable size of screw. In addition, the recently launched double thread screws do not just join together several timber sections to form a compound member, but also enable butt joints between main and secondary members to be achieved without any further means of connection.

Self-drilling dowels render possible multiple-shear dowelled connections without having to drill the timber and sheet metal components first. The short thread just below the head of the dowel serves to secure the position of the member once installed.

4.12 Bolted and dowelled section

4.12.1 Dowel-type fastener joints

EC5 section 8 deals with the ultimate limit state design criteria for joints – with dowel-type fasteners such as nails, screws, staples, bolts and dowels, and connectored-type joints, ie toothed plates, split rings, etc. EC5 provides formulae, based on Johansen's equations, to determine the carrying capacity of timber-to-timber, panel-to-timber and steel-to-timber reflecting all possible failure modes. Load-carrying capacity formulae for steel-to-timber and panel-to-timber joints, with reference to their failure modes and for steel-to-timber joints are compiled in various tables.

Based on Tables 8.4 and 8.5 (EC-5), the minimum bolt and dowel spacings and distances are given in Table 4.23.

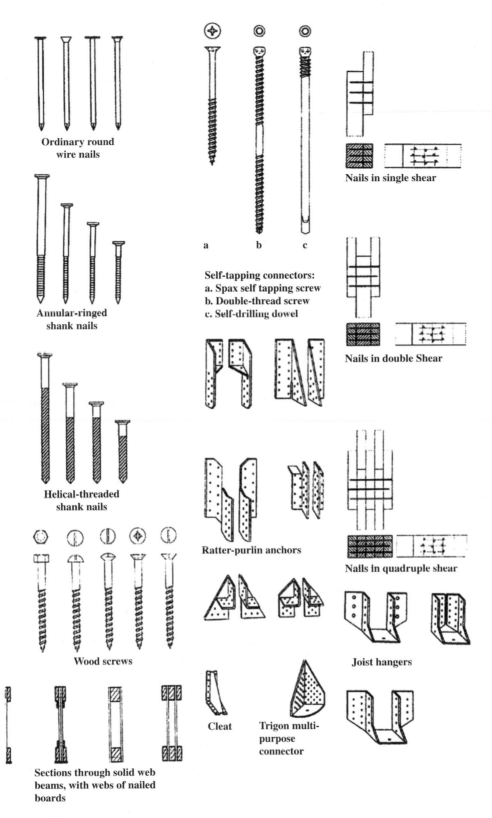

Figure 4.16 *Nails, screws and sheet metal connectors for nails.*

Table 4.23 Minimum bolt and dowel spacings and distances (based on Tables 8.4 and 8.5, EC5)

Spacings and distances	Bolts	Dowels
Spacing parallel (a_1)	$(4+|\cos\alpha|)d$	$(3+2\,|\cos\alpha|)d$
Spacing. Perpendicular (a_2)	$4d$	$3d$
Loaded end distance $(a_{3,t})$		
$-90° \le \alpha \le 90°$	max of [$7d$ and 80 mm]	max of [$7d$ and 80 mm]
Unloaded end distance $(a_{3,c})$		
$150° \le \alpha < 210°$	$4d$	$3d$
$90° \le \alpha < 150°$ and $210° < \alpha \le 270°$	max of $[(1 + 6\,|\sin\alpha|)d$ and $4d]$	max of $[(a_{3,t}\,|\sin\alpha|)d$ and $3d]$
Loaded edge distance $(a_{4,t})$		
$0° \le \alpha \le 180°$	max of $[(2 + 2\,|\sin\alpha|)d$ and $3d]$	max of $[(2 + 2\,|\sin a \,\alpha|)d$ and $3d]$
Unloaded edge distance $(a_{4,c})$		
all other values of α	$3d$	$3d$

d = bolt or dowel diameter in mm and α = angle of force to grain direction.

4.12.2 Comparative study of the values of characteristic embedding strength $f_{h,k}$ based on clauses 8.3 and 8.7 of EC5

It is important at this stage for the designer/detailer to know the comparative values of the embedded strength $f_{h,k}$ of nails, screws and bolts. These are given in Table 4.24.

4.12.3 Joint slip

EC5 clause 7.1 provides a design procedure for the calculation of slip for joints made with metal fasteners such as nails, screws, bolts and dowels and shear plates. In general, the serviceability requirements are to ensure that, for toothed connectors, any deformation caused by slip does not impair the satisfactory functioning of the structure with regard to strength, attached materials and appearance.

The procedure for calculation of slip for joints made with dowel-type fasteners is summarised as follows:

1. The instantaneous slip, u_{inst} in mm, is calculated as:

   for serviceability verification $u_{inst,ser} = \dfrac{F_d}{K_{ser}}$ (4.17)

   for strength verification $u_{inst,u} = \dfrac{3F_d}{2K_{ser}}$ (4.18)

   where $u_{inst.ser}$ = the instantaneous slip at the serviceability limit state.

2. The slip modulus for the ultimate limit state, K_u should be taken as:

$$K_u = \frac{2}{3} K_{ser} \quad (4.18a)$$

Values of K_{ser} are given in Table 4.25.

If the mean densities of the two joined members are $\rho_{m,1}$ and $\rho_{m,2}$, then is calculated as

$$\rho_m = \sqrt{\rho_{m,1}\rho_{m,2}} \quad (4.19)$$

3. The final joint slip, u_{fin}, should be calculated as:

$$u_{fin} = u_{inst}(1 + \psi_2 k_{def}) \quad (4.20)$$

Table 4.24 Values of the characteristic embedding strength, $f_{h,k}$ (based on Clauses 8.3 to 8.7)

	$f_{h,k}$ (N/mm^2)			
Fastener type	Timber-to-limber	Steel-to-timber	Panel-to-timber	
Nails (d ≤ 6 mm) and staples	$0.082\,\rho_\kappa d^{-0.3}$ $0.082(1 - 0.01d)\rho_\kappa$	without pre-drilling pre-drilled	$0.11\,\rho_\kappa d^{-0.3}$ $30t^{0.6}d^{-0.3}$	for plywood for hardboard
Bolls, dowels and nails (d > 6 mm) parallel to grain at an angle α to grain	$0.082(1 - 0.01d)\rho_\kappa$: $f_{h.0.k}/(k_{90}\sin^2\alpha + \cos^2\alpha)$		$0.11(1 - 0.01d)\rho_\kappa$	
Screws (see EC5, clause 8.7.1(3) and (4))	Rules for nails apply		Rules for nails apply	

ρ_κ = the joint member characteristic density, given in Table 1, BS EN 338:1995 (in kg/m^3).
d = the fastener diameter (in mm) characteristic. For screw d is the smooth shank diameter.
t = the panel thickness.
k_{90} = 1.35 + 0.015d for softwoods
     = 0.90 + 0.015d for hardwoods.

Table 4.25 Values of slip moduli K_{ser} per shear plane (based on Table 7.1, EC5)

Fastener Type	K_{ser} (N/mm)	
	Timber-to-timber Panel-to-timber	Steel-to-timber Concrete-to-timber
Bolts (without clearance)* and dowels	$\dfrac{1}{35}\rho_{m}1.5^{1.5}d$	$\dfrac{2}{35}\rho_{m}^{1.5}d$
Screws Nails (pre-drilling) Nails (no pre-drilling)	$\dfrac{1}{45}\rho_{m}^{1.5}d^{0.8}$	$\dfrac{2}{45}\rho_{m}^{1.5}d^{0}$
Staples	$\dfrac{1}{120}\rho_{m}^{1.5}d^{0.8}$	$\dfrac{1}{60}\rho_{m}^{1.5}d^{0}$

ρ_{m} is the joint member mean density, given in Table, BS EN 338: 1995.

* Bolt clearance should be added separately to the joint slip.

where k_{def} is the modification factor for deformation.

4. For a joint constructed from members with different creep properties ($k_{def,1}$ and $k_{def,2}$), the final joint slip should be calculated as:

$$u_{fin} = u_{inst}(1 + \psi_2 \sqrt{k_{def,1}k_{def,2}}) \qquad (4.21)$$

For joints made with bolts, EC5 recommends that the clearance (mm), should be added to joint slip values.

Figure 4.17 gives details of bolted and dowelled joints using European practices in line with EC5.

Dowels are cylindrical rods that are driven into pre-drilled holes. High load and stiffness can be achieved because there is no play through shrinkage or the need for oversized holes. According to recent studies, it is not necessary to stagger the dowels with respect to the line of splitting (DIN 1052 part 2, E 5.7).

In contrast to dowels, bolts require an oversized hole to be pre-drilled. However, bolts can only be employed for locating purposes, or carrying considerably lower loads than dowels when the deformation behaviour has only a small influence on the overall deformation of the structure.

Close tolerance bolts are driven into pre-drilled holes and, as with a dowel, there is no play. When fitted with nut and washer, the short thread at the end of the bolt serves to locate the bolt and the component. Loads and stiffnesses are identical to those of a dowel of the same size.

4.12.4 Halving joints

The various halving joint variations are, first and foremost, structural connections with a low load-carrying capacity involving a considerable weakening of the cross-section. However, they are useful in, for example, roofs and frames.

The use of traditional wood joints is still relevant in the exposed constructions of historical buildings of past centuries and in the reconstruction of historical structures. Beams bearing on masonry, concrete etc must always include a separating pad of material to prevent saturation of the end of the timber.

4.12.5 Craftsman-type connections and connectors

Wood joints made by carpenters (eg tenons, halving joints, oblique dado joints) can now be produced with modern machinery, accurately and economically. Above all, these traditional joints are often a sensible alternative to customary connections using sheet metal parts for fabrication in large, computer-controlled plants. The forces are transferred mostly by contact faces that demand a high level of accuracy. The utmost care should be taken to ensure a low moisture content in the wood. The disadvantages of craftsman-type wood joints are considerable weakening of the members and the (usually) indistinct stress relationships.

Theoretically, only relatively low loads can be carried due to the severe shearing and eccentricity effects.

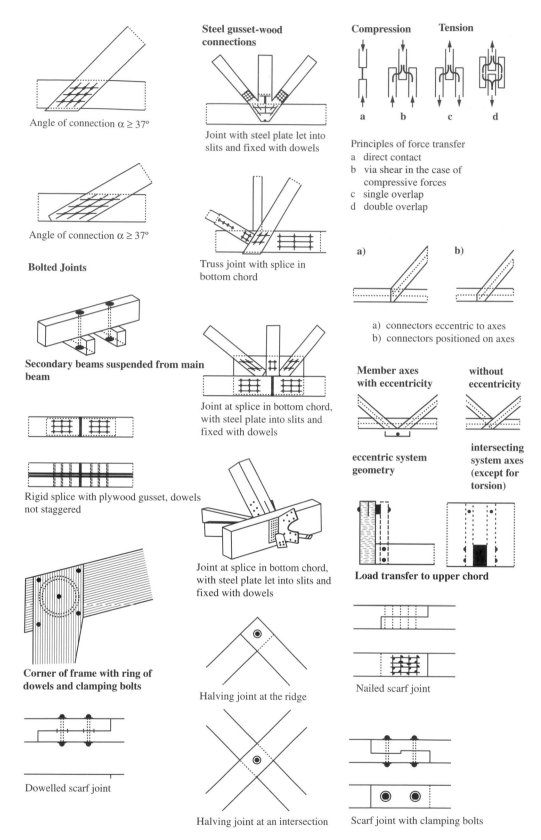

Angle of connection α ≥ 37°

Angle of connection α ≥ 37°

Bolted Joints

Secondary beams suspended from main beam

Rigid splice with plywood gusset, dowels not staggered

Corner of frame with ring of dowels and clamping bolts

Dowelled scarf joint

Steel gusset-wood connections

Joint with steel plate let into slits and fixed with dowels

Truss joint with splice in bottom chord

Joint at splice in bottom chord, with steel plate into slits and fixed with dowels

Joint at splice in bottom chord, with steel plate let into slits and fixed with dowels

Halving joint at the ridge

Halving joint at an intersection

Compression   Tension

a   b   c   d

Principles of force transfer
a   direct contact
b   via shear in the case of compressive forces
c   single overlap
d   double overlap

a)   b)

a) connectors eccentric to axes
b) connectors positioned on axes

Member axes with eccentricity   without eccentricity

eccentric system geometry   intersecting system axes (except for torsion)

Load transfer to upper chord

Nailed scarf joint

Scarf joint with clamping bolts

Figure 4.17 *Bolted and dowelled joints.*

Chapter 5
Structural Detailing of Timber Joints and Components

5.1 Introduction

This chapter deals exclusively with the detailing and fastening of structural elements in timber. Various detailing and fastening practices are introduced and the British, European and American practices are given. Wherever possible, explanatory notes are given to indicate the methods of fastening based on specific codes and practices.

Joints play a decisive role in timber engineering; they are very design-relevant due to the stiffness and strength anisotropy of the material. Timber constructions and elements are neither cast nor welded but consist of assemblies of discrete parts, being columns, members, rafters, studs and sheeting connected by discrete joints. Very often, beam-like elements are not monolithic but are themselves jointed assemblies of smaller parts.

Two main groups of joints are differentiated: joints with mechanical fasteners and glued joints. Both have their advantages and disadvantages; in some areas of application, they offer technical and economical alternative solutions. Several new and promising jointing technologies are likely to enhance the potential of mechanical fastenings: for instance, new types of self-tapping screws, simultaneous steel drilling-bolt setting concepts and hidden joint solutions emanating from furniture technology.

5.2 Joints and fastenings

Figure 5.1 indicates fastening of beam functions based on EC5 and European design practices. As seen, joints are with ledger strip, nailed, screwed and glued. Tenons and screws are provided for joints for tension resistance. In some cases, stiffened angle and joist hangers with fish plates are used to resist tension in joints. For shallow roof pitches, European practices are to use metal framing anchors.

Figures 5.2 and 5.3 are devoted to timber truss joints as practiced in European countries while keeping in mind the strict regulations cited in EC5. Diagonals are nailed to the bottom chord using simple connecters.

If the joints have verticals and diagonals, the verticals are connected with cleats and nails while the diagonals are fully nailed to the bottom chord. Sometimes, struts and diagonals are nailed through metal plates. Dowels are also used on many occasions in pre-drilled plates. Sometimes, diagonals with halving hinge pin,

for beam truss joints, or two part board trussing can be adopted with timber spaces and or glued laminated timber trussing. Sometimes a chord has a block nailed to a chamfered edge or with a strict halving hinge pin.

Figure 5.4 shows truss joints with V-struts based on European practices which are strictly governed by EC5. Dowelled plates and ties are connected with pins. Detail is also given for dowelled web plate welded to timber ties. It is interesting to see a dowelled web plate fastened at 90° to the grain of the strut. Timber joints can also be detailed with external steel gusset plates having steel dowels. This is in line with EC5.

Glued structural built-up beams may take the form of box beams or I-beams as shown in Fig. 5.5. In these components, the web members are plywood and the flanges may be solid pieces of stress-grade lumber or glued laminated timber. Most of the bending stresses are carried by the lumber flanges and the plywood webs carry the shear stresses.

The glued joints between the flanges must be designed with care because they transfer stresses between these two elements of the beam. Figure 5.5c shows vertical stiffeners used to resist web buckling and to aid in distributing concentrated loads; stiffeners are also used in the box beams shown in Fig. 5.5a and 5.5b. The design method for built-up plywood beams is presented in supplement 2 of the *Plywood Design Specification*. The specifications are accepted by ANS and LRFD codes (see Fig. 5.6 for plywood and sheating details).

In addition to its use for subfloors, roof decking and wall sheathing, plywood is fabricated into structural components. These consist of plywood panels and lumber glued together with rigid adhesive to make a single integrated structural element capable of resisting high stresses. One such component is the *stressed-skin panel* which consists of plywood sheets placed above and below relatively small wood stringers or joists. When used as a structural floor element, the flexural stresses are resisted by the plywood and the shear stresses by the lumber stringers. Because the glued joints between the plywood skins and the stringers have no slip, the panel acts as a composite structural assembly. It should be noted that in stressed-skin panels, the direction of the face grain is *parallel* to the stringers; consequently the spacing of stringers may be controlled by the strength of the plywood to the face grain.

Because the stresses developed in the materials in stressed-skin panels may be considerably higher than

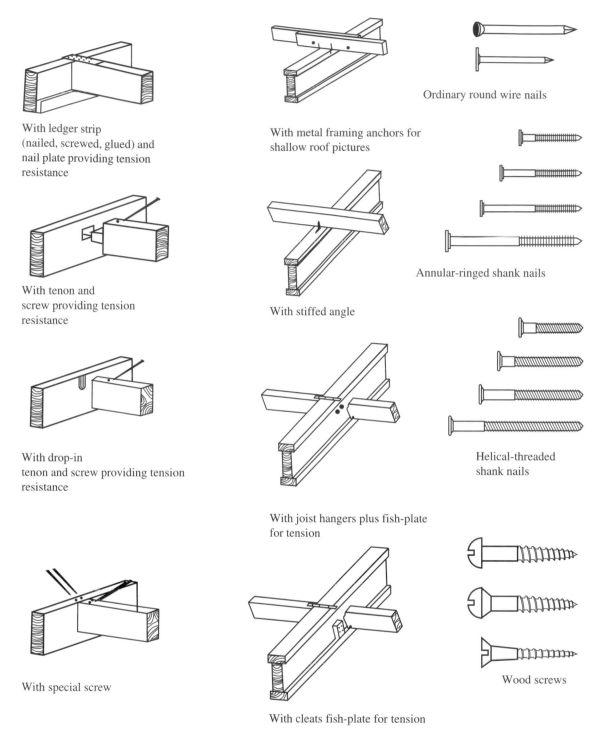

With ledger strip
(nailed, screwed, glued) and
nail plate providing tension
resistance

With tenon and
screw providing tension
resistance

With drop-in
tenon and screw providing tension
resistance

With special screw

With metal framing anchors for
shallow roof pictures

With stiffed angle

With joist hangers plus fish-plate
for tension

With cleats fish-plate for tension

Ordinary round wire nails

Annular-ringed shank nails

Helical-threaded
shank nails

Wood screws

Figure 5.1 *Beam junctions (based on EC5 and European design practices).*

in conventional construction, it is necessary that panels of this type be designed by qualified engineers or architects. They must be fabricated in shops equipped for the purpose where adequate quality control can be maintained throughout the manufacturing process.

Figure 5.6 gives a plywood beam with sheeting details based on LFRD and other American practices.

5.2.1 Stress calculations for plywood

The design aids which enable the required thickness of wood structural panel sheathing to be determined without detailed design calculations have been described in previous sections. For most practical sheathing problems, these methods are adequate to determine the required grade and panel thickness of a wood structural panel.

In the design of structural components such as box beams with wood structural panel webs and lumber flanges, foam-core sandwich panels, or stress-skin panels, it is necessary to use allowable stresses and cross-section panel properties in design calculations. If it becomes necessary to perform structural calculations for wood structural panels, the designer must be familiar

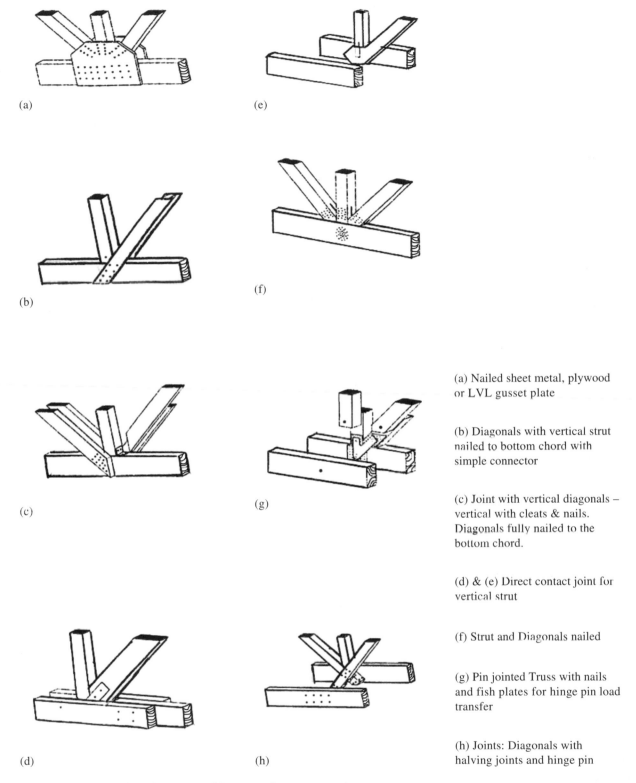

(a)

(b)

(c)

(d)

(e)

(f)

(g)

(h)

(a) Nailed sheet metal, plywood or LVL gusset plate

(b) Diagonals with vertical strut nailed to bottom chord with simple connector

(c) Joint with vertical diagonals – vertical with cleats & nails. Diagonals fully nailed to the bottom chord.

(d) & (e) Direct contact joint for vertical strut

(f) Strut and Diagonals nailed

(g) Pin jointed Truss with nails and fish plates for hinge pin load transfer

(h) Joints: Diagonals with halving joints and hinge pin

Figure 5.2 *Truss joints (based on EC5 and European design practices).*

with a number of factors which interrelate to define the panel's structural capacity. The document which gives effective panel cross-section properties and allowable stresses for PS 1 plywood is the *Plywood Design Specification* (PDS).

It will be helpful to review several factors which are unique to plywood structural calculations. These will provide a useful background to the designer even if structural calculations are not required. Although some of these points have been introduced previously, they are briefly summarised here.

Panel cross-section properties. The panel cross-section properties for plywood are tabulated in the PDS, and

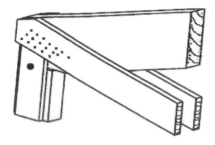

Joint – Consisting two part board trussing

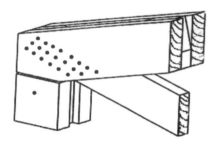

Joint – consisting of two chord having timber spacers and one-piece of glued laminated timber trussing.

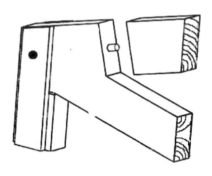

Joint – A chord has a block nailed to a champhered edge

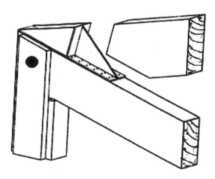

Joint – Chord and street with a hinge pin.

Figure 5.3 *Beam-trussing junction (based on EC5 and European practices).*

UBC Tables 23-2-H and 23-2-1. The variables which affect the properties are

1. Direction of stress
2. Surface condition
3. Species makeup

The *direction of stress* relates to the two-directional behaviour of plywood because of its cross-laminated construction. Because of this type of construction, two sets of cross-section properties are tabulated. One applies when plywood is stressed parallel to the face grain, and

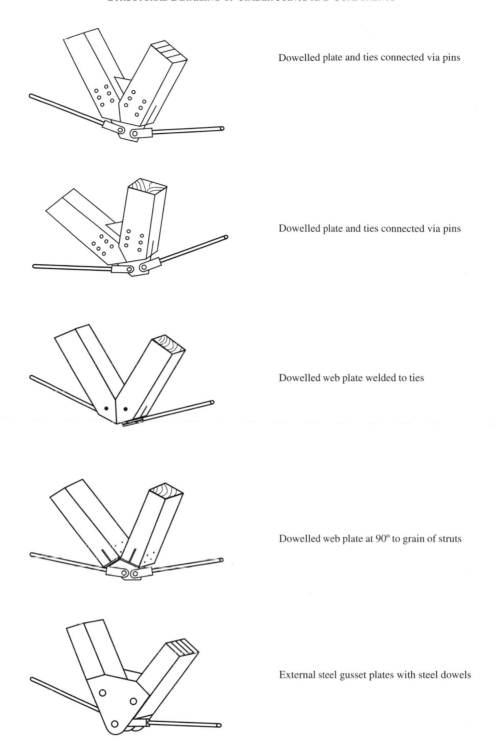

Dowelled plate and ties connected via pins

Dowelled plate and ties connected via pins

Dowelled web plate welded to ties

Dowelled web plate at 90° to grain of struts

External steel gusset plates with steel dowels

Figure 5.4 *Truss joints with V-struts (based on EC5 and European practices).*

the other applies when it is stressed perpendicular to the face grain.

The *basic surface conditions* for plywood are

1. Sanded
2. Touch-sanded
3. Unsanded

Figure 5.7 gives a brief summary of beam grid connections and supports. Here a detail is given using EC5 about a solid web beam strengthened and end dowelled to metal sections. A timber twin beam and a multi-part beam are detailed with vertical metal plates and with a cut-out for building services. A four-part timber connection is

exhibited using dowels, shear plate and toothed plate connectors. This four-part column is again shown with bearing plate and web plates let into slits to secure beams in position. A five-part column with a hardwood block is introduced to strengthen support.

Figure 5.8 is devoted to rigid beam grid joints in timber and beams crossing with and without notch and halving joints. Grid joints in timber with 60 degree angles are introduced. Rails, plates, pins and dowels are examined in 98 degree joints. A grid with an edge-glued member is also shown.

Figure 5.9 is exclusively designed for timber post-truss connections. The timber junction is maintained

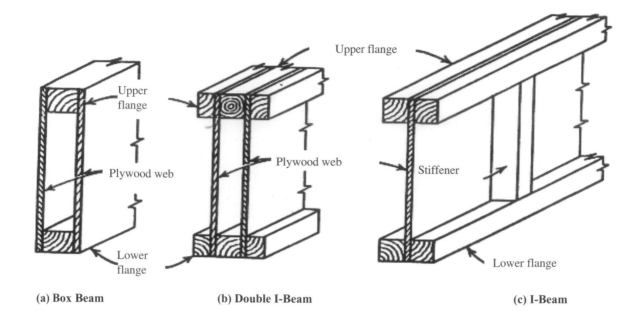

(a) Box Beam   (b) Double I-Beam   (c) I-Beam

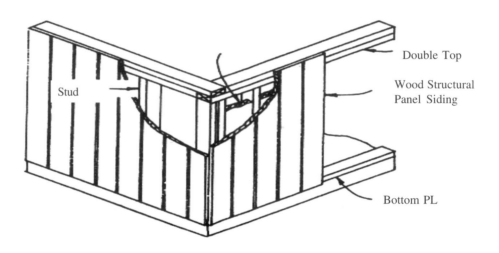

Plywood Combined Sheathing-Siding

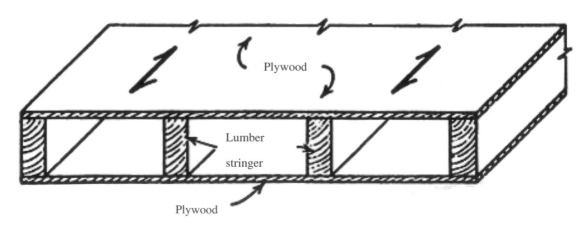

Figure 5.5 *Plywood beam and sheeting details (based on ANS/LFRD and American practices).*

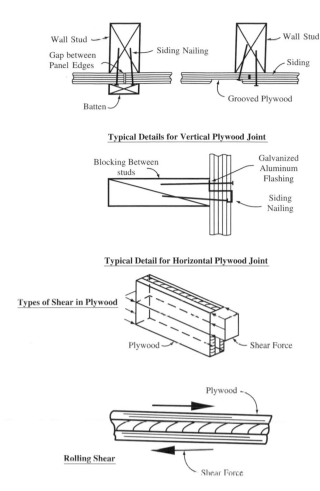

Typical Details for Vertical Plywood Joint

Typical Detail for Horizontal Plywood Joint

Types of Shear in Plywood

Rolling Shear

Figure 5.6 *Plywood beam and sheeting details (based on ANS/LFRD and American practices).*

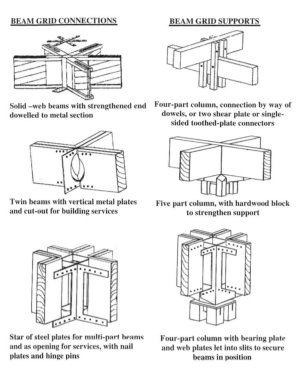

BEAM GRID CONNECTIONS

Solid –web beams with strengthened end dowelled to metal section

Twin beams with vertical metal plates and cut-out for building services

Star of steel plates for multi-part beams and as opening for services, with nail plates and hinge pins

BEAM GRID SUPPORTS

Four-part column, connection by way of dowels, or two shear plate or single-sided toothed-plate connectors

Five part column, with hardwood block to strengthen support

Four-part column with bearing plate and web plates let into slits to secure beams in position

Figure 5.7 *Truss joints with V-struts (based on EC5 and European practices).*

using dowelled plate connected with pins. In one case, dowelled web plate is welded to ties and in another case, pins are used as well.

Figure 5.10 is based on European practice strictly related to EC5 and covers complete tension joints in timber using screws and bolts or with metal plates using pins. Bonded joints with anchors, finger joints and scarf joints are also examined.

Figure 5.11 gives structural details for suspended arch members using pins, plates, steel brackets, lugs and elastomeric bearings.

Sometimes it becomes essential to use splits in timber connections. Figure 5.12 is based on European practices strictly governed by EC5 and shows rigid erection splices and rigid continuous splices over supports. These tension splices can be notched or with scarf joint with fish plates with and without slits. For the first time, nail plates with channels and fish plates are used in combinations using pins. Over continuous supports, round and squared-section purlins are used either screwed to the main beams or nailed using splice plates.

Figure 5.13 gives typical details of rigid timber functions using EC5. Structural details are given for joints with hanger's spacer blocks with two posts, wood wedges, connectors, nailed plywood gussets, nailed sheet metals and halving joints and hardwood blocks.

Figure 5.14 using European practices governed by EC5, shows roof frame corners, edges and joints. Columns with base details using pins and from EC5 are given in Fig. 5.15.

People have been building with logs for more than two thousand years, but in the last 25 years the craft of log building has changed. This is because there is a concentration of fine and innovative log builders in Canada and in the United States, and because they belong to an association that encourages all.

Log homes are beautiful, durable, energy efficient, non-toxic, quiet, and strong. The organic shapes of natural logs have a rhythmic pattern that is attractive. Log homes use excellent building materials excellent with excellent R-values. Log homes are energy efficient in cold climates and have good air infiltration. These homes are renewable and have longevity when they are properly maintained and are built with natural materials. Roofs are made with log common rafters, log trusses, timber frame trusses hybrid trusses and mitred log trusses. Purlins and ridge tie beams are also manufactured from logo. A typical layout is given in Figs. 5.16–5.18 for log construction profile, edge-sawn and square logs involved. Connections cover stave wall construction.

Figure 5.19 shows three-dimensional timber joints and base details based on European practices and EC5. Three-dimensional joints are shown with post connection and spreader plates, using single, multiple and knee-braced multiple struts. In a similar manner, 3D base details are given for such joints.

Based on the American practices related to LRFD, Fig. 5.20 shows the timber joints using nails, bolts and split-ring connectors under tensile loads. It is interesting

Beam Grid

Beam Grid

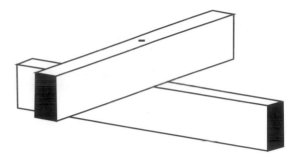

Horizontal metal plates in star
arrangement let in to slits in beams
and dowelled

Beams crossing

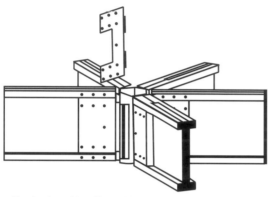

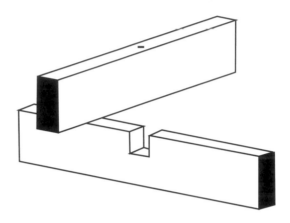

Steel tube with rails welded on and
beams hung on rails

with notch in one member

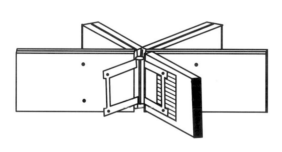

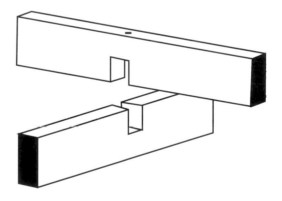

Twin beams with hinge pins

with halving joint

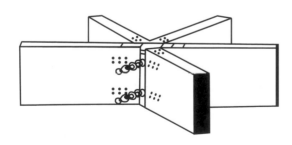

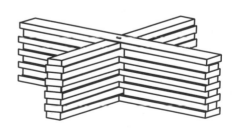

Steel tube with threaded bars or
dowels bonded in

Using edge-glued members

Figure 5.8 *Rigid beam grid joints.*

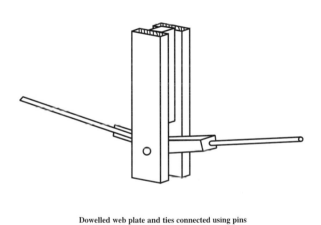

Dowelled plate connected using pins

Dowelled web plate welded to ties

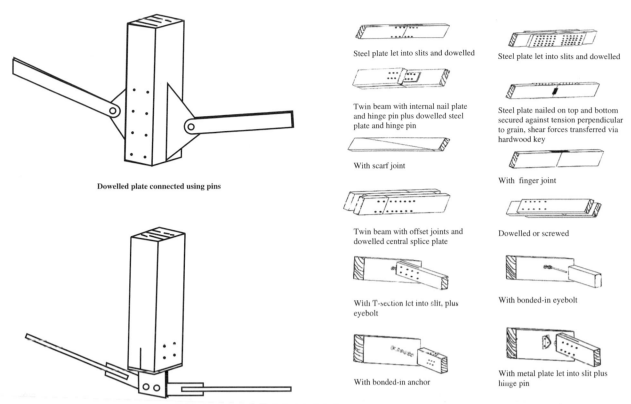

Figure 5.10 *Tension joints in timber (based on EC5 and European practices).*

Dowelled web plate and ties connected using pins

Figure 5.9

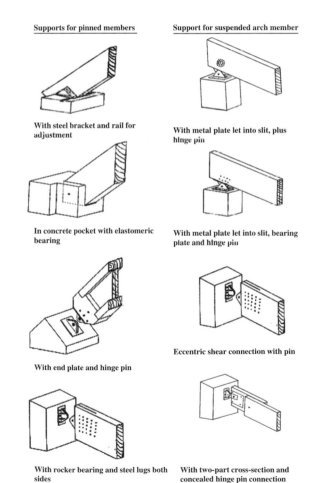

Figure 5.11 *Supports for pinned and suspended arch members (based on EC5 and European practices).*

to note that, like America, Europe has been the traditional home of log construction. Based on EC5, step joints in timber logs are shown in Fig. 5.21. Details are given for normal form step joints and double-step joints in logs which are currently used in triangular trusses. Compared with other codes of practice, it is essential to look at connectors and connector joints as practiced in the European countries. Figures 5.22 and 5.23 show isometric views and sections of different types of joint, the range of size of claw plates and an example of the use of washers with connectors of the dowel type.

The lower washer of the safety bolt is supported on the sloping chamfer cut on the end of the bolster, which is fastened to the lower chord of the truss by nails. In

Rigid Erection Splices   Rigid continuous splices over supports

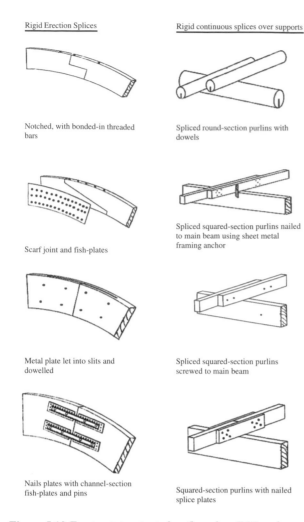

Notched, with bonded-in threaded   Spliced round-section purlins with
bars   dowels

Scarf joint and fish-plates   Spliced squared-section purlins nailed
                                          to main beam using sheet metal
                                          framing anchor

Metal plate let into slits and   Spliced squared-section purlins
dowelled   screwed to main beam

Nails plates with channel-section   Squared-section purlins with nailed
fish-plates and pins   splice plates

Figure 5.12 *Tension joints in timber (based on EC5 and European practices).*

the most simple forms of triangular truss carrying a distributed load the force R, acting on the joint from the side of the upper chord, usually has an angle of slope to the centre line of the lower chord of about 60–70°. In these supporting joints with the usual step joint, the danger of shearing is extremely slight and even timber-logs could serve as safety connectors.

If a simple step joint cannot be used in a round timber truss owing to the inadequate sectional area of the available logs, a more complicated notched butt joint is used, the *double-step joint*, which allows a slight reduction in the section of the chords. Any increase in the strength of a joint of this type depends on the simultaneous stressing of both steps. The design calculation for bearing stress is carried out assuming the distribution of the compression stresses N_C between both bearing surfaces is in proportion to their areas. Of the surfaces is not dangerous from the point of view of resistance to bearing, since, if there should be any overstressing, the other surface comes into play and the uneven distribution of the stresses is reduced. In the case of shearing, the matter is not satisfactory.

In accordance with the general requirements set out above for the basic types of joint in structural members, double-step joints can only be recommended when the governing factor is not shearing, but bearing stresses, ie

with large values of the angle γ, and especially when a load is applied to the upper chords of round or squared-timber trusses away from the joint.

In the USA, connectors are manufactured by various companies. TECO connectors are the most popular ones and are described in this section. It is generally advisable to use the same size connector and the same size bolt throughout a design. This simplifies the fabrication procedure, lessens the chance of fabrication error and leads to greater economy. It is advisable not to specify lumber of a higher grade than is required. The loads, spacing and other design data published here are based on the use of TECO timber connectors installed with TECO tools. Therefore, when using data, TECO products should be specified to protect the integrity of the design (see www.clevelandsteel.com/documentation. html for design manual for connectors.

The following is offered as a guide for the selection of size and type of connector or connectors to be used in a design and to point out the respective advantages of using the different sizes and types under various conditions. Figure 5.24 shows the types of TECO connectors.

5.2.2 TECO Wedge-Fit

2 1/2″ and 4″ split rings are placed in pre-cut conforming grooves made with TECO Wedge-Fit grooving tools. The tongue and groove "split" in the ring permits simultaneous bearing of the inner surface of the ring against the core left in grooving and the outer face of the ring against the outer wall of the groove. The special wedge shape of the ring section provides maximum tolerance for easy insertion but a tight-fitting joint when the ring is seated in the conforming groove.

TECO Wedge-Fit split rings are the most efficient mechanical devices used for joints in timber construction. The 2½″ split rings are widely used in trussed rafter construction in spans up to 50ft and in all wood framing involving lumber of 2″ nominal dimension. The 4″ split rings are used in heavy beam and girder construction, towers, bridges and moderate and long span roof trusses, generally in members 3″ or more in nominal thickness. TECO Wedge-Fit grooving tools are precision-made, high-speed woodworking tools, designed for power operation only.

As to choice of size, the 2 6/8″ shear plate compares with the ring and the 4″ split ring.

The TECO circular grid is used between the top of a pile and cap to prevent lateral movement. It is usually installed for this purpose with several light blows from the pile driver hammer. It is also used in trestle and other heavy timber construction in much the same manner as the flat and single curve grids. It is also used as a tie space.

End distance is the distance from a connector to the square cut measured parallel to the grain from the centre of the end of the member. If the end of the member is not square cut, the end distance shall be taken as the distance from any point on the centre half of the connector diameter drawn perpendicular to the centre line of the piece to the nearest point on the end of the member

Simply Supported Trusses   Pinned Type

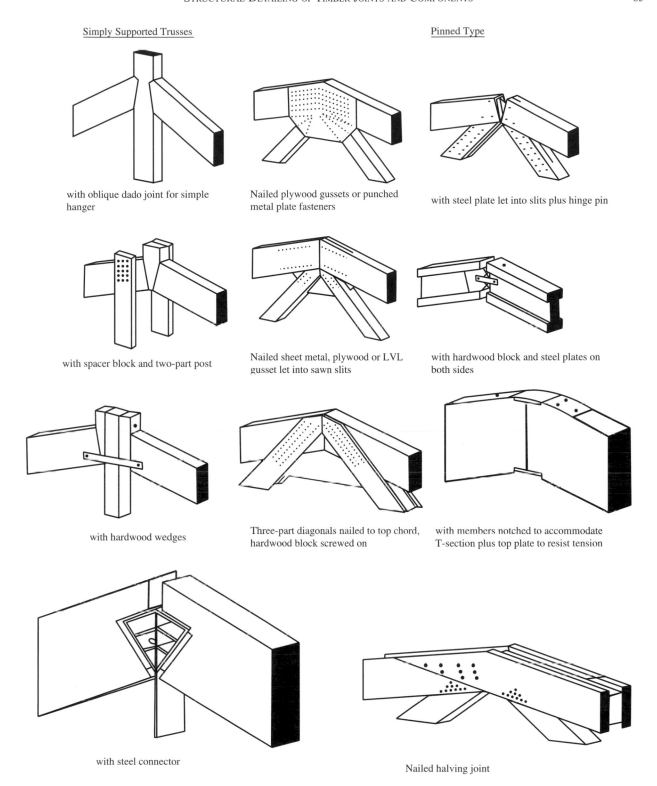

with oblique dado joint for simple hanger

Nailed plywood gussets or punched metal plate fasteners

with steel plate let into slits plus hinge pin

with spacer block and two-part post

Nailed sheet metal, plywood or LVL gusset let into sawn slits

with hardwood block and steel plates on both sides

with hardwood wedges

Three-part diagonals nailed to top chord, hardwood block screwed on

with members notched to accommodate T-section plus top plate to resist tension

with steel connector

Nailed halving joint

Figure 5.13 *Ridge junction (based on EC5 and other European practices).*

measured parallel to the grain. The distance measured perpendicular to the end cut to the centre of the connector shall never be less than the required edge distance.

The end distance chart in the TECO manual is divided into two sections, depending on whether the member is in tension or compression. If in tension, project vertically on the chart from the end distance to the curve, then horizontally to obtain the percentage of full load allowable. This process can be reversed of course by going from percent of full load required to

spacing required. In compression, there is an additional variable of angle of load to grain. The operation is the same except the curve for the proper angle of load to grain should be used. For intermediate angles, interpolate between the curves on a straight line. On some of the distance dimension markings, this cut-off is the minimum permissible and gives the full load for an angle of load to grain of zero degrees. Reductions in load for edge distance or spacing are not additive to end distance reductions but are coincident.

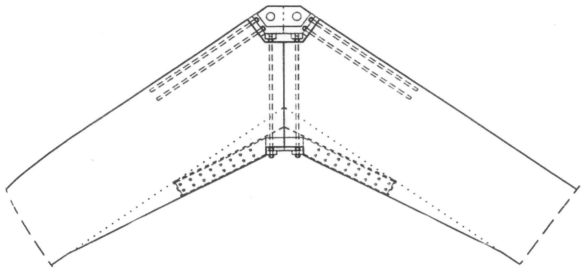

(a) Frame corners two roof ridge members joined at corners with steel bolts, nuts, and plates

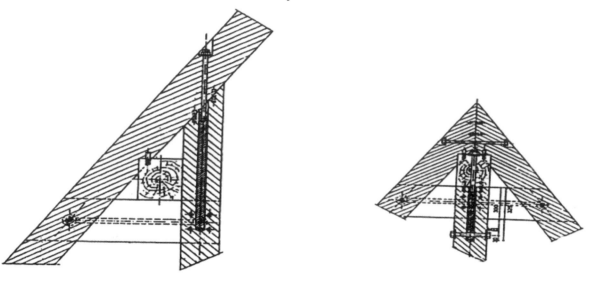

(b) Joint: Stud, beam and rafter

(c) Joint: Stud with two inclined struts and middle pole

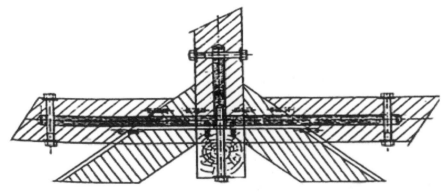

(d) Ridge Details: Joint between ridge pole, beam and two rafters.

Figure 5.14 *Roof frame corners edges and joints (based on EC5 and other European practices).*

Column base details

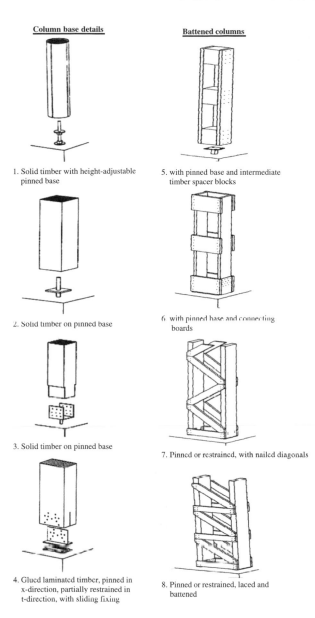

1. Solid timber with height-adjustable pinned base

2. Solid timber on pinned base

3. Solid timber on pinned base

4. Glued laminated timber, pinned in x-direction, partially restrained in t-direction, with sliding fixing

Battened columns

5. with pinned base and intermediate timber spacer blocks

6. with pinned base and connecting boards

7. Pinned or restrained, with nailed diagonals

8. Pinned or restrained, laced and battened

Figure 5.15 *Column with base details (based on EC5 and European practices).*

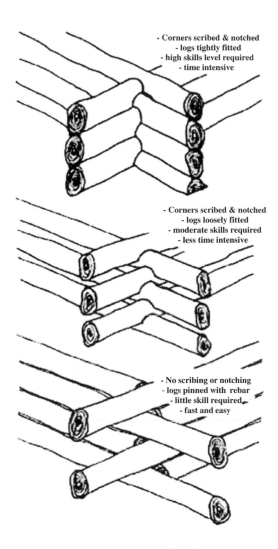

- Corners scribed & notched
- logs tightly fitted
- high skills level required
- time intensive

- Corners scribed & notched
- logs loosely fitted
- moderate skills required
- less time intensive

- No scribing or notching
- logs pinned with rebar
- little skill required
- fast and easy

Figure 5.16 *LDG construction – walls and corner joints (based on EC5 and European practices).*

5.2.2.1 Angle of load to grain

The angle of load to grain is the angle between the resultant load exerted by the connector acting on the member and the longitudinal axis of the member.

5.2.2.2 Angle of axis to grain

The angle of axis to grain is the angle of the connector axis formed by a line joining the centres of two adjacent connections located in the same face of a member in a joint and the longitudinal axis of the member.

5.2.2.3 Load charts

The load charts show the allowable normal loads for one connector unit and bolt in single shear. The connector unit consists of one split ring or toothed ring, a pair of shear plates, or a single shear plate used with a steel sideplate.

The charts are divided vertically into the three species groups. Select the group from page 4 according to the species of lumber specified and use the portion of the chart applying to this group. Within each group there are several curves, each representing a thickness of lumber and the number of loaded faces. Use the curve conforming to the condition existing in the joint. Each curve is plotted according to the Hackinson Formula with load in pounds and angle of load to grain as the variables. Select the proper angle at the bottom or top of the chart, proceed vertically to the selected curve and proceed horizontally to read the allowable normal load. Lumber thicknesses less than those shown on the load data charts for the corresponding number of loaded faces are not recommended. For more than one connector unit, multiply the connector load by the number of units.

5.2.2.4 Connector data

The data given cover the dimensions of the connectors, minimum lumber sizes, recommended bolt and bolt hole diameters, recommended washer sizes and similar self explanatory information. Lag screws of the same diameter

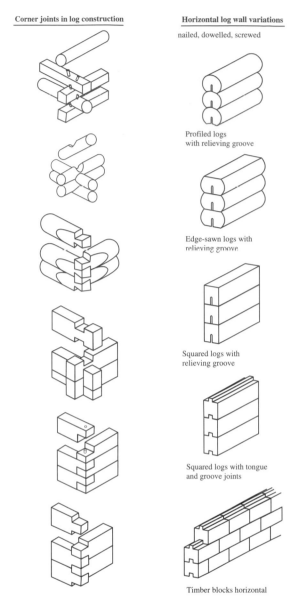

Figure 5.17 *LDG construction – walls and corner joints (based on EC5 and European practices).*

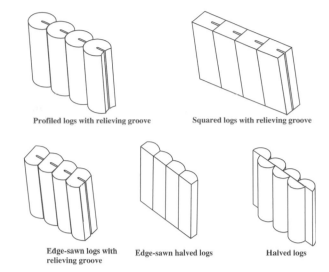

Figure 5.18 *LDG construction – walls and corner joints (based on EC5 and European practices).*

as the recommended bolt sizes may be used instead of bolts in accordance with provisions set forth in the 'National Design Specification for Wood Construction'.

5.2.2.5 Connector specifications

The design data and other information in the publication are based on the use of standard TECO connectors and tools. It is recommended that the specifications be closely followed to ensure the use of products for which the data were prepared.

5.2.2.6 Adjustments for load and duration

The allowable connector loads given on the charts are for normal duration. Normal duration of loading contemplates the joint being fully loaded for approximately 10 years, either continuously or cumulatively, by the maximum allowable normal design load shown on the charts and/or that 90% of this load is applied throughout the remainder of the life of the structure. Factors are given with the

data on each connector which, when applied to the load values in the charts, give the allowable loads for other durations of loading from those of impact character to those applied permanently.

Permanent loading for which a factor of 90% is given, contemplates the connector will be fully loaded to that percentage of the normal design values on the chart applied permanently or for many years.

5.2.2.7 Design of eccentric joints and of beams supported by fastenings

Eccentric connector joints and beams supported by connectors shall be designed so that the computed stress does not exceed the unit stresses in horizontal shear in which h (with connectors) = the depth of the member less the distance from the unloaded edge of the member to the nearest edge of the nearest connector.

5.2.2.8 Maximum permissible loads on shear plates

Due to the type of test failure, there are limits beyond which shear plates should not be loaded. These limits vary with the bolt size and are given with the shear plate shape data.

5.2.2.9 Spacing charts

Spacing is the distance between centres of connectors measured along a line joining their centres. "R" is the spacing between the rings shown. On the right-hand page of the data for each connector, is the spacing chart. Each chart has five parabolic curves representing recommended spacing for a full load at the particular angle of load to grain noted on the curve. For intermediate angles of load, straight-line interpolation may be used. If the spacing for full load is designated, select the proper angle of load to grain curve and find where it intersects the radial lines representing angle

Three-Dimensional Timber Joints

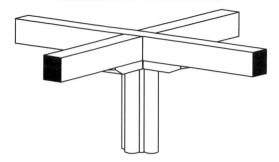

Post connection with spreader plates at support

Three-dimensional Base Details

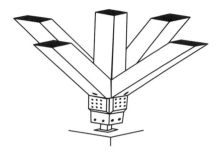

Base with cleats

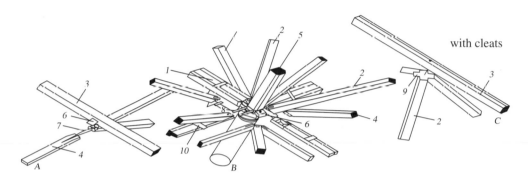

1. Bottom chord, 160 × 100 mm + 2 No. 80 × 200 mm, 2. Strut to pyramid, 160 × 160 mm
3. Purlin, 160 × 260 mm, 4. Strut, 160 × 160 mm, 5. Column, 2 No. 120 × 240 mm +
2 No. 40 × 160 mm, 6. Hardwood block 11 cm thk, 7. Plywood gusset, 11 cm thk,
8. Steel hanger, 9. Metal plate 10 mm thk, 10. Perforated metal strap, 2 mm thk.

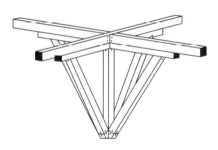

Multiple struts with spreader plates
and oblique dado joints

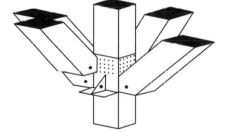

Base with steel angles let into slits

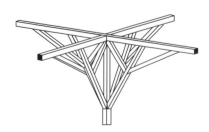

Kneebraced multiple struts

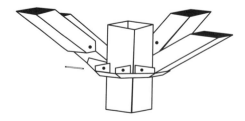

Base connected to steel stanchion

Figure 5.19 *Three-dimensional joints and base details (based on EC5 and European practices).*

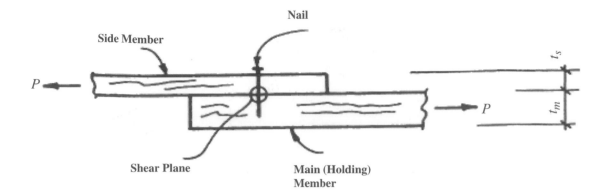

Nail Connection

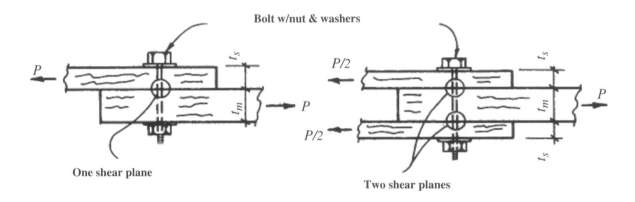

Bolt Connections

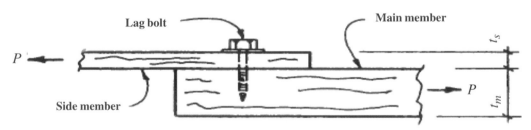

Lag Bolt

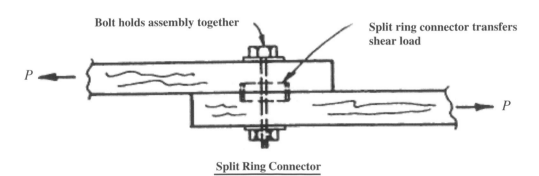

Split Ring Connector

Figure 5.20 *Nail, screw and split-ring connectors used in timber joints (based on American practices).*

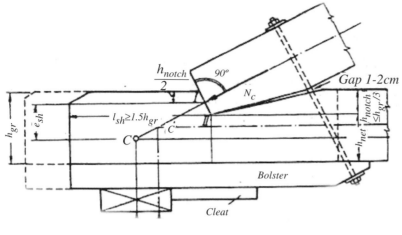

(a) Normal Form of Step Joint

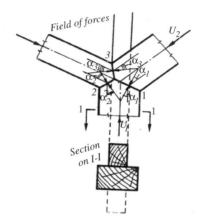

(d) Three-way Butt joint

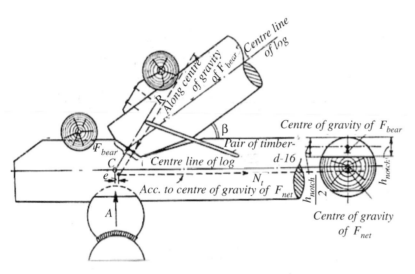

(b) The nominal form of step joint in the support joint of a simple triangular truss of logs when the load is distributed along the principal rafter (dot-dash lines indicate axial alignment, accurate enough in practice for round-timber structures)

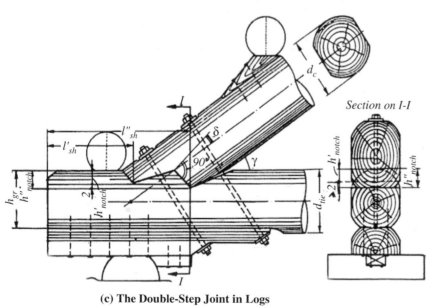

(c) The Double-Step Joint in Logs

Figure 5.21 *Step joints in timber logs (based on EC5 and Russian practices).*

92   STRUCTURAL DETAILING IN TIMBER

Type of joint (the facing piece is omitted for clearness)

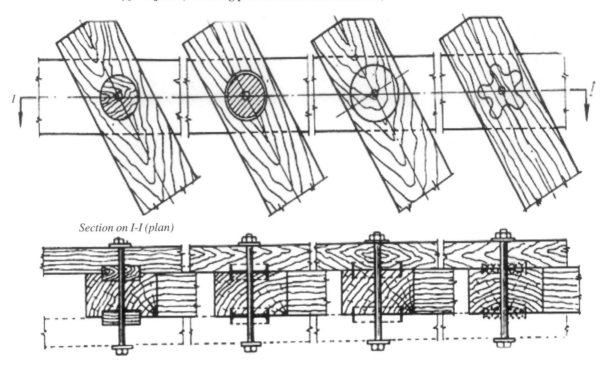

Section on I-I (plan)

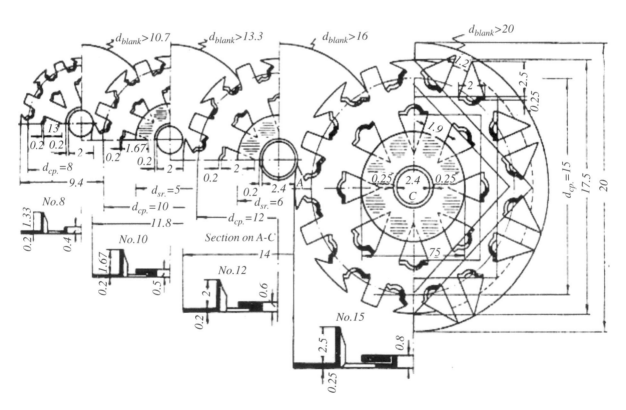

Figure 5.22 *Connectors and connector joints (based on EC5 and Russian practices).*

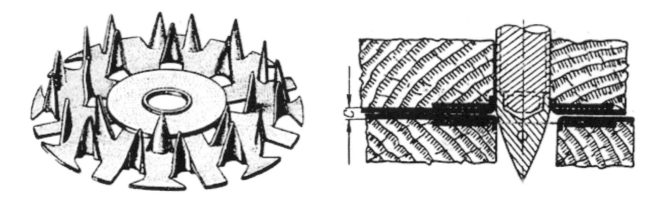

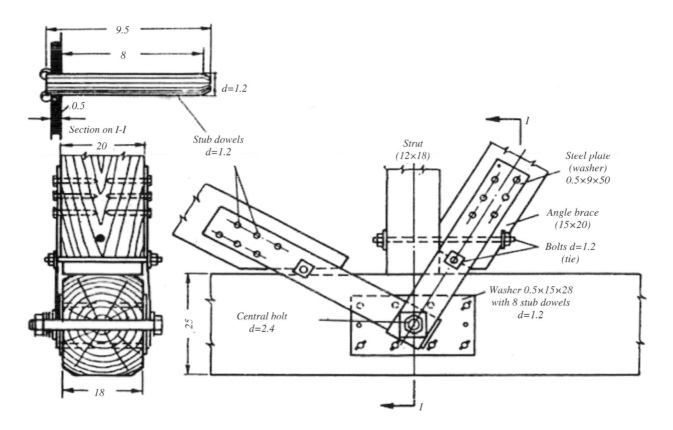

An example of the use of washers (plates) with connectors of the dowel type for a
single-bolt knock-down joint between square-sawn timber struts and chord

Figure 5.23 *Connectors and connector joints (based on EC5 and European practices).*

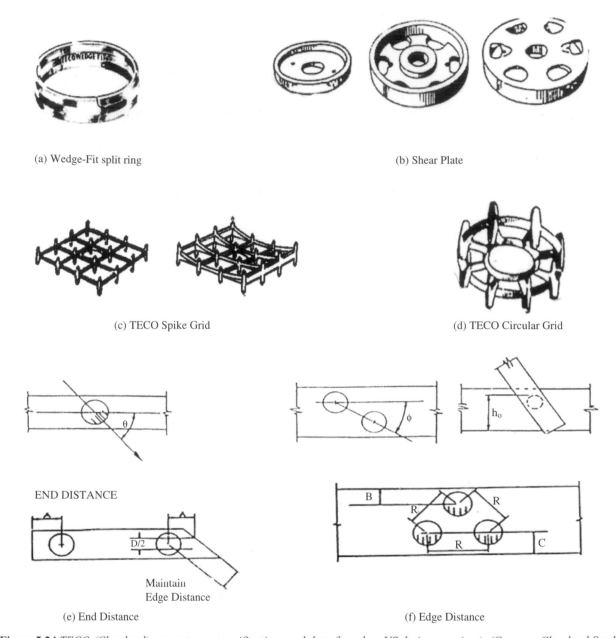

(a) Wedge-Fit split ring   (b) Shear Plate

(c) TECO Spike Grid   (d) TECO Circular Grid

END DISTANCE

Maintain
Edge Distance

(e) End Distance   (f) Edge Distance

Figure 5.24 *TECO (Cleveland) connectors – specifications and data (based on US design practices). (Courtesy Cleveland Steel Specialty Company, USA).*

of axis to grain. The distance from the point to the lower left-hand corner is the spacing. It is probably more convenient, however, in laying out this to use the parallel to grain and perpendicular to grain components for measurements of the spacing. The parallel to grain component may be read at the bottom of the chart by projecting downward from the point on the curve. The perpendicular component of the spacing may be read at the left-hand side of the chart by projecting horizontally from the point on the curve.

The sixth curve on the chart is a quarter-circle. This curve represents the spacing for 50% of full load for any angle of load to grain and also the minimum spacing permissible. For percentages between 50% and 100% of full load for an angle of load to grain, interpolate radially on a straight line between the 50% curve and the curve corresponding to the proper angle of load to grain.

Reductions in load for edge distance and end distance are not additive to spacing reductions but are coincident.

5.2.2.10 Inter-relationship of thickness, distance and spacing

Loads reduced because of thickness do not permit any reduction of edge distance, end distance or spacing without further reduction of load and conversely loads reduced for edge distance, end distance or spacing do not permit reduction of thickness.

When allowed load is reduced due to reduced edge distance, end distance or spacing, the reduced allowable load for each shall be determined separately and the lowest allowable load so determined for any one connector shall apply to this and all connectors resisting a common force joint. Such load reductions are not cumulative.

Conversely, if the allowable load is reduced because of a reduced distance or spacing, the other distances or spacings may be reduced to those resulting in the same reduced allowable load.

5.2.2.11 Net section – tension or compression

The net section of a timber in a connector joint is usually adequate to transmit the full strength of the timber which can be developed outside of the joint when lower grades of lumber are used. However, it may be desirable to check the strength of the net section of timbers when they are of the minimum size recommended herein for a given connector, particularly if a high stress-grade of lumber is used.

The critical "net" section of a timber joint, which will generally pass through the centre line of a bolt and connector, occurs at the plane of maximum stress. The next cross-section at the plane is equal to the full cross-sectional area of the timber minus the projected area of that portion of the bolt hole, not within the connector projected area, located at this point.

Due to wood being able to support loads of short-term duration greatly in excess of permanently applied loads, computation of the required net section involves consideration of the different types and amounts of loadings.

The net cross-sectional area in square inches required at the critical section may be determined by multiplying the total load in pounds, which is transferred through the critical section of the member, by the appropriate constant given in the table. Conversely, the total working load capacity in pounds of a given net area may be determined by dividing the net area in square inches by the appropriate constant.

5.2.2.12 Design and load data for TECO (USA) connectors

Figures 5.26 to 5.34 give the complete design data recommended by TECO. These sheets can easily be adopted by all those who are familiar with the American practices. Every sheet gives relevant specifications and data geared to specific subjects. British, European and any other codes can produce similar design data sheets with minimum alterations required. These sheets and their contents can easily be adjusted where differences exist owing to rules established by other codes. These sheets give the layouts which can be useful for designing similar design datasheets required by other codes and practices.

CONSTANTS FOR USE IN DETERMINING REQUIRED NET SECTION IN SQUARE INCHES

Type of Loading	Thickness of Wood Member in Inches	Constants for Each Connector Load Group		
		Group A	Group B	Group C
Normal	4″ or less	.00043	.00050	.00061
	over 4″	.00054	.00063	.00077
Permanent	4″ or less	.00048	.00055	.00067
	over 4″	.00059	.00069	.00083
Snow	4″ or less	.00037	.00044	.00053
	over 4″	.00047	.00054	.00067
Wind or Earthquake	4″ or less	.00032	.00038	.00045
	over 4″	.00040	.00047	.00057

PROJECTED AREA OF CONNECTORS AND BOLTS

(For Use In Determining Net Sections)

Connectors		Bolt Diam.	Placement of Connectors	Total Projected Area in Square Inches of Connectors & Bolts in Lumber Thickness of				
No.	Size			1½″	2½″	3½″	5½″	7½″
SPLIT RINGS		½	One Face	1.71	2.27	2.84	3.89	5.02
1	2 ½	½	Two Faces	2.60	3.16	3.73	4.78	5.91
2	4	¼	One Face	3.01	3.86	4.64	6.16	7.79
		¾	Two Faces	4.85	5.66	6.47	8.00	9.62
SHEAR PLATES		¼	One Face	2.00	2.81	3.62	4.14	6.77
1	2⅝	¾	Two Faces	2.81	2.68	4.43	5.96	7.58
1	2⅛ LG	¾	One Face	1.87	2.68	3.50	5.02	6.65
		¾	Two Faces	2.56	3.38	4.19	5.71	7.34
2	4	¼	One Face	3.24	4.05	4.87	6.39	8.01
		¾	Two Faces	–	6.11	6.93	8.45	10.07
2-A	4	⅞	One Face	3.33	4.27	5.21	6.97	8.84
		⅞	Two Faces	–	6.25	7.19	8.95	10.82

Figure 5.25

CLEVELAND DESIGN AND LOAD DATA FOR TECO CONNECTORS

2½" TECO SPLIT RINGS

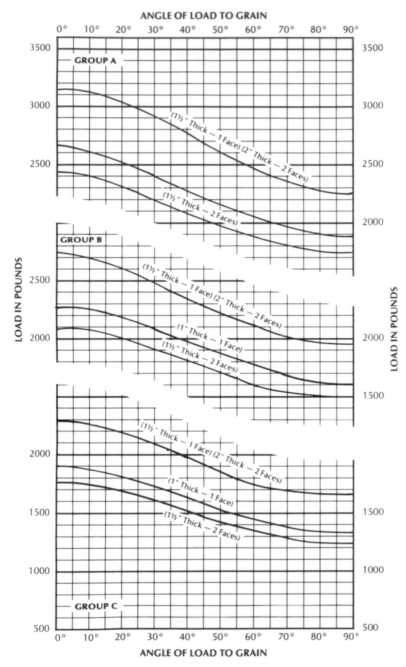

LOAD CHART FOR NORMAL LOADING
ONE 2½" SPLIT RING AND BOLT IN SINGLE SHEAR

2½" SPLIT RING DATA

Split Ring — Dimensions	
Inside Diameter at center when closed	2½"
Inside diameter at center when installed	2.54"
Thickness of ring at center	0.163"
Thickness of ring at edge	0.123"
Depth	¾"
Lumber, Minimum dimensions allowed	
Width	3½"
Thickness, rings in one face	1"
Thickness, rings opposite in both faces	1½"
Bolt, diameter	½"
Bolt hole, diameter	%₁₆"
Projected Area for portion of one ring within a member, square inches	1.10
Washers, minimum	
Round, Cast or Malleable Iron, diameter	2⅛"
Square Plate	
Length of Side	2"
Thickness	⅛"
(For trussed rafters and similar light construction standard wrought washers may be used.)	

SPLIT RING SPECIFICATIONS

Split rings shall be TECO split rings as manufactured by CSS, Cleveland, OH. Split rings shall be manufactured from hot rolled S. A. E. — 1010 carbon steel. Each ring shall form a closed true circle with the principal axis of the cross section of the ring metal parallel to the geometric axis of the ring. The ring shall fit snugly in the prepared groove. The metal section of each ring shall be beveled from the central portion toward the edges to a thickness less than that at mid-section. It shall be cut through in one place in its circumference to form a tongue and slot.

PERCENTAGES FOR DURATION OF MAXIMUM LOAD

Two Months Loading, as for snow	115%
Seven Days Loading	125%
Wind or Earthquake Loading	133⅓%
Impact Loading	200%
Permanent Loading	90%

DECREASES FOR MOISTURE CONTENT CONDITIONS

	Condition when Fabricated	Condition when Used	
	Seasoned	Unseasoned	Unseasoned
	Seasoned	Seasoned	Unseasoned or Wet
Split Rings	0%	20%	33%

Figure 5.26 *Charts for normal loading-I. (Courtesy Cleveland Steel Specialty Company, USA)*

CLEVELAND — DESIGN AND LOAD DATA FOR TECO CONNECTORS

4″ TECO SPLIT RINGS

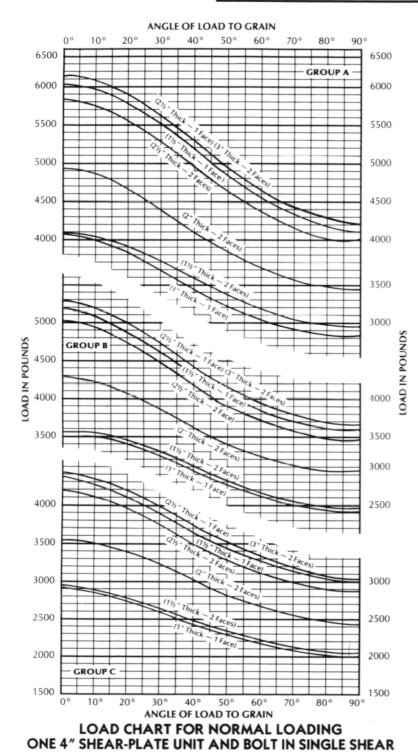

ANGLE OF LOAD TO GRAIN

LOAD CHART FOR NORMAL LOADING
ONE 4″ SHEAR-PLATE UNIT AND BOLT IN SINGLE SHEAR

4″ SPLIT RING DATA

Split Ring — Dimensions	
Inside Diameter at center when closed	4″
Inside diameter at center when installed	4.06″
Thickness of ring at center	0.193″
Thickness of ring at edge	0.133″
Depth	1″
Lumber, Minimum dimensions allowed	
Width	5½″
Thickness, rings in one face	1″
Thickness, rings opposite in both faces	1½″
Bolt, diameter	¾″
Bolt hole, diameter	13/16″
Projected Area for portion of one ring within a member, square inches	2.24
Washers, minimum	
Round, Cast or Malleable Iron, diameter	3″
Square Plate	
Length of Side	3″
Thickness	7/16″
(For trussed rafters and similar light construction standard wrought washers may be used.)	

SPLIT RING SPECIFICATIONS

Split rings shall be TECO split rings as manufactured by CSS, Cleveland, OH. Split rings shall be manufactured from hot rolled S. A. E. — 1010 carbon steel. Each ring shall form a closed true circle with the principal axis of the cross section of the ring metal parallel to the geometric axis of the ring. The ring shall fit snugly in the prepared groove. The metal section of each ring shall be beveled from the central portion toward the edges to a thickness less than that at mid-section. It shall be cut through in one place in its circumference to form a tongue and slot.

PERCENTAGES FOR DURATION OF MAXIMUM LOAD

Two Months Loading, as for snow	115%
Seven Days Loading	125%
Wind or Earthquake Loading	133⅓%
Impact Loading	200%
Permanent Loading	90%

DECREASES FOR MOISTURE CONTENT CONDITIONS

Condition when Fabricated	Seasoned	Unseasoned	Unseasoned
Condition when Used	Seasoned	Seasoned	Unseasoned or Wet
Split Rings	0%	20%	33%

Figure 5.27 *Charts for normal loading-II. (Courtesy Cleveland Steel Specialty Company, USA)*

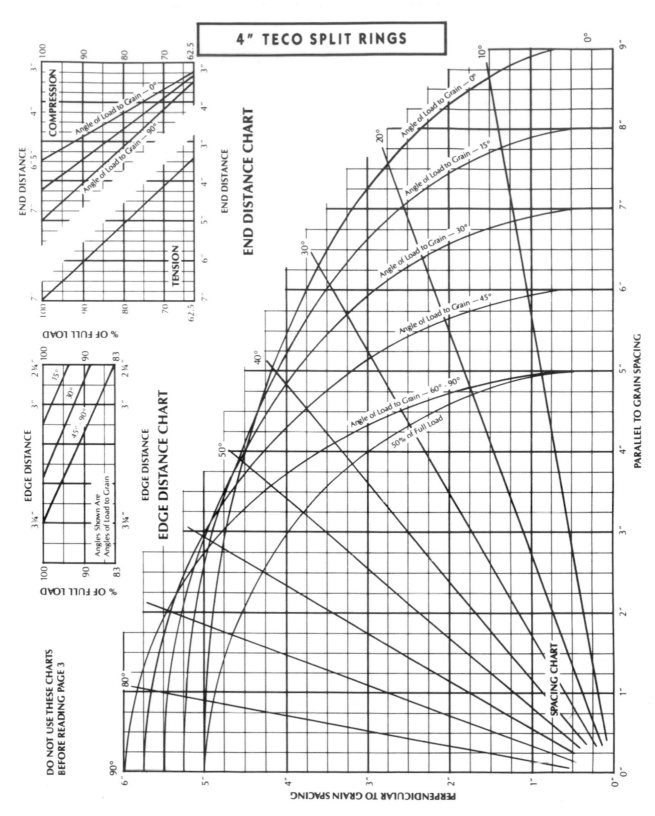

Figure 5.28 *Charts for normal loading-II (contd.). (Courtesy Cleveland Steel Specialty Company, USA)*

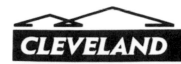

CLEVELAND DESIGN AND LOAD DATA FOR TECO CONNECTORS

2⅝" TECO SHEAR PLATES

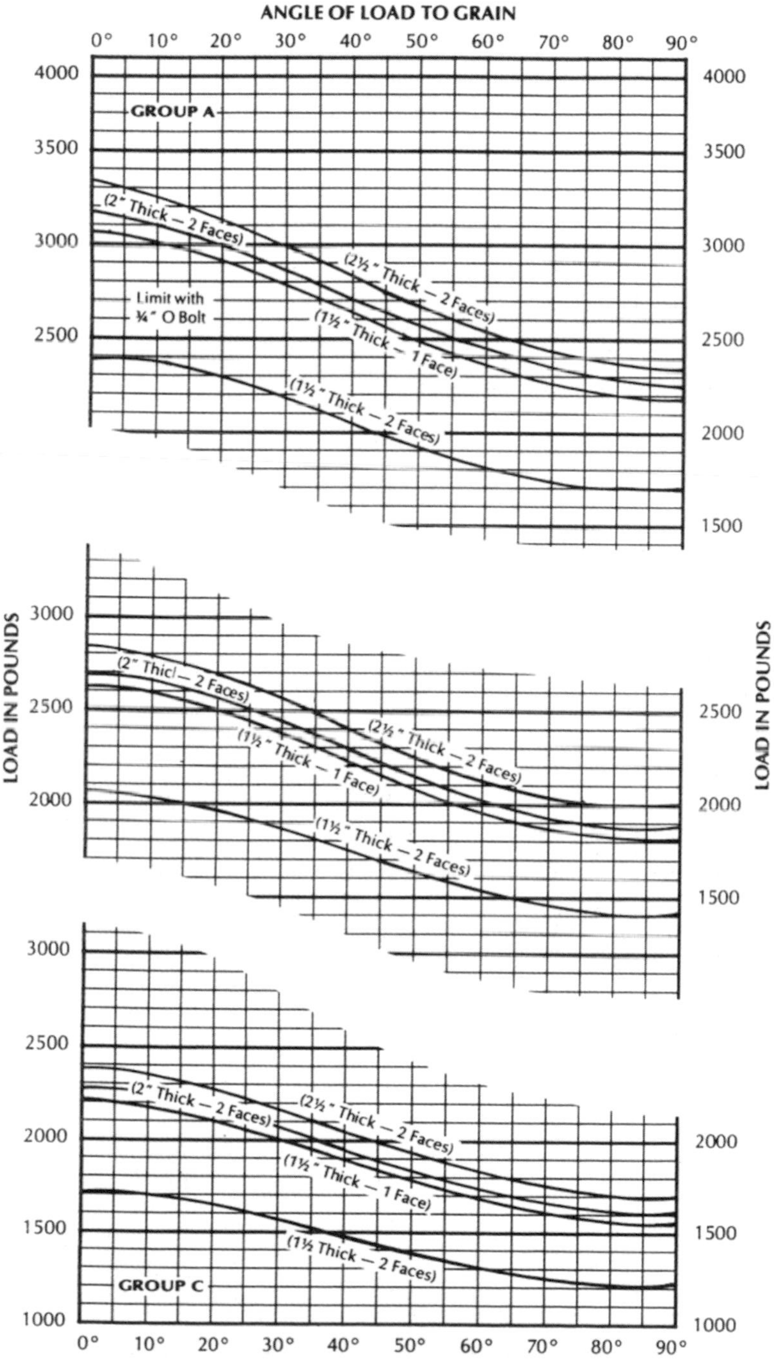

LOAD CHART FOR NORMAL LOADING
ONE 2⅝" SHEAR-PLATE UNIT AND BOLT IN SINGLE SHEAR

2⅝" SHEAR PLATE DATA

Shear Plates, Dimensions		
Material	Pressed Steel	
	Reg.	Lt. Ga.
Diameter of plate	2.62"	2.62"
Diameter of bolt hole	.81"	.81"
Depth of plate	.42"	.35"
Lumber, Minimum Dimensions		
Face, width	3½"	3½"
Thickness, plates in one face only	1½"	1½"
Thickness, plates opposite in both faces	1½"	1½"
Steel Shapes or Straps (Thickness required when used with shear plates) Thickness of steel side plates shall be determined in accordance with A.I.S.C. recommendations.		
Hole, diameter in steel straps or shapes	13/16"	13/16"
Bolt, diameter	¾"	¾"
Bolt Hole, diameter in timber	13/16"	13/16"
Washers, standard, timber to timber connections only		
Round, cast or malleable iron, dia.	3"	3"
Square Plate		
Length of side	3"	3"
Thickness	¼"	¼"
(For trussed rafters and other light structures standard wrought washers may be used.)		
Projected Area, for one shear plate, square inches	1.18	1.00

SHEAR PLATE SPECIFICATIONS

Shear plates shall be CSS shear plates as manufactured by CSS, Cleveland, OH. Malleable Iron Types — Malleable iron shear plates shall be manufactured according to current A. S. T. M. Standard Specifications A 47, Grade 32510, for malleable iron castings. Each casting shall consist of a perforated round plate with a flange around the edge extending at right angles to the face of the plate and projecting from one face only, the plate portion having a central bolt hole reamed to size with an integral hub concentric to the bolt hole and extending from the same face as the flange.

PERCENTAGES FOR DURATION OF MAXIMUM LOAD

Two Months Loading, as for snow	*115%
Seven Days Loading	*125%
Wind or Earthquake Loading	*133⅓%
Impact Loading	*200%
Permanent Loading	90%

*Do not exceed limitations for maximum allowable loads for shear plates given elsewhere on this page.

DECREASES FOR MOISTURE CONTENT CONDITIONS

Condition when Fabricated	Seasoned	Unseasoned	Unseasoned
Condition when Used	Seasoned	Seasoned	Unseasoned or Wet
Shear Plates	0%	20%	33%

MAXIMUM PERMISSIBLE LOADS ON SHEAR PLATES

The allowable loads for all loadings except wind shall not exceed 2900 lbs. for 2⅝" shear plates with ¾" bolts. The allowable wind load shall not exceed 3870 lbs. If bolt threads bear on the shear plate, reduce the preceding values by one-ninth.

Figure 5.29 *Charts for shear plates-I. (Courtesy Cleveland Steel Specialty Company, USA)*

CLEVELAND   DESIGN AND LOAD DATA FOR TECO CONNECTORS

4" TECO SHEAR PLATES (WOOD-TO-WOOD)

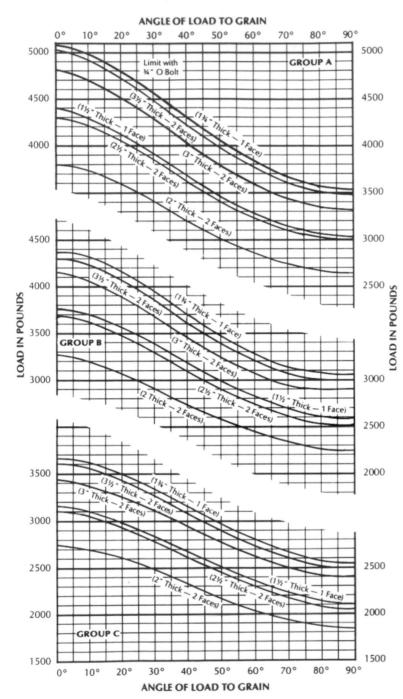

4" SHEAR PLATE DATA

Shear Plates, Dimensions	Melleable Iron	Melleable Iron
Material	4.03″	4.03″
Diameter of plate	4.03″	4.03″
Diameter of bolt hole	.81″	.94″
Depth of plate	.64″	.64″
Lumber, Minimum Dimensions		
Face, width	5½″	5½″
Thickness, plates in one face only	1½″	1½″
Bolt, diameter	¾″	⅞″
Bolt Hole, diameter in timber	1³⁄₁₆″	1⁵⁄₁₆″
Washers, standard, timber to timber connections only		
Round, cast or malleable iron, dia.	3″	3½″
Square Plate		
Length of side	3″	3″
Thickness	¼″	¼″
(For trussed rafters and other light structures standard wrought washers may be used.)		
Projected Area, for one shear plate, square inches	2.58	2.58

SHEAR PLATE SPECIFICATIONS

Shear plates shall be TECO shear plates as manufactured by CSS, Cleveland, OH. Malleable Iron Types — Malleable iron shear plates shall be manufactured according to current A. S. T. M. Standard Specifications A 47, Grade 32510, for malleable iron castings. Each casting shall consist of a perforated round plate with a flange around the edge extending at right angles to the face of the plate and projecting from one face only, the plate portion having a central bolt hole reamed to size with an integral hub concentric to the bolt hole and extending from the same face as the flange.

PERCENTAGES FOR DURATION OF MAXIMUM LOAD

Two Months Loading, as for snow	*115%
Seven Days Loading	*125%
Wind or Earthquake Loading	*133⅓%
Impact Loading	*200%
Permanent Loading	90%

*Do not exceed limitations for maximum allowable loads for shear plates given elsewhere on this page.

DECREASES FOR MOISTURE CONTENT CONDITIONS

Condition when Fabricated	Seasoned	Unseasoned	Unseasoned
Condition when Used	Seasoned	Seasoned	Unseasoned or Wet
Shear Plates	0%	20%	33%

MAXIMUM PERMISSIBLE LOADS ON SHEAR PLATES

The allowable loads for all loadings except wind shall not exceed 4970 lbs. for 4″ shear plates with ¾″ bolts and 6760 lbs. for 4″ shear plates with ⅞″ bolts. The allowable wind load shall not exceed 6630 lbs. when used with a ¾″ bolt and 9020 lbs. when used with a ⅞″ bolt. If bolt threads bear on the shear plate, reduce the preceding values by one-ninth.

LOAD CHART FOR NORMAL LOADING
ONE 4" SHEAR-PLATE UNIT AND BOLT IN SINGLE SHEAR

Figure 5.30 *Charts for shear plates-II. (Courtesy Cleveland Steel Specialty Company, USA)*

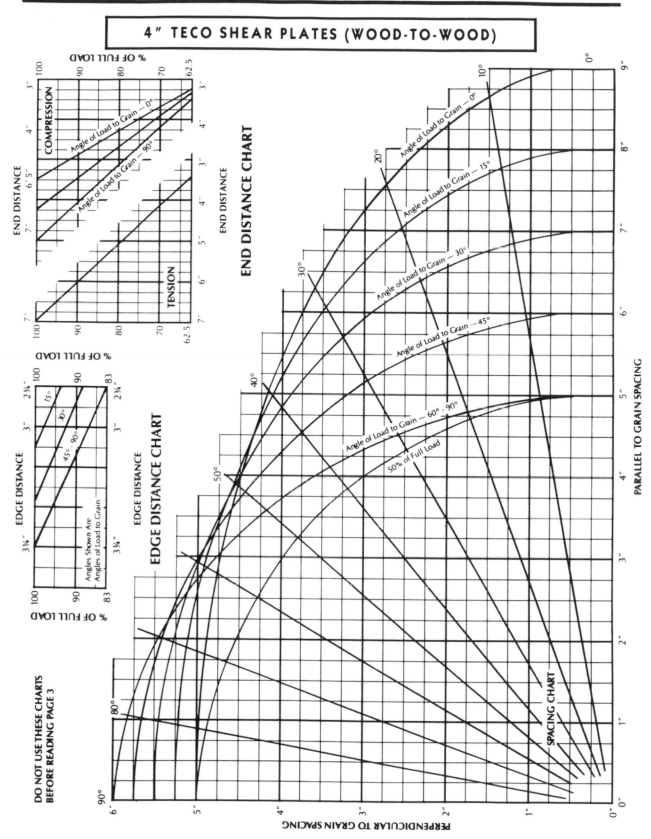

Figure 5.31 *Charts for shear plates-III. (Courtesy Cleveland Steel Specialty Company, USA)*

CLEVELAND   DESIGN AND LOAD DATA FOR TECO CONNECTORS

4" TECO SHEAR PLATES (WOOD-TO-STEEL)

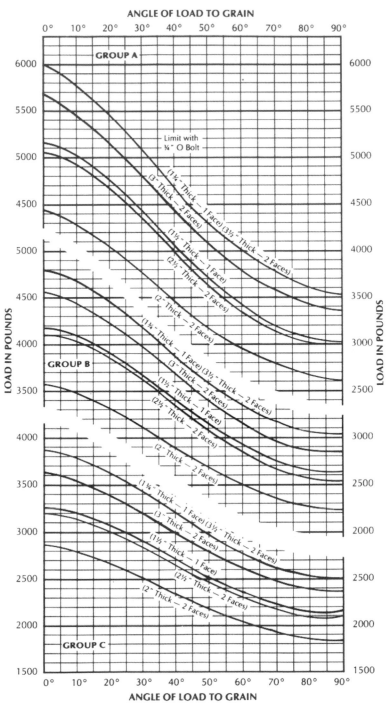

LOAD CHART FOR NORMAL LOADING
ONE 4" SHEAR-PLATE UNIT AND BOLT IN SINGLE SHEAR

4" SHEAR PLATE DATA

Shear Plates, Dimensions	Melleable Iron	Melleable Iron
Material		
Diameter of plate	4.03"	4.03"
Diameter of bolt hole	.81"	.94"
Depth of plate	.64"	.64"
Lumber, Minimum Dimensions		
Face, width	5½"	5½"
Thickness, plates in one face only	1½"	1½"
Steel Shapes or Straps (Thickness required when used with shear plates) Thickness of steel side plates shall be determined in accordance with A.I.S.C. recommendations.		
Hole, diameter in steel straps or shapes	1³⁄₁₆"	1⁵⁄₁₆"
Bolt, diameter	¾"	⅞"
Bolt Hole, diameter in timber	1³⁄₁₆"	1⁵⁄₁₆"
Projected Area, for one shear plate, square inches	2.58	2.58

SHEAR PLATE SPECIFICATIONS

Shear plates shall be TECO shear plates as manufactured by CSS, Cleveland, OH. Malleable Iron Types — Malleable iron shear plates shall be manufactured according to current A. S. T. M. Standard Specifications A 47, Grade 32510, for malleable iron castings. Each casting shall consist of a perforated round plate with a flange around the edge extending at right angles to the face of the plate and projecting from one face only, the plate portion having a central bolt hole reamed to size with an integral hub concentric to the bolt hole and extending from the same face as the flange.

PERCENTAGES FOR DURATION OF MAXIMUM LOAD

Two Months Loading, as for snow *115%
Seven Days Loading *125%
Wind or Earthquake Loading *133⅓%
Impact Loading *200%
Permanent Loading 90%
*Do not exceed limitations for maximum allowable loads for shear plates given elsewhere on this page.

DECREASES FOR MOISTURE CONTENT CONDITIONS

Condition when Fabricated	Seasoned	Unseasoned	Unseasoned
Condition when Used	Seasoned	Seasoned	Unseasoned or Wet
Shear Plates	0%	20%	33%

MAXIMUM PERMISSIBLE LOADS ON SHEAR PLATES

The allowable loads for all loadings except wind shall not exceed 4970 lbs. for 4" shear plates with ¾" bolts and 6760 lbs. for 4" shear plates with ⅞" bolts. The allowable wind load shall not exceed 6630 lbs. when used with a ¾" bolt and 9020 lbs. when used with a ⅞" bolt. If bolt threads bear on the shear plate, reduce the preceding values by one-ninth.

Figure 5.32 *Charts for shear plates-IV. (Courtesy Cleveland Steel Specialty Company, USA)*

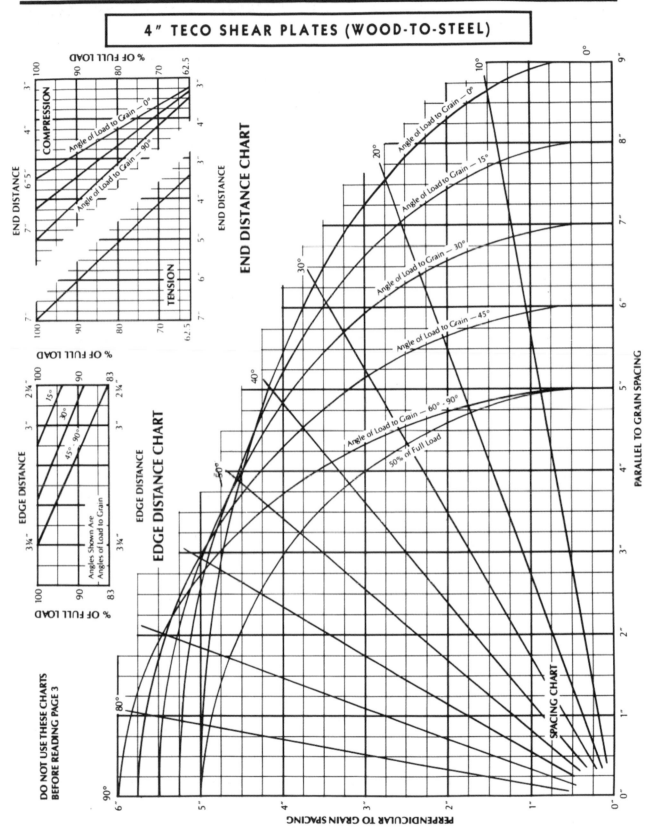

Figure 5.33 *Charts for shear plates-V. (Courtesy Cleveland Steel Specialty Company, USA)*

CLEVELAND DESIGN AND LOAD DATA FOR TECO CONNECTORS

TECO SPIKE-GRIDS

TECO SPIKE-GRIDS

TECO SPIKE-GRID Type	Flat	Single Curve	Circular
Size, square	4⅛″	4⅛″	3¼″
Total depth of grids, maximum	1″	1.38″	1.20″
Diameter of bolt hole	1.06″	1.06″	1.33″
Weight, per 100 grids, lbs.	50	75	26
LUMBER DIMENSIONS, minimum recommended for installation of flat grids			
Face, width	5½″	5½″	5½″
Thickness			
Grids one face only	1½″	1½″	1½″
Grids opposite in both faces	2½″		2½″
Minimum diameter of pile for curved grids		10″	
BOLT, diameter	¾″ or 1″	¾″ or 1″	¾″ or 1″
BOLT HOLE, diameter in timber	¹³⁄₁₆″ or 1¹⁄₁₆″	¹³⁄₁₆″ or 1¹⁄₁₆″	¹³⁄₁₆″ or 1¹⁄₁₆″
WASHERS			
Round, cast or malleable iron	Standard Size for Bolt Diameter Used.		
Square plate	3″x3″x⅜″ Punched for Bolt Diameter Used.		
SPACING OF GRIDS, minimum, center to center			
0°-30° angle of load to grain			
Spacing parallel to grain	7″	7″	7″
Spacing perpendicular to grain	5½″	5½″	5½″
30°-90° angle of load to grain			
Spacing parallel or perpendicular to grain	5½″	5½″	5½″
END DISTANCES, center of grid to end of piece (tension or compression members)			
Standard	7″	7″	7″
Minimum, reduce loads 15%	5″	5″	5″
EDGE DISTANCES, center of grid to edge of piece			
Load applied at any angle to grain			
Standard	3⅝″	3⅝″	3⅝″
Minimum, reduce loads 15%	2¾″	2¾″	2¾″
PROJECTED AREA for portion of one grid within member, square inches	2.06	2.06	1.95

SPIKE-GRID SPECIFICATIONS

Spike-grids shall be TECO spike-grids as manufactured by CSS, Cleveland, OH. Spike-grid timber connectors shall be manufactured according to current A. S. T. M. Standard Specifications A47, Grade 32510, for malleable iron castings. They shall consist of rows of opposing spikes which are held in place by fillets. Fillets for the flat and circular grid in cross section shall be diamond shaped. Fillets for the single curve grids shall be increased in depth to allow for curvature.

WIND AND EARTHQUAKE LOADS

For wind or earthquake loads alone or a combination of wind or earthquake with dead or live loads or both, the safe loads on spike-grids may be taken as 120% of the Design Loads provided the resulting size and number of connectors is not less than required for the dead and live loads alone.

IMPACT

When using Design Loads, the load on a spike-grid due to a force producing impact shall be taken as 115% of the sum of the force as a static load and the load due to its impact.

LOADS IN RELATION TO DISTANCES AND SPACINGS

Standard Design Loads are for standard distances and spacings. Standard and minimum distances and spacings with load reduction factors are given in the table. Loads for end and edge distances and spacings intermediate of standard and minimum may be determined by interpolation.

DESIGN LOADS* FOR ONE SPIKE-GRID AND BOLT IN SINGLE SHEAR

GROUP A			GROUP B			GROUP C		
Type of Grid	Bolt Diameter	Allowable Load	Type of Grid	Bolt Diameter	Allowable Load	Type of Grid	Bolt Diameter	Allowable Load
FLAT	¾″	3900#	FLAT	¾″	3500#	FLAT	¾″	3000#
	1″	4200#		1″	3800#		1″	3300#
SINGLE CURVE	¾″	4200#	SINGLE CURVE	¾″	3800#	SINGLE CURVE	¾″	3200#
	1″	4500#		1″	4100#		1″	3500#
CIRCULAR	¾″	3500#	CIRCULAR	¾″	3100#	CIRCULAR	¾″	2600#
	1″	3800#		1″	3400#		1″	2900#

*Allowable loads on spike-grids same for all angles of load to grain.

Figure 5.34 *Results for spike grids. (Courtesy Cleveland Steel Specialty Company, USA)*

Chapter 6
Roofs and Roofing Elements

6.1 Introduction

A vast amount of literature exists on the analysis, design, manufacture and construction of structural elements forming roofs and other associated structures. Roofs of various shapes have been designed and constructed. The most commonly known roofs are:

- Symmetrical Pitched Roof
- Mono-Pitch Roof
- Coupled Roof
- Close Coupled Roof
- Collar Roof
- Purlin Roof and Trussed Rafter Roof
- Hipped Roof
- Multi-pitch and Multi-bay Roofs
- Mansard Roof
- Shell Roof
  - Butterfly Type
  - Barrel Vault
  - Hyperbolic Paraboloid
  - Domed Roof

In general, roofs have three principal shapes: gabled, hipped and flat, but each has many variations. For instance, the gambrel roof is in the gabled family while the mansard roof is in the hipped family. The specific material should be appropriate for the architectural style of the building. Tile roofs are compatible with many architectural styles and capable of long life spans. Various codes give some properties and characteristics that should be considered in the selection of the tile. Specification of a symmetrical shape, in which the curve of the tile is an arc less than half round, is generally encouraged. The slight arc allows the tile to lay flatter on the roof than a tile that is half round.

Details of the installation are crucial components of successful tile roofs. Elements that require specific attention include:

- the eave conditions
- the gable end
- the rip ridge
- the primary ridge
- the detailing around penetrations such as walls and projections.

Timber roofs with slates are compatible with many architectural styles and with good maintenance are capable of a long life span.

Eaves are critical components of the functions between the wall plane and the roof plane. Some of the building elements to consider in detailing of the eave of a house include:

- Overhang dimensions
- Correct scale of overhang dimension to building
- Exposed rafters and facia treatment
- Rain gutter placement and shape
- Correct scale of decorative elements

Coming to lofts and mansard roofs, this chapter gives a complete design of a loft. As far as mansard roofs are concerned, they have two slopes on each side, like a hipped structure, the lower face being more steeply pitched than the upper thus providing more space for accommodation than would be with a conventional roof of single slope. The exact angles and proportions of the slopes may vary according to the depth of the house and other circumstances. Generally, tiles require a steeper pitch, slates a lower one. Roofs with an upper slope of pitch below 5° or totally flat are not usually acceptable.

For correct appearance, the lower slope of a domestic mansard should ride from a point sufficiently behind the main parapet wall of the house at both front and back. It should not rest on a parapet at all and should normally be separated from the wall by a substantial greeter. Bulk heads and other protrusions rising out of the slopes of the roof are to be avoided. Only party walls with their chimney stack, if any, and windows of traditional-type should interrupt the roof planes.

Particular problems in design can arise in the construction of roof extensions. It may be necessary for the sake of good appearance, to construct roofs with proper tipped ends. There are several ways of setting out mansard roofs. Two of the most suitable solutions are based on semi-circles with the span of the roof as the diameter. The upper slope should generally not be greater than 30° depending on roofing materials and other factors. The lower slope should be in the region of 70°.

The coping should always fall towards the gutter and the party wall line should start behind the back line of the coping. One example would be where the circumference of the semi-circle is divided into five equal parts. The junction between the two slopes (knee) is located between the lower two parts. The other example is where the height from the diameter to

106   STRUCTURAL DETAILING IN TIMBER

the ridge is divided into two equal parts and the knee is located on the circumference of this level. Dormer windows in the lower roof slopes should be modest in size and should generally project from the roof slopes rather than be inset.

In this chapter, a number of roofs have been discussed which include pitched roofs, coupled and close coupled roofs, collar roofs, purlin roofs, trussed roofs and shell-type roofs. Other related elements and their design drawings are also discussed.

6.2 Description of roofing structures and structural timber detailing

6.2.1 Symmetrical pitched roof

Figure 6.1 gives the geometrical layout of all components of a symmetrically pitched roof. All areas have been duty identified on the various figures.

Timber pitched roof structures are used for two main reasons: first, to ensure adequate weather-proofing and second, if required, to provide roof-space utilisation such as additional storage. Relatively low pitch angles such as 10 degrees are acceptable for roofs surfaced with multi-layered bituminous felts or similar sheet materials. In situations where a roof is tiled, slated or covered with overlapping profile sheeting, the pitch must be sloped sufficiently to permit rapid drainage of water and prevent

the ingress of water by wind-driven rain. The minimum slope necessary is dependent on the type of covering, particularly the degree of overlap between individual units. Typically, large slates require approximately 22.5 degree slopes whilst small slates and single lap tiles may require nearer 35 degree slopes. Plain tiles require at least a 40 degree slope to inhibit water penetration.

The construction of a timber pitched roof can take one of many structural forms. Parapets reduce the high suction in the edge zones around the periphery of the roof and the effects of parapets should be taken into account.

6.2.1.1 Roof overhangs

Where the roof overhangs the walls by an amount less than 6/10 inches (15 mm), pressure coefficients should be assessed. Larger overhangs should be treated as open-sided buildings, with internal pressure coefficients determined.

6.2.1.2 Small overhangs

The net pressure across a small roof overhang should be calculated taking the pressure coefficient on the upper surface as appropriate, and the pressure coefficient on the lower surface as that on the adjacent wall.

6.2.1.3 Canopies, grandstands and open-sided buildings

Free-standing canopies

Net pressure coefficients C_p for free-standing canopy roofs are given in many standard tables of many codes which take account of the combined effect of the wind on both upper and lower surfaces of the canopy for all wind directions.

A comparative study of triangular roof design and construction with typical wall section and ground floor with foundations based on European, British and American codes and practices is given in Figs 6.2–6.4, respectively. These are treated as constructed facilities.

6.2.2 Mono-pitch roof

Mono-pitch roofs consist of series of single rafters supported on wall plates directly carried onto the walls, or on a wall plate directly carried on a wall at the lower end and another wall plate fastened to, or corbelled from, a wall at the upper level. The latter method is often referred to as a lean-to construction.

Since there is a tendency for an inclined rafter to induce a horizontal thrust at the support, provision must be made to transfer this force to the wall and prevent sliding. Clearly the wall plates and the supporting wall must be capable of resisting the outward thrust from the rafter. The transfer of thrust is achieved by creating a notch (bird's mouth) in the rafter at the location where it meets the wall.

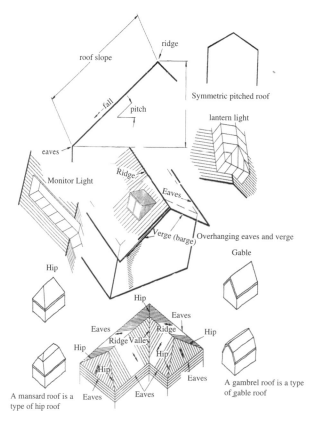

Figure 6.1 *Timber roof configuration for symmetric pitched roof.*

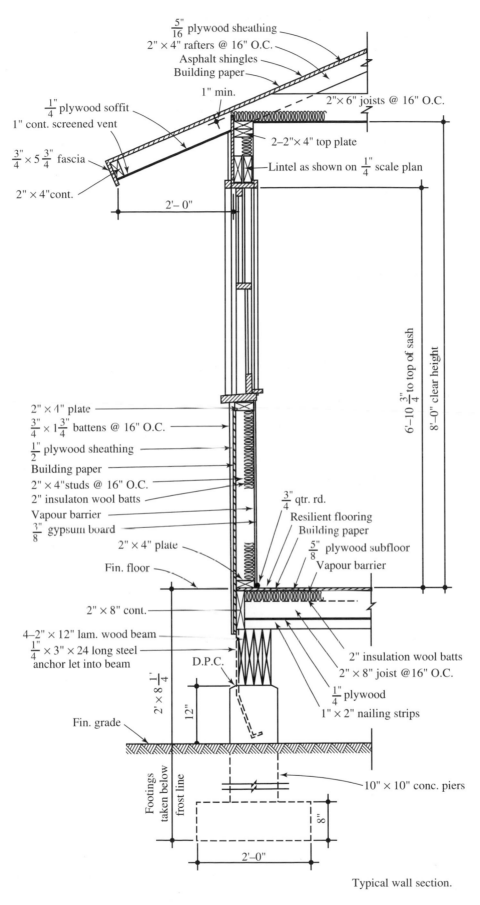

$\frac{5"}{16}$ plywood sheathing

2" × 4" rafters @ 16" O.C.

Asphalt shingles

Building paper

1" min.

2"× 6" joists @ 16" O.C.

$\frac{1"}{4}$ plywood soffit

1" cont. screened vent

2–2"× 4" top plate

$\frac{3"}{4}$ × 5 $\frac{3"}{4}$ fascia

Lintel as shown on $\frac{1"}{4}$ scale plan

2" × 4"cont.

2'– 0"

6'–10 $\frac{3"}{4}$ to top of sash

8'–0" clear height

2" × 4" plate

$\frac{3"}{4}$ × 1 $\frac{3"}{4}$ battens @ 16" O.C.

$\frac{1"}{2}$ plywood sheathing

Building paper

2" × 4"studs @ 16" O.C.

2" insulaton wool batts

Vapour barrier

$\frac{3"}{8}$ gypsum board

2" × 4" plate

Fin. floor

$\frac{3"}{4}$ qtr. rd.

Resilient flooring

Building paper

$\frac{5"}{8}$ plywood subfloor

Vapour barrier

2" × 8" cont.

4–2" × 12" lam. wood beam

$\frac{1"}{4}$ × 3" × 24 long steel anchor let into beam

D.P.C.

2' × 8 $\frac{1'}{4}$

12"

Fin. grade

Footings taken below frost line

2" insulation wool batts

2" × 8" joist @16" O.C.

$\frac{1"}{4}$ plywood

1" × 2" nailing strips

10" × 10" conc. piers

8"

2'–0"

Typical wall section.

Key: 1" = 25.4 mm   1 ft = 0.3048 m

Figure 6.2 *Pitched roof based on American practice.*

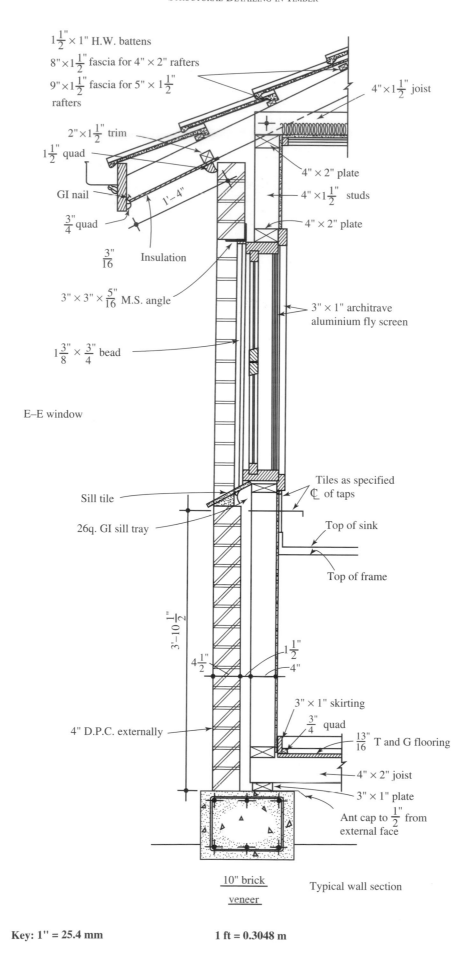

$1\frac{1}{2}$" × 1" H.W. battens

8" × $1\frac{1}{2}$" fascia for 4" × 2" rafters

9" × $1\frac{1}{2}$" fascia for 5" × $1\frac{1}{2}$"
rafters

4" × $1\frac{1}{2}$" joist

2" × $1\frac{1}{2}$" trim

$1\frac{1}{2}$" quad

GI nail

$\frac{3}{4}$" quad

$\frac{3}{16}$"

Insulation

1'–4"

4" × 2" plate

4" × $1\frac{1}{2}$" studs

4" × 2" plate

3" × 3" × $\frac{5}{16}$" M.S. angle

$1\frac{3}{8}$" × $\frac{3}{4}$" bead

3" × 1" architrave
aluminium fly screen

E–E window

Sill tile

26q. GI sill tray

Tiles as specified
℄ of taps

Top of sink

Top of frame

$3'-10\frac{1}{2}$"

$4\frac{1}{2}$"

$1\frac{1}{2}$"

4"

3" × 1" skirting

$\frac{3}{4}$" quad

$\frac{13}{16}$" T and G flooring

4" D.P.C. externally

4" × 2" joist

3" × 1" plate

Ant cap to $\frac{1}{2}$" from
external face

10" brick
veneer

Typical wall section

Key: 1" = 25.4 mm   **1 ft = 0.3048 m**

Figure 6.3 *Pitched roof based on British codes and practices.*

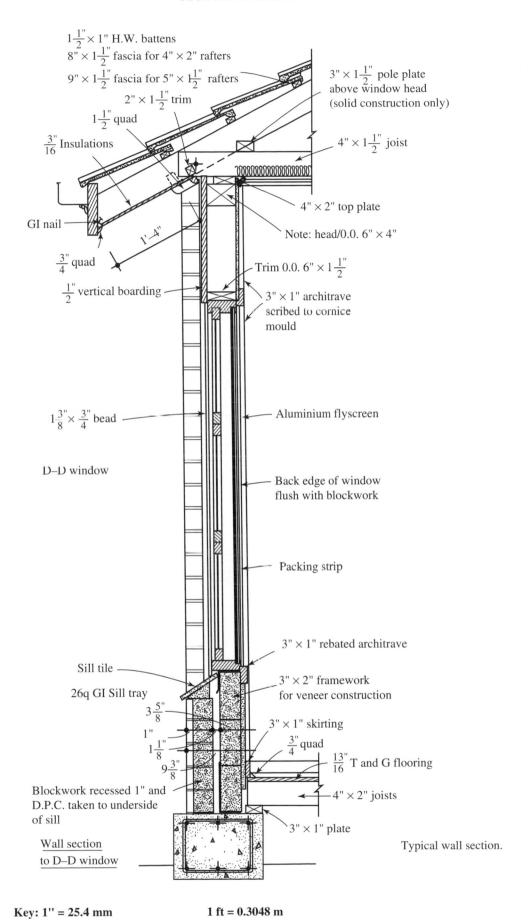

$1\frac{1}{2}" \times 1"$ H.W. battens
$8" \times 1\frac{1}{2}"$ fascia for $4" \times 2"$ rafters
$9" \times 1\frac{1}{2}"$ fascia for $5" \times 1\frac{1}{2}"$ rafters
$2" \times 1\frac{1}{2}"$ trim
$1\frac{1}{2}"$ quad
$\frac{3}{16}"$ Insulations

GI nail

$\frac{3}{4}"$ quad

$\frac{1}{2}"$ vertical boarding

$1\frac{3}{8}" \times \frac{3}{4}"$ bead

D–D window

Sill tile

26q GI Sill tray

$3\frac{5}{8}"$
$1"$
$1\frac{1}{8}"$
$9\frac{3}{8}"$

Blockwork recessed 1" and
D.P.C. taken to underside
of sill

Wall section
to D–D window

$3" \times 1\frac{1}{2}"$ pole plate
above window head
(solid construction only)

$4" \times 1\frac{1}{2}"$ joist

$4" \times 2"$ top plate

Note: head/0.0. $6" \times 4"$

Trim 0.0. $6" \times 1\frac{1}{2}$

$3" \times 1"$ architrave
scribed to cornice
mould

Aluminium flyscreen

Back edge of window
flush with blockwork

Packing strip

$3" \times 1"$ rebated architrave

$3" \times 2"$ framework
for veneer construction

$3" \times 1"$ skirting

$\frac{3}{4}"$ quad

$\frac{13}{16}"$ T and G flooring

$4" \times 2"$ joists

$3" \times 1"$ plate

Typical wall section.

Key: 1" = 25.4 mm   **1 ft = 0.3048 m**

Figure 6.4 *Pitched roof based on American practice.*

The wall plate is located at the top of the wall in a mortar bed to provide a level bearing for the rafters and distribute the end reaction evenly over the wall. The rafters are normally skew-nailed to a plate. A similar detail is used at the upper level. The covering material can be sheeted or tiled. When using sheeted materials, care must be taken to ensure that appropriate rafter spacing is provided accurately to accommodate the standard sheet sizes. In tiled roofs, softwood battens are provided, the size of which depends upon the spacing of the rafters which normally range from 400 mm to 600 mm centres. When using larger spacings, care must be taken to ensure that the battens have adequate stiffness to enable tiling or slating nails to be driven in. The economic spans of mono-pitch roofs of this type are limited to approximately 3 m to 4 m.

Figure 6.5 gives an illustration of the mono-pitch roof in dictating the pitch angle and wind directions. External pressure coefficients, C_{pe}, for mono-pitch roofs of buildings and net pressure coefficients C_p for free standing mono-pitch canopies roofs can be found in BS 6399-2:1997.

6.2.3 Duopitch roofs

Like mono-pitch roofs, a duopitch roof is also with a gable end. Figures 6.6 and 6.7 show elevations and plan of a duopitch roof indicating pitch angles and wind angles.

External pressure coefficients, C_{pe}, for duopitch roofs should be obtained from BS 6399-2:1997. Values

are given for two wind directions: wind normal to the low eaves ($\theta = 0°$) and wind normal to the gable ($\theta = 90°$). These coefficients are appropriate to duopitch faces of equal pitch but may be used without modification provided the upwind and downwind pitch angles are within 5° of each other.

The code provides tables of pitch change vs. C_{pe} for mono-pitch roofing.

When $\alpha < 7°$ and $W < b_L$, zone C load cases $\theta = 0°$ should be considered to extend for a distance $<b/2$ downwind from the windward eave replacing ridge zones E and F and part of zone G.

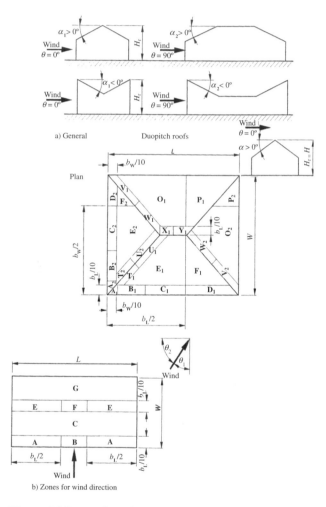

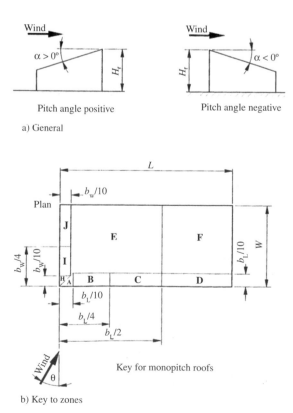

Figure 6.6 *Duopitch roofs.*

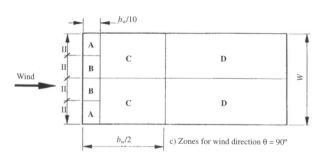

Figure 6.5 *Monopitch roof.*

Figure 6.7 *Duopitch roofs (contd.).*

6.2.4 Coupled and close coupled roofs

6.2.4.1 Coupled roof

A coupled roof is assembled by nailing rafters to a ridge board at the upper level and to wall plates at the lower level as shown in Fig. 6.8(a). The horizontal thrust which is produced at the wall plate normally limits the economic clear span to approximately 3 m. Spans above this level require a significant increase in the depth of the rafter to control deflection, and in addition buttressing of the support walls may be necessary to resist the lateral thrust. The ridge board locates the rafters in opposite alignment on each side preventing lateral movement. This form of construction is sometimes used with very steeply pitched roofs in church buildings utilising solid or glued members and relatively low support walls.

6.2.4.2 Close coupled roof

Close coupled roofs are a development of coupled roofs, in which the horizontal thrust at the lower rafter ends is contained internally within a triangulated framework as shown in Fig. 6.8(b) and (c).

The additional tie member, which is securely nailed at its junctions with the rafters, may also be used as a ceiling joist. In this instance, a heavier section than is required simply for tying is usually required. Economic spans for this type of construction are approximately 5.0 m to 6.0 m. In spans of this value, the section size adopted for the ceiling joist can be minimised by the introduction of additional supports such as binders and hangers as shown in Fig. 6.8(d).

A brief calculation is given in Fig. 6.8(e) for the triangular roof system based on a coupled roof with a vertical strut in the centre.

6.2.5 Collar roof

Close coupled roofs in which the tie is placed at a level above the supporting walls, typically one-third to one-half of the rise as shown in Fig. 6.9(a) are referred to as collar roofs. This form of construction is less efficient in resisting the spread of the main rafters. A disadvantage in this type of roof is that the section of rafter between the wall plate and the collar is subject to considerable binding. In addition, a bolted/connected joint between the collar and the rafter is normally required to adequately resist the rafter thrust. Collar roofs are economic for relatively short spans not exceeding 5.0 m.

Explanatory calculation notes for Fig. 6.9 are now given based on the idealised diagrams adopted in European practices.

The thrust H at the level of the support is found from the condition $M_c = 0$:

$$H = \frac{M_c^0 + Xf_1}{f} \qquad (6.1)$$

where M_c^0 = moment in the middle of the span in a simple beam of span l on the two supports A and E

(Fig. 6.9(a)). Thus when the collar is brought in, the thrust H in the support joints of a couple truss is increased in comparison with that in a triangular system having a tie rod at the level of the supports, where $H = \frac{M_c^0}{f}$.

In the statically determinate triangular system with a raised tie, working according to the diagram in Fig. 6.9(b), the support reactions are found as for the usual simply supported beams. The force in the tie is equal to

$$N_{\text{tie}} = \frac{M_c^0}{f_1} \qquad (6.2)$$

where $M_c^0 = M_{l/2}^0$ is the moment in the middle of a beam of span l.

When the entire span is loaded with a uniformly distributed load q

$$N_{\text{tie}} = \frac{ql^2}{8f_1}.$$

If the rafter in a triangular system with a collar and a uniform distribution of the load over the entire span is regarded as a double-span continuous beam on rigid supports, the compression stress in the collar can be determined according to the formula

$$\upsilon = \frac{ql^2}{32} \times \frac{f_1^2 \times 3f_1f_2 + f_2^2}{f_1f_2f} \qquad (6.3)$$

When half the span is loaded, the stress in the collar is equal to $\sigma/2$.

6.2.6 Purlin roofs and trussed rafter roofs

Timber pitched roof structures are used for two main reasons: first to ensure adequate weather-proofing and second, if required, to provide roof-space additional storage. Relatively low pitch ingles such as 10° are acceptable for roofs surfaced with multi-layered bituminous felts or similar sheet materials. In situations where a roof is tiled, slated or covered with overlapping profile sheeting, the pitch must be sufficient to permit rapid drainage of water to prevent the ingress of water by wind-driven rain. The minimum slope necessary is dependent on the type of covering, particularly the degree of overlap between individual units. Typically, large slates require approximately 22.5° slopes whilst small slates and single lap tiles may require nearer 35° slopes. Plain tiles require at least a 40° slope to inhibit water penetration.

The construction of a timber pitched roof can take one of many structural forms, such as purlin roofs and trussed rafter roofs. All the roof types with the exception of the purlin roof and elements of the trussed rafter roofs are essentially assembled on site. Trussed rafters are normally prefabricated by a specialist manufacturer and are made using timber members of the same thickness

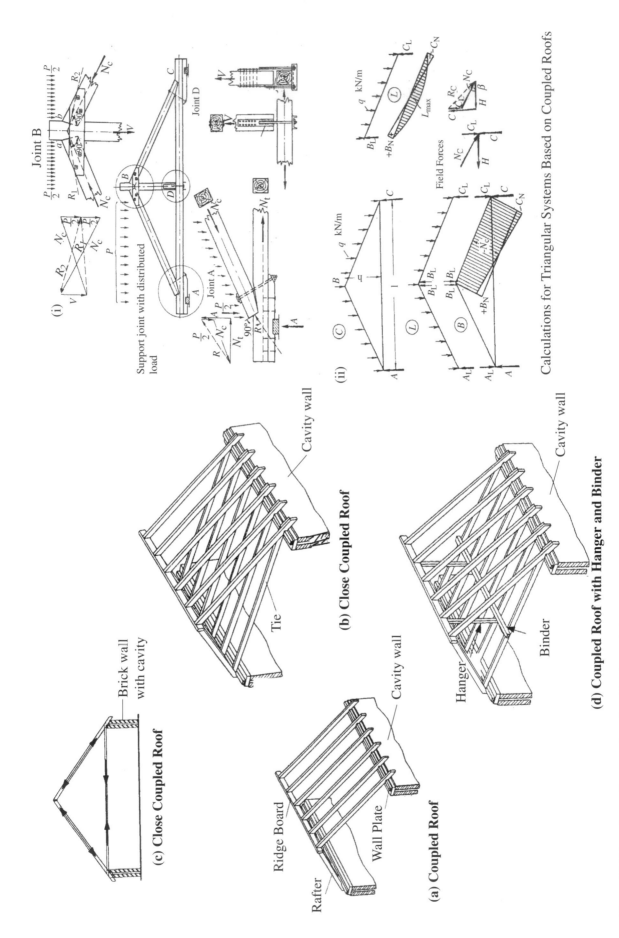

Figure 6.8 *Triangular roof systems (based on Russian codes).*

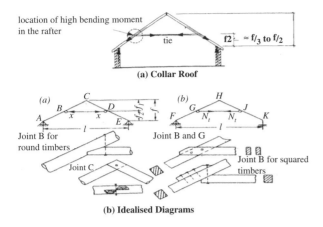

Figure 6.9 *Triangular roof systems – collar roofs.*

fastened together in one plane by metal plate fasteners or plywood gussets.

A complete drawing based on the American Standards for a trussed rafter design is shown in Fig. 6.10, giving specifications, dimensions and design methodology. Figures 6.11 to 6.14 give a comparative study of the timber trusses based on the Canadian design practices and the Forest Products Laboratories of Canada. Figure 6.15 shows a trussed roof with longitudinal and diagonal braces and ceiling ties. This kind of trussed roof is designed by the American practices and has been a very popular choice.

The standard Canadian and US roof truss designs given are suited to spans of 20 ft to 32 ft. For the convenience of the designers/detailers, a quick force polygon diagram can also be given in the design. For a comprehensive design, various new methods of analysis such as finite element should also be considered. The truss rafters are preferred to be placed at spaces. Where no specifications are available, the normal load on roof shall be taken as 35 lb/ft2, with the normal load for ceiling taken as 10 lb/ft2. All other details and variations based on span and plates are given in Figs 6.16 to 6.18.

6.2.7 Hipped roof

The provisions in BS 6399 apply to conventional hipped roofs on rectangular-plan buildings, where the main pitch of the main ridged faces have pitch angle $\acute{\alpha}_1$ and the triangular side faces have pitch angle $\acute{\alpha}_2$. Zones of external $\theta_2 = 90° - \theta_1$.

Thus, for the main ridged faces the pitch is $\acute{\alpha}_1$, the wind direction is θ_1 and the zones are A_1 to Y_1, and for the triangular side faces the pitch is $\acute{\alpha}_2$, the wind direction is θ_2 and zones are A_2 to Y_2. The reference height H_r is the height above ground of the ridge.

External pressure coefficients for zones A to E on the upwind faces are given in any code. External pressure coefficients for zones O and P on the downwind faces are given in Table 35. The size of each of these zones is given in Fig. 6.19.

External pressure coefficients for the additional zones T to W along the hip ridges and for zones X and Y along

the main ridge are given in Table 36. The width of each of these additional zones is shown in plan. The boundary between each pair of additional zones, T–U, V–W and X–Y, is the mid-point of the respective hip or main ridge.

6.2.7.1 Mixed gables and hipped roofs

Roofs with a standard gamble at one end and a hip at the other are a frequent occurrence. In such cases, the governing criterion is the form of the upwind corner for the wind direction being considered.

External pressure coefficients for conventional hipped roofs on cuboidal-plan buildings, where all faces of the roof have the same pitch angle and are in the range $\alpha = -45°$ to 75°, are given in Fig. 6.20. The definitions of loaded zones and the pitch angles are given in the table in Fig. 6.20. The data may be applied to hipped roofs where the main faces and hipped faces have different pitch angles, provided the pitch angle of the upwind face is used for each wind direction (Fig. 6.19). Negative pitch angles occur when the roof is a hipped-trough form.

1. Upwind main and hip faces. For the part of the roof below the top of the parapet, external pressure coefficients should be determined, ie of the top of the parapets, irrespective of actual roof pitch. For any part of the roof that is above the top of the parapet, ie if the top of the parapet is below the level of the ridge, external pressure coefficients should be determined.
2. Downwind main and hip faces. External pressure coefficients should be determined as if the parapet did not exist. The reduction factors should not be applied to any zone.

6.2.8 Multi-pitch, multi-bay and mansard roofs

Pitch roofs have already been discussed in this chapter. Multiples of such roofs are then known as multi-pitch roofs. Any one of them can be a mansard roof. In a similar manner, multi-bay roofs can be designed in timber. Timber and wood can easily be chosen for these types of roofs using specific dead and imposed loads. Regarding wind loads, external pressure coefficients for mansard roofs and other-pitch roofs can be derived for each plane face of the roof with cable verges or hipped verges, using the pitch angle for each plane face. Figure 6.21 indicates where edge zones should be omitted.

External pressure coefficients should be determined in accordance with BS 6399 using the appropriate zones for the pitched roofs.

6.2.8.1 Friction induced loads on pitched roofs

Friction forces should be determined for long pitched roofs when the wind is parallel to the eaves or ridge, ie $\theta = 90°$. The frictional drag coefficients should be assumed to act over zones F and P only of such roofs, with the values given by the code. The resulting

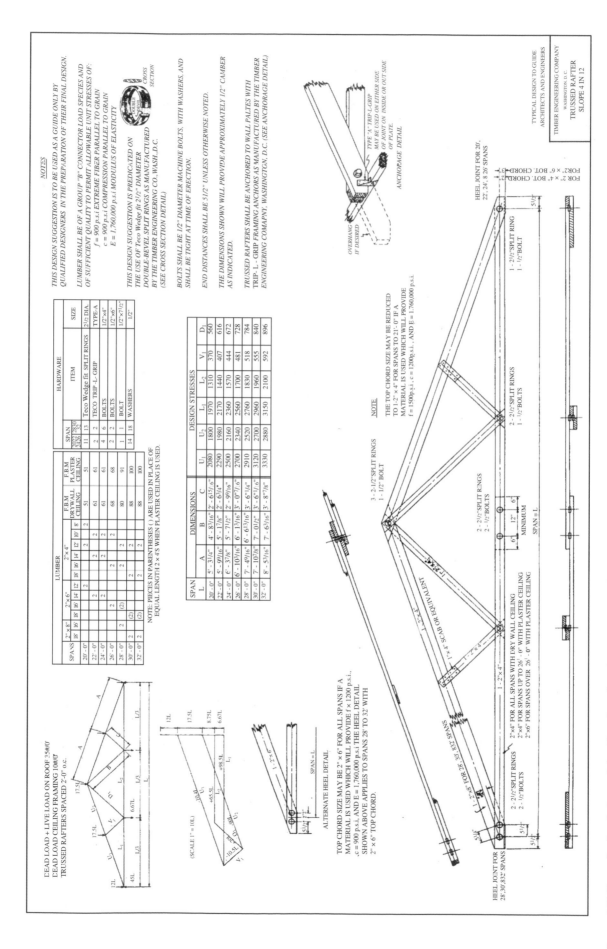

Figure 6.10 *Trussed rafter design and detailing (based on American practice).*

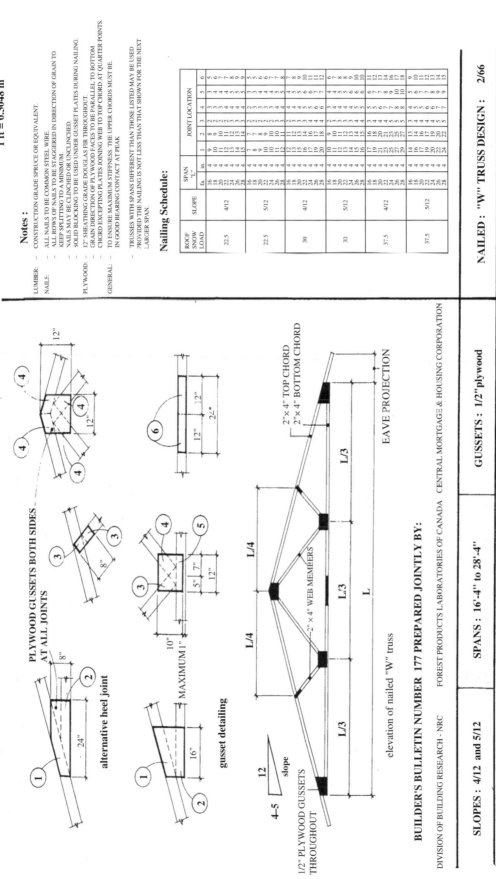

Figure 6.11 *Design detailing for a truss roof with gussets and nailing schedule (based on Canadian practices) (Source: Canada Mortgage and Housing Corporation (CMHC). Builders' Bulletin 177, 1966 All rights reserved. Reproduced with consent of CMHC. All other uses and reproductions of this material are expressly prohibited.)*

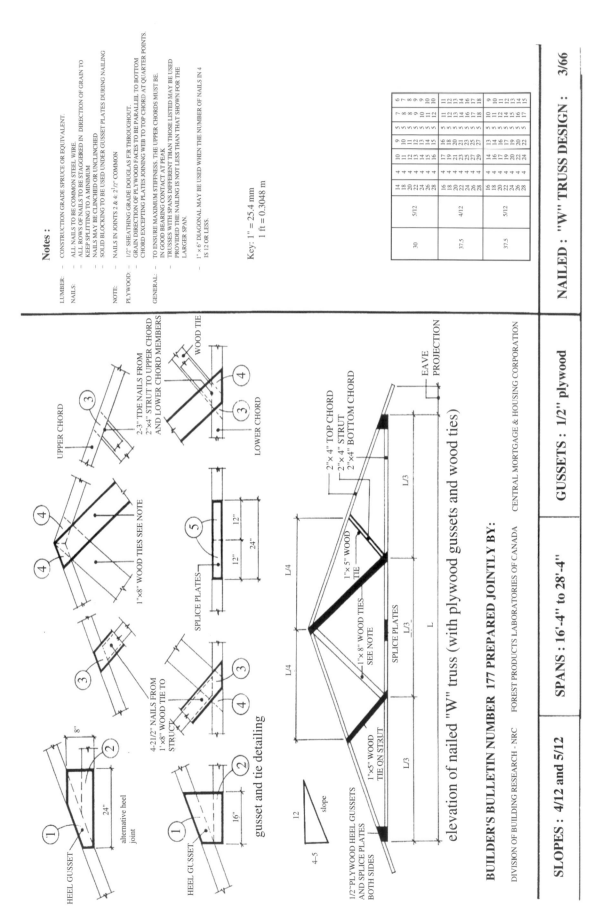

Figure 6.12 *Nailed "W" truss design with plywood gussets and wood ties (based on Canadian practices) (Source: Canada Mortgage and Housing Corporation (CMHC). Builders' Bulletin 177,*

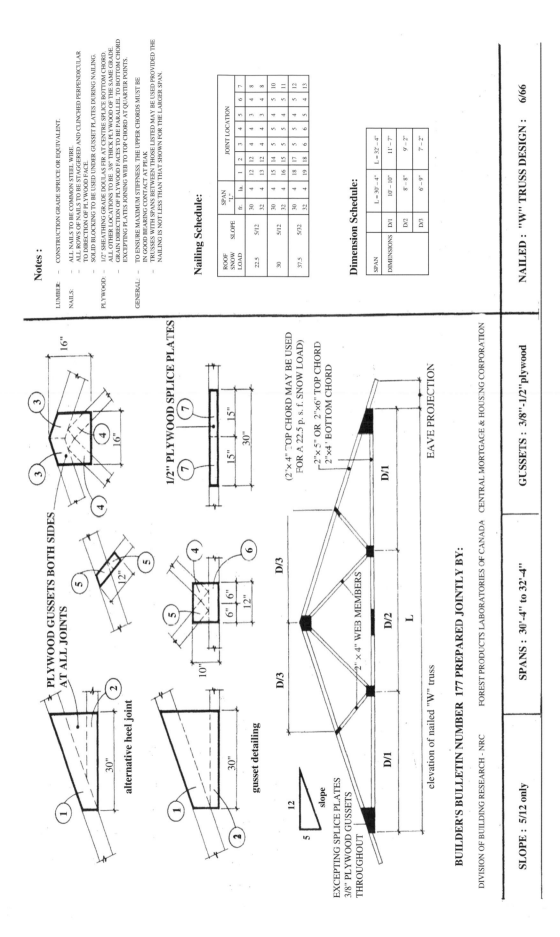

Figure 6.13 *Nailed "W" truss with plywood gussets (based on Canadian practices) (Source: Canada Mortgage and Housing Corporation (CMHC). Builders' Bulletin 177, 1966 All rights reserved. Reproduced with consent of CMHC. All other uses and reproductions of this material are expressly prohibited.)*

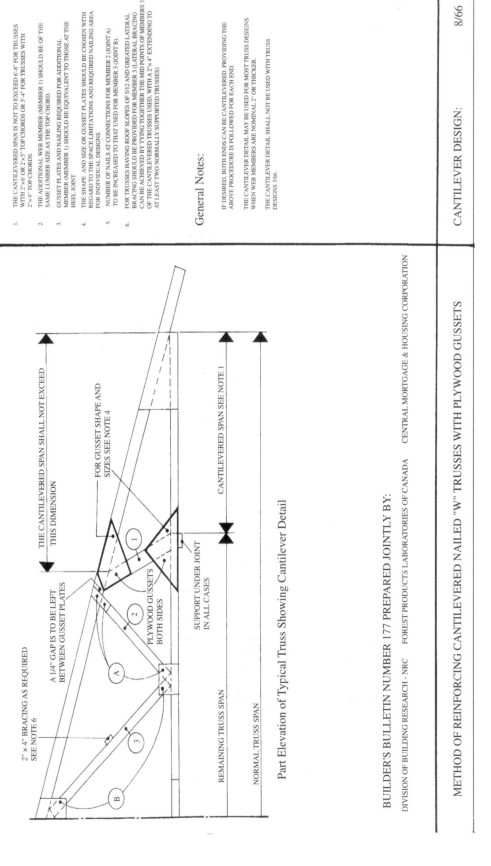

Reinforcing Requirements:

1. THE CANTILEVERED SPAN IS NOT TO EXCEED 6'-8" FOR TRUSSES WITH 2"×6" OR 2"×5" TOP CHORDS OR 5'-4" FOR TRUSSES WITH 2"×4" TOP CHORDS.

2. THE ADDITIONAL WEB MEMBER (MEMBER 1) SHOULD BE OF THE SAME LUMBER SIZE AS THE TOP CHORD.

3. GUSSET PLATES AND NAILING REQUIRED FOR ADDITIONAL MEMBER (MEMBER 1) SHOULD BE EQUIVALENT TO THOSE AT THE HEEL JOINT.

4. THE SHAPE. AND SIZE OF GUSSET PLATES SHOULD BE CHOSEN WITH REGARD TO THE SPACE LIMITATIONS AND REQUIRED NAILING AREA FOR INDIVIDUAL DESIGNS.

5. NUMBER OF NAILS AT CONNECTIONS FOR MEMBER 2 (JOINT A) TO BE INCREASED TO THAT USED FOR MEMBER 3 (JOINT B).

6. FOR TRUSSES HAVING ROOF SLOPES OF 5/12 AND GREATER LATERAL BRACING SHOULD BE PROVIDED FOR MEMBER 3 (LATERAL BRACING CAN BE ACHIEVED BY TYING TOGETHER THE MID POINTS OF MEMBERS 3 OF THE CANTILEVERED TRUSSES USED, WITH A 2"×4" EXTENDING TO AT LEAST TWO NORMALLY SUPPORTED TRUSSES)

General Notes:

IF DESIRED, BOTH ENDS CAN BE CANTILEVERED. PROVIDING THE ABOVE PROCEDURE IS FOLLOWED FOR EACH END.

THE CANTILEVER DETAIL MAY BE USED FOR MOST TRUSS DESIGNS WHEN WEB MEMBERS ARE NOMINAL 2" OR THICKER.

THE CANTILEVER DETAIL SHALL NOT BE USED WITH TRUSS DESIGNS 396.

2" × 4" BRACING AS REQUIRED SEE NOTE 6

A 1/4" GAP IS TO BE LEFT BETWEEN GUSSET PLATES

THE CANTILEVERED SPAN SHALL NOT EXCEED THIS DIMENSION

FOR GUSSET SHAPE AND SIZES SEE NOTE 4

PLYWOOD GUSSETS BOTH SIDES

SUPPORT UNDER JOINT IN ALL CASES

CANTILEVERED SPAN SEE NOTE 1

REMAINING TRUSS SPAN

NORMAL TRUSS SPAN

Part Elevation of Typical Truss Showing Cantilever Detail

BUILDER'S BULLETIN NUMBER 177 PREPARED JOINTLY BY:

DIVISION OF BUILDING RESEARCH - NRC   FOREST PRODUCTS LABORATORIES OF CANADA   CENTRAL MORTGAGE & HOUSING CORPORATION

METHOD OF REINFORCING CANTILEVERED NAILED "W" TRUSSES WITH PLYWOOD GUSSETS	CANTILEVER DESIGN:	8/66

Figure 6.14 Cantilever truss design (based on Canadian practices) (Source: Canada Mortgage and Housing Corporation (CMHC). Builders' Bulletin 177, 1966 All rights reserved. Reproduced with consent of CMHC. All other uses and reproductions of this material are expressly prohibited.)

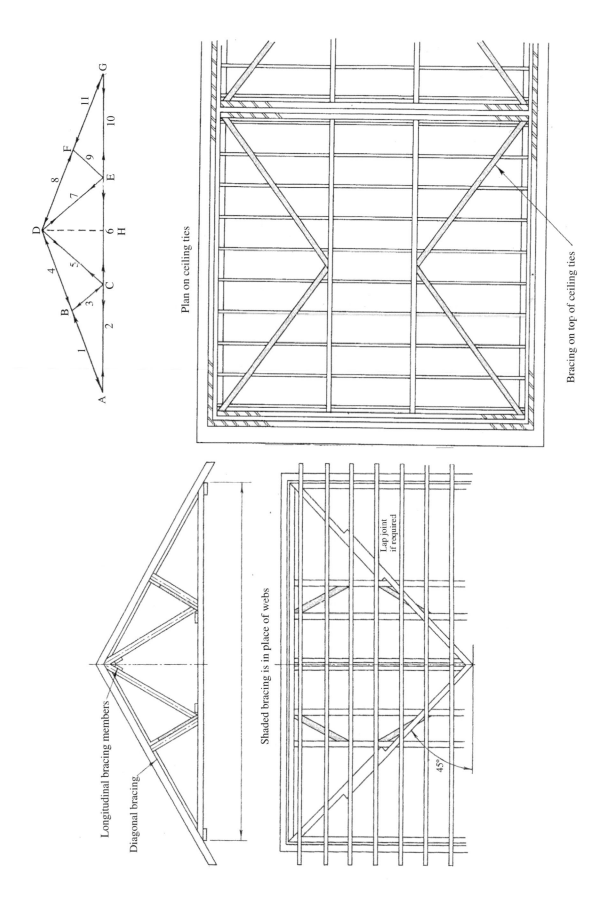

Figure 6.15 *Trussed roof with longitudinal and diagonal braces, webs and ceiling ties (based on American practice).*

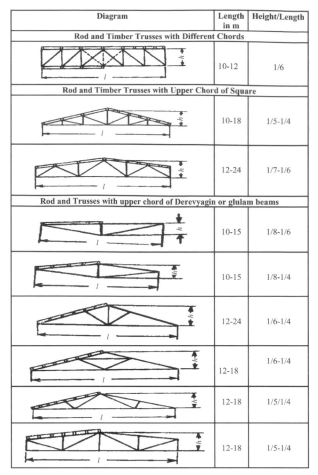

Diagram	Length in m	Height/Length
Rod and Timber Trusses with Different Chords		
	10-12	1/6
Rod and Timber Trusses with Upper Chord of Square		
	10-18	1/5-1/4
	12-24	1/7-1/6
Rod and Trusses with upper chord of Derevyagin or glulam beams		
	10-15	1/8-1/6
	10-15	1/8-1/4
	12-24	1/6-1/4
	12-18	1/6-1/4
	12-18	1/5/1/4
	12-18	1/5-1/4

Figure 6.16 *Rod and timber trusses (based on Russian practice).*

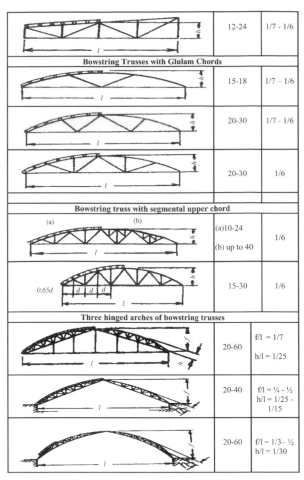

	12-24	1/7 - 1/6
Bowstring Trusses with Glulam Chords		
	15-18	1/7 – 1/6
	20-30	1/7 - 1/6
	20-30	1/6
Bowstring truss with segmental upper chord		
	(a)10-24 (b) up to 40	1/6
	15-30	1/6
Three hinged arches of bowstring trusses		
	20-60	f/l = 1/7 h/l = 1/25
	20-40	f/l = ¼ - ½ h/l = 1/25 - 1/15
	20-60	f/l = 1/3– ½ h/l = 1/30

Figure 6.17 *Bow-string wooden trusses (based on Russian practice).*

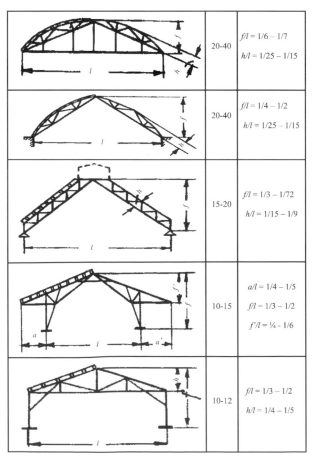

	20-40	f/l = 1/6 – 1/7 h/l = 1/25 – 1/15
	20-40	f/l = 1/4 – 1/2 h/l = 1/25 – 1/15
	15-20	f/l = 1/3 – 1/72 h/l = 1/15 – 1/9
	10-15	a/l = 1/4 – 1/5 f/l = 1/3 – 1/2 f'/l = ¼ - 1/6
	10-12	f/l = 1/3 – 1/2 h/l = 1/4 – 1/5

Figure 6.18 *Trussed frames made in wood (based on Russian practice).*

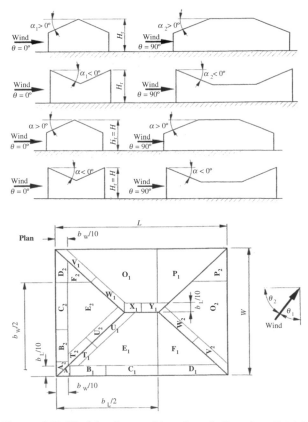

Figure 6.19 *Wind loading on hipped roofs (based on British codes and practices).*

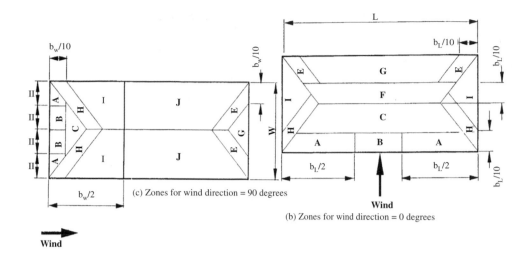

Figure 6.20

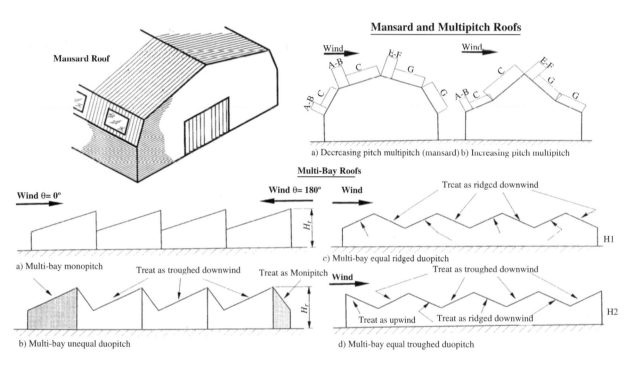

Figure 6.21 *Multi-pitch, multi-bay and mansard roof.*

frictional forces should be applied in accordance with the code.

6.2.8.2 Multi-pitch and multi-bay roofs

Multi-pitch roofs

Multi-pitch roofs are defined as a roof in which each span is made up of two or more pitch angles for the standard method. The form is commonly known as a mansard roof. External pressure coefficients for each pitch should be derived in accordance with the code, according to the form of the verges, but omitting the eaves edge zones along the change in slope where indicated.

1. the eaves edge zones on the bottom edge of all windward faces should be included when the pitch angle of that face is less than that of the pitch below, including the lowest face forming the actual eaves of the windward side. The eaves zones should be excluded when the pitch angle is greater than that of the pitch below.
2. The ridge zones for gambled roofs should be included only on the highest downwind face along the actual ridge. Ridge zones on all other downwind faces should be excluded.
3. Verge zones H to J on gambled roofs should be included for all faces.

6.2.8.3 Multi-bay roofs

Multi-bay roofs are defined as roofs made up of a series of monopitch, duopitch, hipped or similar spans. Pressure coefficients on the first span, ie the upwind pitch of multi-bay monopitch roofs and the upwind pair

Table 6.1 Reduction factors for free-standing multi-bay canopy roofs

Location	Factors for all solidity ratio ζ	
	On maximum	On minimum
End bay	1.00	0.81
Second bay	0.87	0.64
Third bay subsequent bays	0.68	0.63

of pitches of duopitch roofs, may be taken to be the same as for single span roofs. However, these pressures are reduced in value for the downwind spans.

Account may be taken of this reduction by following the procedure in the code of the standard method to define regions of shelter at any given wind direction.

Reduction factors for free-standing multi-bay canopy roofs are available in the code.

When the wind is normal to the gables, ie $\theta = 90°$, there are no regions of shelter. When the wind is normal to the eaves, $\theta = 0°$.

For winds from $\theta = 0°$ and $\theta = 180°$, in all of the above cases, a further reduction in external pressure may be obtained by applying the reduction factors to the second and subsequent downwind bays as given in the code.

6.2.8.4 Timber framing design and construction

6.2.8.4.1 Introduction

This section is devoted to different type of framing and assembly design to assist designers and manufacturers in their own designs.

6.2.8.4.2 Timber framing

Figure 6.22 shows four different methods of timber floor framing. A 3D view of timber framing with bay window is shown in Fig. 6.23, which also illustrates second floor framing with exterior wall platform and framing over bearing platform. Figure 6.24 provides 3D views of timber door and window assembly with structural details, stud assembly with and without partitions meeting, and with and without corner details.

Figure 6.25 covers the following:

a) Nailing of built-up beam
b) Glue laminated beam with steel dowel and metal strap connecting it with wooden posts
c) Plywood cladding and sheeting for walls
d) Woodstrip flooring and resilient flooring
e) Plywood roof sheathing

Table 6.2 Reduction factors

Pitch angle α	Zone for $\theta = 0°$ and $\theta = 90°$								
	A	B	C	E	F	G	H	I	J
−45°	−1.4	−1.0	−1.0	−0.7	−0.4	−0.7	1.1	−1.0	−0.9
−30°	−2.3	−1.2	−1.0	−1.3	−1.3	−0.8	−0.7	−1.0	−0.8
−15°	−2.6	−1.0	0.9	1.4	−1.3	−0.6	−0.9	−0.9	−0.8
−5°	−2.3	−1.1	−0.8	−0.8	−0.6	−0.6	−1.1	−0.8	−0.8
+5°	−1.8	−1.2	−0.6	−0.8	−0.6	−0.6	−1.1	−0.6	−0.6
	+0.0	+0.0	+0.0	−0.8	−0.6	−0.6	+0.0	+0.0	+0.0
+15°	−1.3	−0.8	−0.5	−1.4	−1.3	−0.6	−0.9	−0.6	−0.4
	+0.2	+0.2	+0.2	−1.4	1.3	−0.6	+0.0	+0.0	+0.0
+30°	−0.5	−0.5	−0.2	−1.3	−0.8	−0.6	−1.0	−0.6	−0.5
	+0.8	+0.5	+0.4	−1.3	−0.8	−0.6	+0.0	+0.0	+0.0
+45°	−0.0	−0.0	+0.0	−0.7	−0.4	−0.4	−1.1	−1.15	−0.4
	+0.8	+0.6	+0.7	−0.7	−0.4	−0.4	+0.0	+0.0	+0.0
+60°	+0.8	+0.8	+0.8	−0.6	−0.3	−0.7	−1.2	−0.7	−0.6
+75°	+0.8	+0.8	+0.8	−0.6	−0.3	−1.2	−1.2	−0.5	−0.6

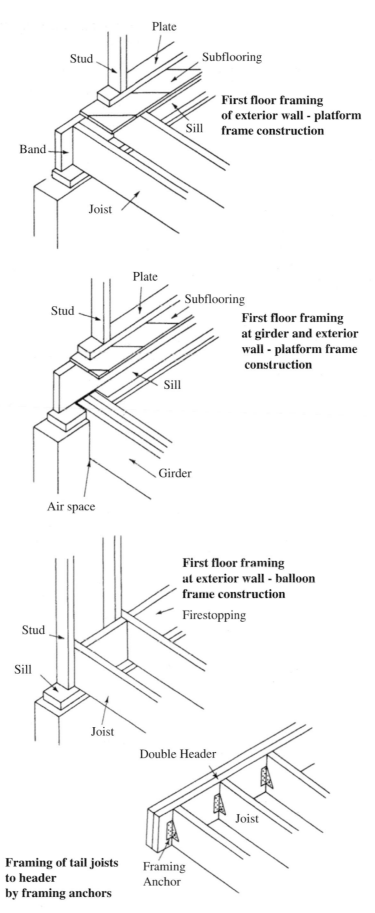

Plate

Stud

Subflooring

**First floor framing
of exterior wall - platform
frame construction**

Sill

Band

Joist

Plate

Stud

Subflooring

**First floor framing
at girder and exterior
wall - platform frame
construction**

Sill

Girder

Air space

**First floor framing
at exterior wall - balloon
frame construction**

Firestopping

Stud

Sill

Joist

Double Header

Joist

**Framing of tail joists
to header
by framing anchors**

Framing
Anchor

Figure 6.22 *Timber floor framing (based on American practice).*

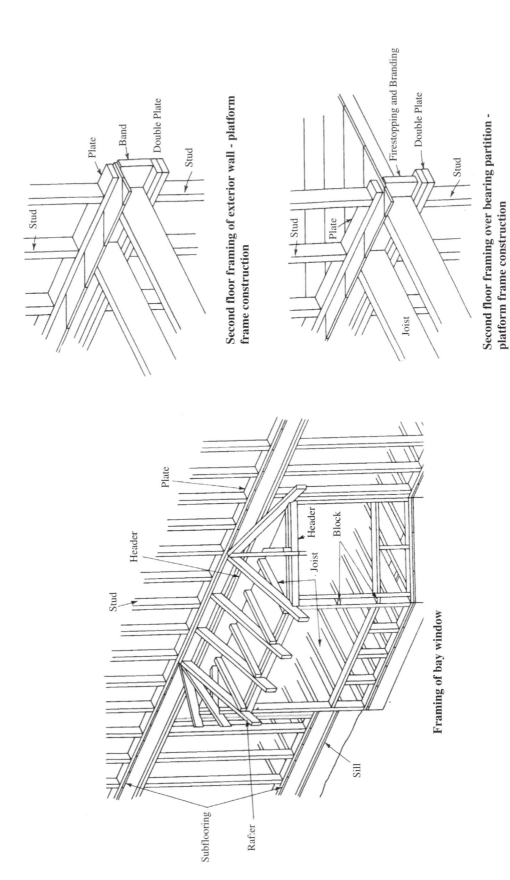

Figure 6.23 *Flooring and timber structural assemblies (based on American practice).*

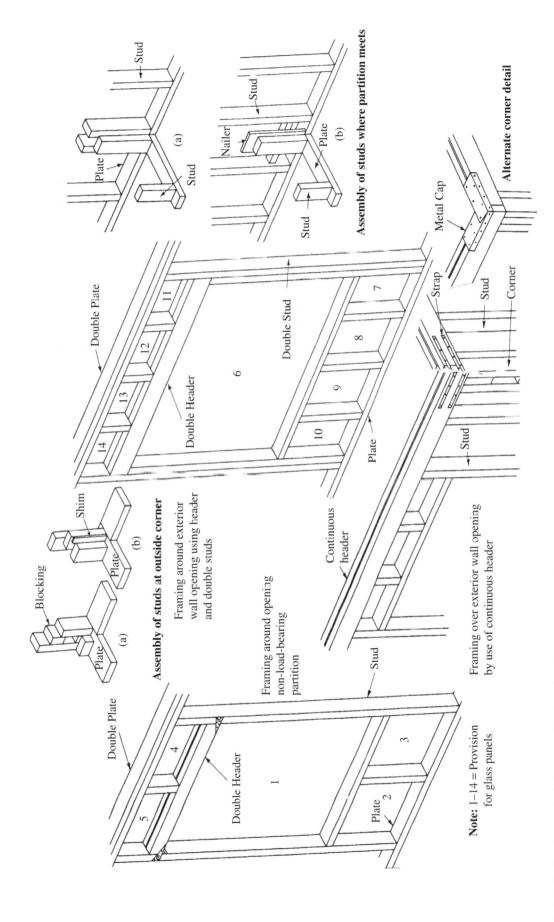

Fig 6.24 *Timber door and window assembly (based on American practice).*

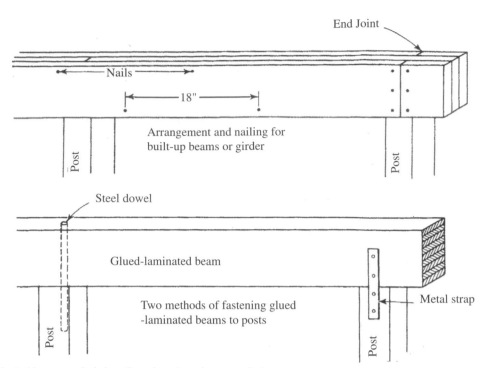

(a) Laminated beams and girders (based on American practice).

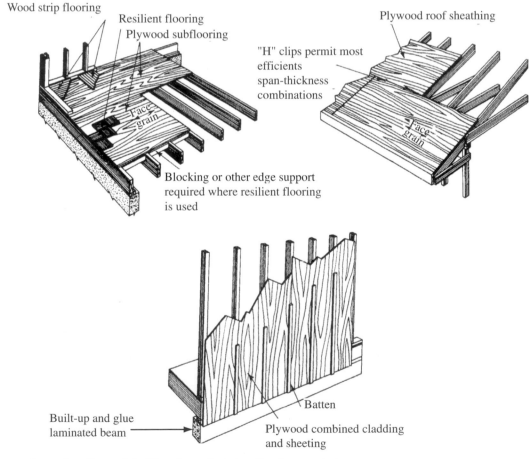

(b) Plywood roofing and cladding sheets (based on Russian practice).

Figure 6.25 *Built-up and glue laminated beams; plywood cladding and roof sheeting.*

Chapter 7

Timber Structures – Case Studies

7.1 Introduction

The case studies discussed in this chapter are constructed facilities in timber. The individual designs have also been checked using Eurocode EC5. The following major case studies are included together with their respective structural timber details:

a) Lofts
b) Roof Glued Beams
c) Steel Cover Plates Glued to Timber Facility
d) Nailed I-Beams with Diagonally Boarded Webs
e) Triangular Timber Truss with Collar Beams and Struts
f) Arches, Portals and Frames
g) Timber Grid Work
h) Timber Shell Roofs
i) Folded Plates
j) Vehicle Bridges in Timber
k) Footbridges

Wherever possible, a detailed explanation is given for each case study supported by relevant structural details. Although the individual designs are based on their respective codes, these have been checked using EC5.

7.2 Case studies

7.2.1 Lofts and attic roofs in timber (British design and practices)

7.2.1.1 Introduction

For the purposes of this section, an attic will be defined as a roof void used for living, bedroom, bathroom and kitchen, playroom or studio use, whilst a loft is a roof void intended for storage only.

One of the disadvantages of the normal fan trussed rafter roof is that many web members obstruct easy access within the roof void, a space traditionally used for storage of household items. The traditional purlin and common rafter roof described in Chapter 6 gives a relatively unobstructed loft space, although the collars may be well below a desirable head roof height.

The trussed rafter roof can provide either a loft or an attic. If the roof is designed with a loft space which the designer knows, or because of the shape of the internal members of the truss, can provide a room, then he/she should allow for domestic floor loading.

There are four main types of loft or attic roof:

1. Loft void – trussed rafters carrying roof loads with separate floor joists for storage load.
2. Loft void – trussed rafters carrying roof and floor loads.
3. Attic – trussed rafters carrying most of the roof load with separate joists for floor loads.
4. Attic – trussed rafters carrying floor and roof loads.

For special trusses designed to provide an unobstructed roof void for some parts of the roof, the truss loading is as for standard trusses. In between the trusses, independent joists are installed, taking support on the same wall plate as the trussed rafter and on some intermediate load-bearing wall.

The timber sections used in the trussed rafters will be standard and not the heavier sections required with the full attic, thus the cost of the truss is not greatly increased. Care must be taken to add necessary herringbone strutting, and the strutting indicated on the intermediate wall plate to brace the infilling floor joists.

7.2.1.2 Loft conversion and design (UK practice)

A typical loft conversion for which the structural design and detailing were carried out by SC Consulting at High Wycombe, Bucks, UK is given as a typical example. The drawings in Figs 7.1 to 7.11 summarise the design/detailing and construction of a loft.

7.2.2 Roof glued beams (Russian practice)

Glued beams are built up with boards laid flat or with boards and plywood (Fig. 7.12). Glued members, jointed together in the form of assemblies of boards, are also used for the upper chords of trusses and arches. The thickness of the boards should not be more than 50 mm, while the moisture content should not exceed 15%. With greater thicknesses and higher moisture content, the shrinkage deformation across the grain and the warping of the timber increases, resulting in fracture of the glue line.

Glued beams have a number of advantages over other forms of compound beam. Sections of greater height and more efficient shape (such as I-beams) can be made with a very high load-carrying capacity. Splicing is possible with no strength reduction in the section by means of

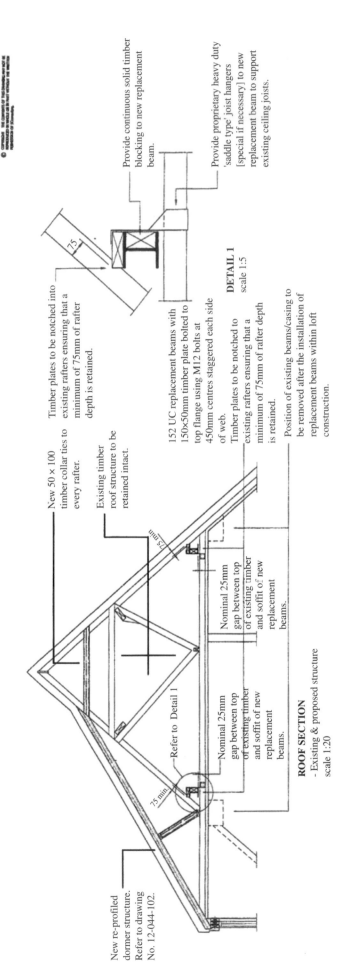

Provide continuous solid timber blocking to new replacement beam.

Provide proprietary heavy duty 'saddle type' joist hangers [special if necessary] to new replacement beam to support existing ceiling joists.

Timber plates to be notched into existing rafters ensuring that a minimum of 75mm of rafter depth is retained.

75

DETAIL 1
scale 1:5

New 50 × 100 timber collar ties to every rafter.

Existing timber roof structure to be retained intact.

152 UC replacement beams with 150×50mm timber plate bolted to top flange using M12 bolts at 450mm centres staggered each side of web.

Timber plates to be notched to existing rafters ensuring that a minimum of 75mm of rafter depth is retained.

Position of existing beams/casing to be removed after the installation of replacement beams within loft construction.

75 min.

Nominal 25mm gap between top of existing timber and soffit of new replacement beams.

Refer to Detail 1

Nominal 25mm gap between top of existing timber and soffit of new replacement beams.

75 min.

ROOF SECTION
- Existing & proposed structure
scale 1:20

New re-profiled dormer structure. Refer to drawing No. 12-044-102.

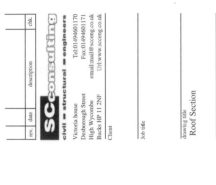

rev.	date	description	chk.

SCconsulting
civil ▪ structural ▪ engineers

Victoria house   Tel:01494601170
Desborough Street   Fax:01494601171
High Wycombe   email:mail@scceng.co.uk
Bucks HP 11 2NF   Url:www.scceng.co.uk
Client

Job title

drawing title
Roof Section

code   date   drawn by: checked by:
job code   August 2004

Figure 7.1 *Roof section (Courtesy SC Consulting, High Wycombe, UK).*

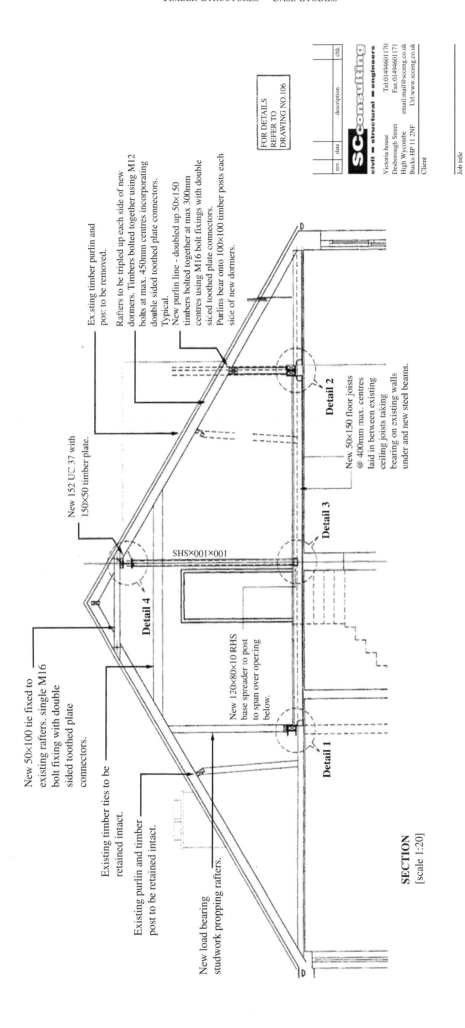

SC consulting
civil ■ structural ■ engineers

rev.	date	description	chk

Victoria house   Tel:01494601170
Desborough Street   Fax:01494601171
High Wycombe   email:mail@scceng.co.uk
Bucks HP 11 2NF   Url:www.scceng.co.uk
Client

Job title

drawing title
Structural Alterations
section through Roof

code: date: drawn by: checked by:
          Jan 2004   MRH

Existing timber purlin and
post to be removed.

Rafters to be tripled up each side of new
dormers. Timbers bolted together using M12
bolts at max. 450mm centres incorporating
double sided toothed plate connectors.
Typical.

New purlin line - doubled up 50×150
timbers bolted together at max 300mm
centres using M16 bolt fixings with double
sided toothed plate connectors.

Purlins bear onto 100×100 timber posts each
side of new dormers.

New 50×100 tie fixed to
existing rafters. single M16
bolt fixing with double
sided toothed plate
connectors.

Existing timber ties to be
retained intact.

Existing purlin and timber
post to be retained intact.

New load bearing
studwork propping rafters.

New 152 UC 37 with
150×50 timber plate.

New 50×150 floor joists
@ 400mm max. centres
laid in between existing
ceiling joists taking
bearing on existing walls
under and new steel beams.

100×100×SHS

New 120×80×10 RHS
base spreader to post
to span over opering
below.

Detail 1

Detail 2

Detail 3

Detail 4

SECTION
[scale 1:20]

Figure 7.2 *Structural alterations: section through roof (Courtesy SC Consulting, High Wycombe, UK).*

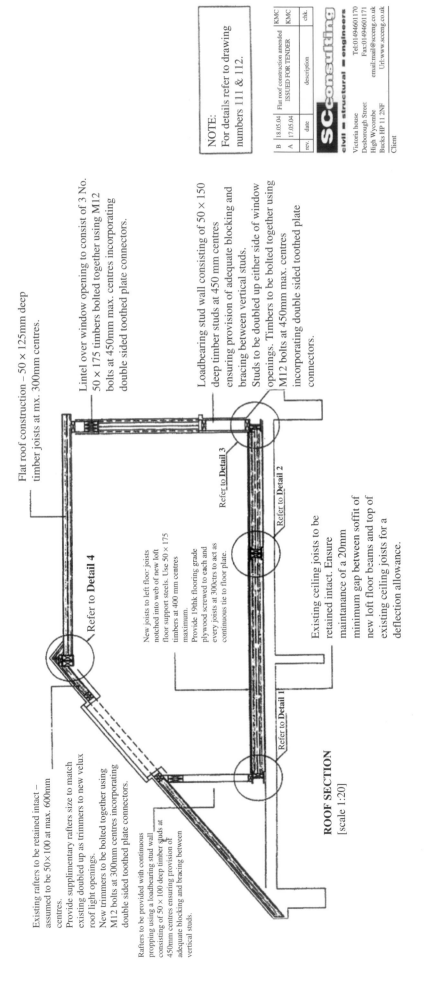

Flat roof construction – 50 × 125mm deep timber joists at mx. 300mm centres.

Lintel over window opening to consist of 3 No. 50 × 175 timbers bolted together using M12 bolts at 450mm max. centres incorporating double sided toothed plate connectors.

Loadbearing stud wall consisting of 50 × 150 deep timber studs at 450 mm centres ensuring provision of adequate blocking and bracing between vertical studs.
Studs to be doubled up either side of window openings. Timbers to be bolted together using M12 bolts at 450mm max. centres incorporating double sided toothed plate connectors.

Existing rafters to be retained intact – assumed to be 50×100 at max. 600mm centres.
Provide supplementary rafters size to match existing doubled up as trimmers to new velux roof light openings.
New trimmers to be bolted together using M12 bolts at 300mm centres incorporating double sided toothed plate connectors.

Rafters to be provided with continuous propping using a loadbearing stud wall consisting of 50 × 100 deep timber studs at 450mm centres ensuring provision of adequate blocking and bracing between vertical studs.

New joists to left floor joists notched into web of new loft floor support steels. Use 50 × 175 timbers at 400 mm centres maximum.
Provide 19thk flooring grade plywood screwed to each and every joists at 300ctrs to act as continuous tie to floor plate.

Existing ceiling joists to be retained intact. Ensure maintanance of a 20mm minimum gap between soffit of new loft floor beams and top of existing ceiling joists for a deflection allowance.

Refer to **Detail 4**

Refer to **Detail 3**

Refer to **Detail 2**

Refer to **Detail 1**

ROOF SECTION
[scale 1:20]

NOTE:
For details refer to drawing numbers 111 & 112.

B	18.05.04	Flat roof construction amended	KMC
A	17.05.04	ISSUED FOR TENDER	KMC
rev.	date	description	chk.

SCconsulting
civil ■ structural ■ engineers

Victoria house   Tel:01494601170
Desborough Street   Fax:01494601171
High Wycombe   email:mail@sceng.co.uk
Bucks HP 11 2NF   Url:www.sceng.co.uk
Client

Job title

drawing title
Loft Alterations
Typical Section

date   drawn by: checked by:
May 2004   MRH

Figure 7.3 *Loft alterations: typical section (Courtesy SC Consulting, High Wycombe, UK).*

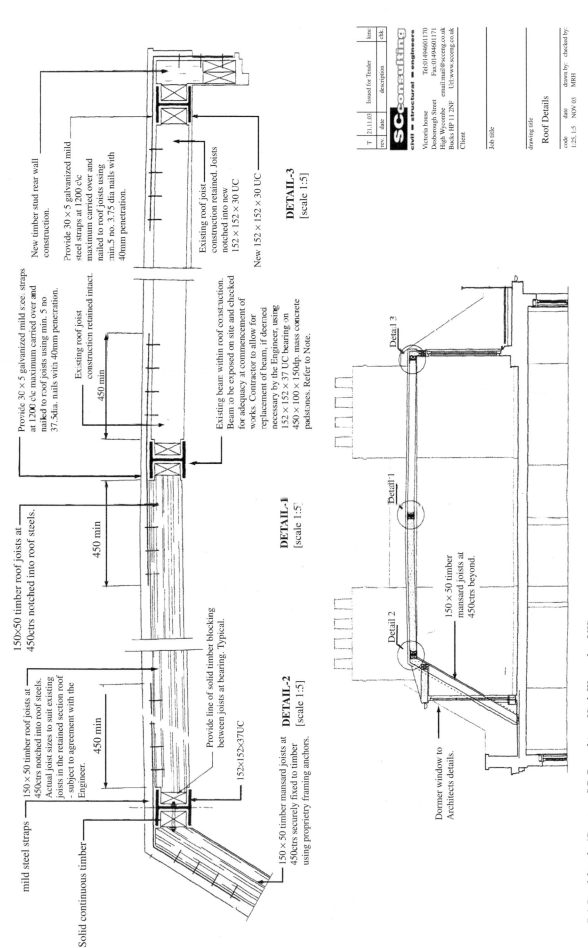

Figure 7.4 *Roof details (Courtesy SC Consulting, High Wycombe, UK).*

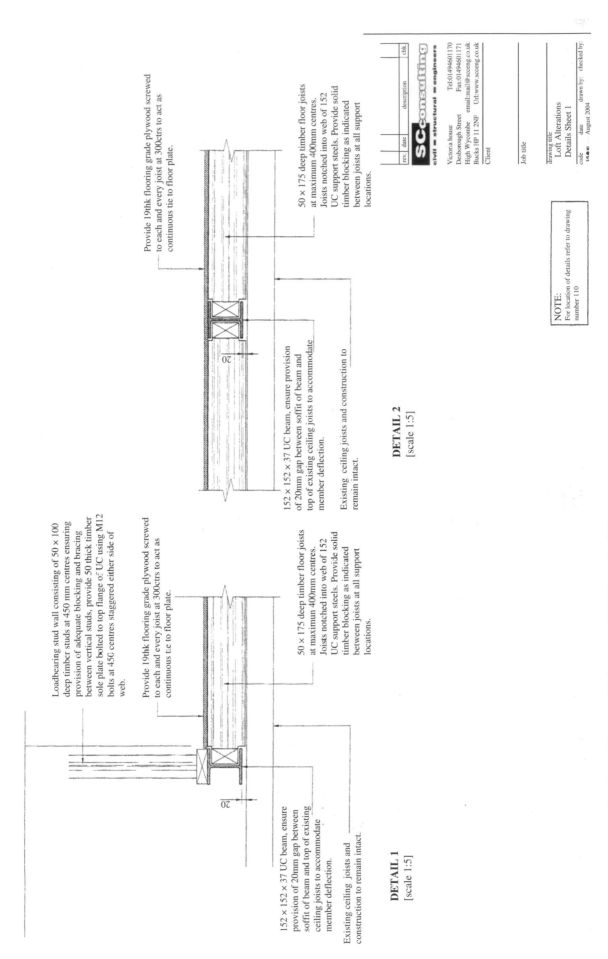

Figure 7.5 *Loft alterations details – sheet 1 (Courtesy SC Consulting, High Wycombe, UK).*

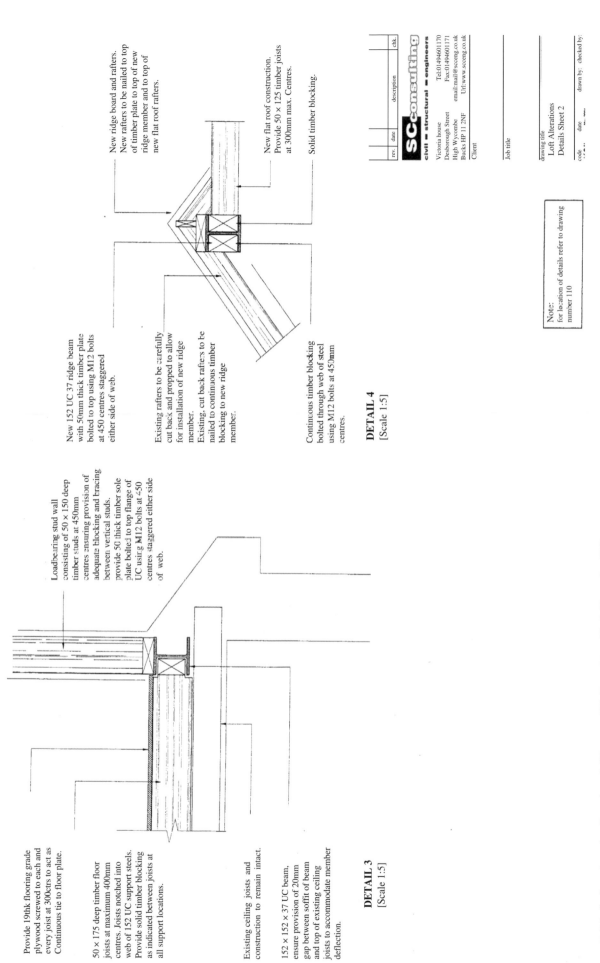

New ridge board and rafters.
New rafters to be nailed to top of timber plate to top of new ridge member and to top of new flat roof rafters.

New flat roof construction.
Provide 50 × 125 timber joists at 300mm max. Centres.

Solid timber blocking.

New 152 UC 37 ridge beam with 50mm thick timber plate bolted to top using M12 bolts at 450 centres staggered either side of web.

Existing rafters to be carefully cut back and propped to allow for installation of new ridge member.
Existing, cut back rafters to be nailed to continuous timber blocking to new ridge member.

Continuous timber blocking bolted through web of steel using M12 bolts at 450mm centres.

DETAIL 4
[Scale 1:5]

Provide 19thk flooring grade plywood screwed to each and every joist at 300ctrs to act as Continuous tie to floor plate.

50 × 175 deep timber floor joists at maximum 400mm centres. Joists notched into web of 152 UC support steels. Provide solid timber blocking as indicated between joists at all support locations.

Existing ceiling joists and construction to remain intact.

152 × 152 × 37 UC beam, ensure provision of 20mm gap between soffit of beam and top of existing ceiling joists to accommodate member deflection.

Loadbearing stud wall consisting of 50 × 150 deep timber studs at 450mm centres ensuring provision of adequate blocking and bracing between vertical studs. provide 50 thick timber sole plate bolted to top flange of UC using M12 bolts at 450 centres staggered either side of web.

DETAIL 3
[Scale 1:5]

SC consulting
civil ■ structural ■ engineers
Victoria house   Tel:01494601170
Desborough Street   Fax:01494601171
High Wycombe   email:mail@sceng.co.uk
Bucks HP 11 2NF   Url:www.sceng.co.uk
Client

Job title

drawing title
Loft Alterations
Details Sheet 2

Note:
for location of details refer to drawing number 110

Figure 7.6 *Loft alterations details – sheet 2 (Courtesy SC Consulting High Wycombe, UK).*

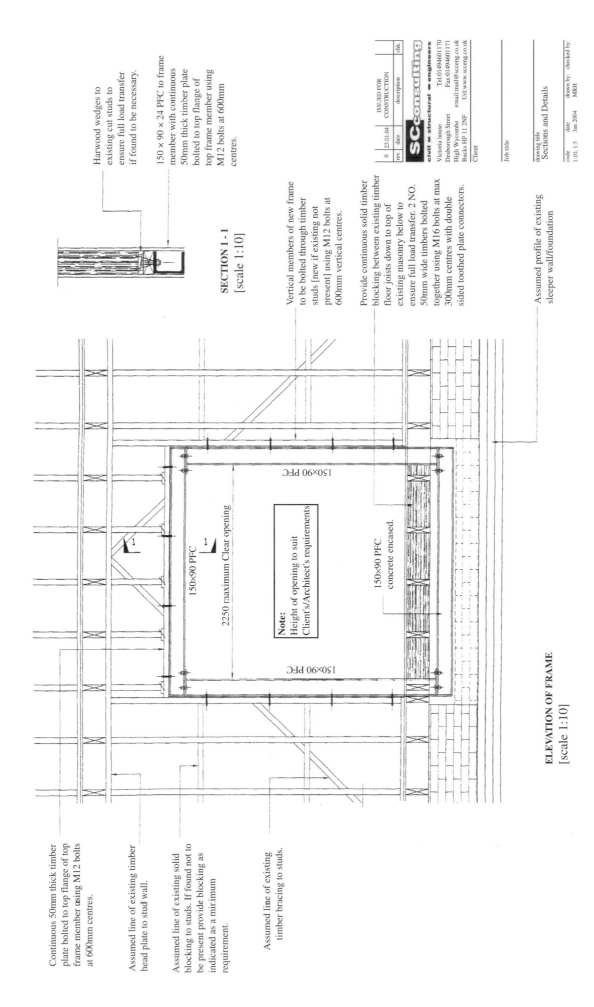

Figure 7.7 *Sections and details (Courtesy SC Consulting, High Wycombe, UK).*

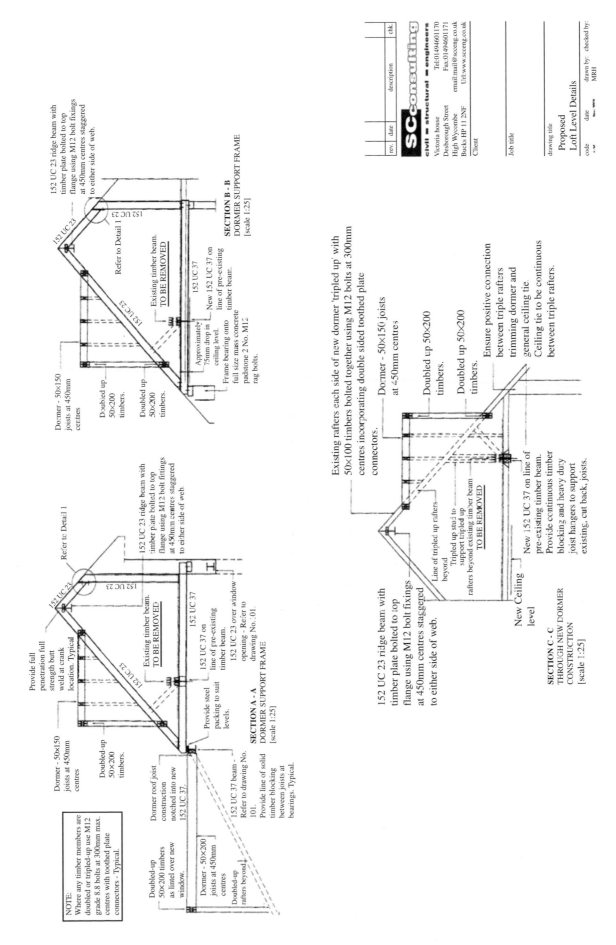

Figure 7.8 *Proposed loft level details (Courtesy SC Consulting, High Wycombe, UK).*

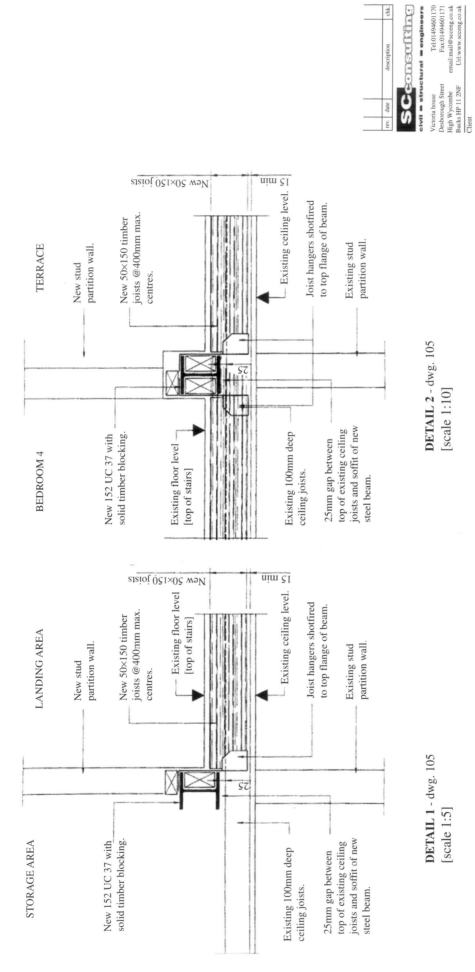

Figure.7.9 *Structural alterations details – sheet 1 (Courtesy SC Consulting, High Wycombe, UK).*

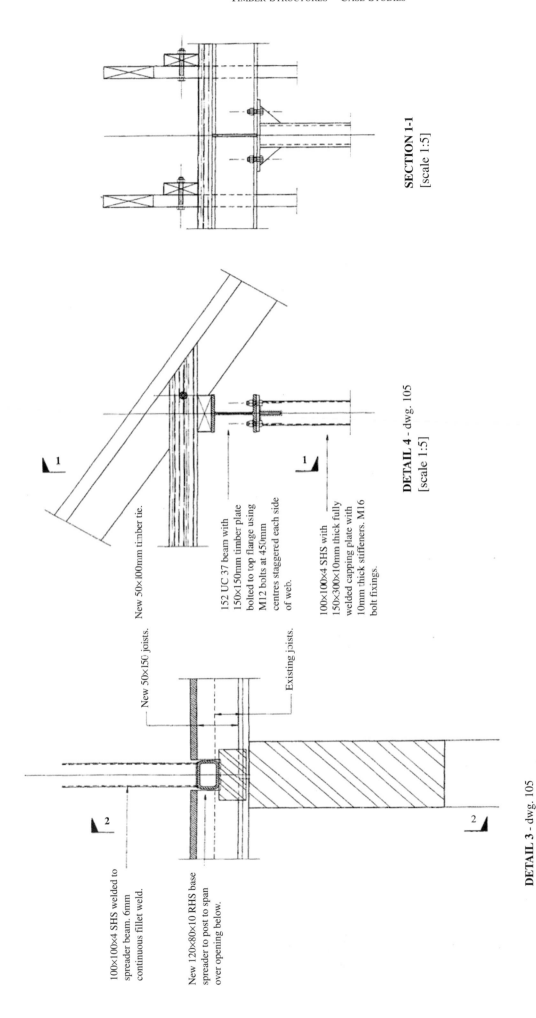

SECTION 1-1
[scale 1:5]

DETAIL 4 - dwg. 105
[scale 1:5]

New 50×100mm timber tie.

152 UC 37 beam with 150×150mm timber plate bolted to top flange using M12 bolts at 450mm centres staggered each side of web.

100×100×4 SHS with 150×300×10mm thick fully welded capping plate with 10mm thick stiffeners. M16 bolt fixings.

New 50×150 joists.

Existing joists.

DETAIL 3 - dwg. 105
[scale 1:5]

100×100×4 SHS welded to spreader beam. 6mm continuous fillet weld.

New 120×80×10 RHS base spreader to post to span over opening below.

Figure 7.10 *Structural alterations details – sheet 2 (Courtesy SC Consulting, High Wycombe, UK).*

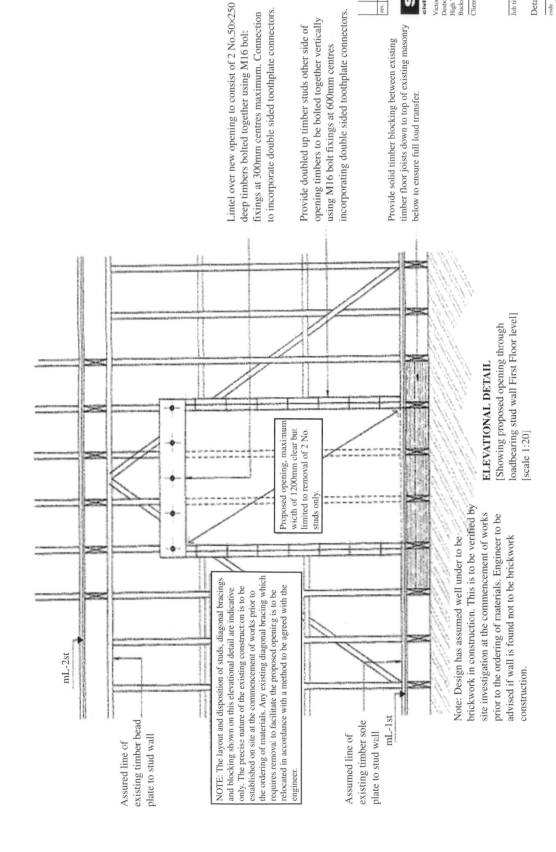

Lintel over new opening to consist of 2 No.50×250 deep timbers bolted together using M16 bolt fixings at 300mm centres maximum. Connection to incorporate double sided toothplate connectors.

Provide doubled up timber studs other side of opening timbers to be bolted together vertically using M16 bolt fixings at 600mm centres incorporating double sided toothplate connectors.

Provide solid timber blocking between existing timber floor joists down to top of existing masonry below to ensure full load transfer.

mL-2st

Assured line of existing timber head plate to stud wall

NOTE: The layout and disposition of studs, diagonal bracings and blocking shown on this elevational detail are indicative only. The precise nature of the existing construction is to be established on site at the commencement of works prior to the ordering of materials. Any existing diagonal bracing which requires removal to facilitate the proposed opening is to be relocated in accordance with a method to be agreed with the engineer.

Proposed opening, maximum width of 1200mm clear but limited to removal of 2 No. studs only.

Assured line of existing timber sole plate to stud wall

mL-1st

Note: Design has assumed well under to be brickwork in construction. This is to be verified by site investigation at the commencement of works prior to the ordering of materials. Engineer to be advised if wall is found not to be brickwork construction.

ELEVATIONAL DETAIL
[Showing proposed opening through loadbearing stud wall First Floor level]
[scale 1:20]

SC consulting
civil ■ structural ■ engineers

Victoria house   Tel:01494601170
Desborough Street   Fax:01494601171
High Wycombe   email:mail@sceng.co.uk
Bucks HP 11 2NF   Url:www.sceng.co.uk

Client

rev. date   description   chk.

Job title

Details

code   date   drawn by: checked by:
                            MRH

Figure 7.11 *Elevation details (Courtesy SC Consulting, High Wycombe, UK).*

finger joints (the best method) or bevelled scarfing, and members of any required size can be made irrespective of the lengths of the available sawn stock.

The glue line is sufficiently rigid under shear and consequently the member acts under load in the same manner as that of a solid section. Boards of varying quality can be efficiently used according to the height of the section: the better grades in the more highly-stressed upper and lower chords and the inferior grades in the lower-stressed middle section. In this way, both small-dimensioned and low-grade timber can be utilised in these beams.

When deciding whether to adopt glued construction, it must be borne in mind that the strength of a glued member depends entirely on the quality rules of the technique of gluing and careful adherence to the processes of production.

Small-scale preparation of glued structures by hand is inadmissible. It is only possible under factory conditions, in specially heated workshops and using the necessary equipment: drying kilns for seasoning the timber, machines for sheeting and trimming and processing the splices to increase the length of the material, glue-mixers and rollers for spreading the glue, presses and so on. The work must be carried out by specially-trained personnel. All this calls for extra expenditure, but experience has shown that if the conditions required are not complied with, the desired quality of the glued members cannot be guaranteed and consequently their preparation cannot proceed.

I-beams with a web of boards on edge with spans of 3–7 m are mainly used for inter-storey and attic floors and are jointed with water-resistant phenol formaldehyde (p.f.) glues. In order to utilize short lengths of sawn timber, the splices in the double web can be arranged in any position; in the upper chord, in the end thirds of the length, as long as they do not coincide with the web joints. It is desirable to leave the lower chord without splices.

Beams glued together with boards laid on the flat are used as floor girders under heavy loading for spans of 4–8 m and as load carrying structures for roofs with spans of 6–12 m. In order to provide the required stiffness of the beam out of its own plane (sideways), the thickness of the web and the relationship between the height of the beam and its width must be limited. The splicing of all of the boards should be carried out as a rule with finger joints (in the workpiece).

The relative arrangement of the rings in the boards and the beam should be "in agreement" with the code in order to avoid the development of rupturing stresses across the grain in the glue lines when the timbers continue to shrink in service.

Beams with plywood webs are used in roofs with spans of up to 15 m. They consist of boarded chords and a plywood web, using building plywood with a thickness of not less than 10 mm. Water-resistant plywoods made with p.f. resin adhesives are used as a rule for the web. The use of ordinary plywood for glued structures is not permitted.

To increase the stiffness of the thin plywood web and prevent it from buckling, strengthening ribs are fitted

at intervals of about one-eighth of the span, and at the application points of concentrated loads. In the support panel, where the web is most highly stressed, it is usual to fit a bracing strut. The chord of the beam consists of two vertical layers of boards on each side of the web. The first layer, glued directly to the plywood, is composed of two narrow boards with a gap between them, the second layer is composed of one wide board glued to the narrow ones.

7.2.3 Glued steel cover plates

A glued steel cover plate is a steel plate or sleeve with a hole for a bolt usually arranged in the middle of the plate, glued to the wooden or plywood member with all-purpose water-resistant glue. Some examples of glue steel connectors are shown in Fig. 7.13.

Boring the holes in the wooden members and gluing the plates must be done in jigs; this ensures the interchange ability of the separate members and makes it easy to assemble and take them apart. Various types of glued-steel connectors are available in the market:

1. tension splice between boards with glue-steel plates;
2. joint of boarded truss with glue-steel plates;
3. connecting glued plywood tubes with glue-steel sleeve;
4. joint in plywood members working under bending stresses.

7.2.4 Nailed I-beams with diagonally boarded webs

Nailed beams with diagonally crossed webs (Fig. 7.14) are used for roofs under a sloping roof-covering for spans of not more than 12 m. During the Second World War, beams of this type with squared timber chords were used successfully for bridges with spans of up to 40 m. These beams are structures of the site-built type. Long and reliable service can only be guaranteed when the lower chords under tension are made with high-quality (first grade) timber, and when the wood is given preservative treatment, since the multi-layer construction of the beams is extremely vulnerable to decay. For this reason, the roof-covering should always be separated from the upper chord to allow ventilation. These beams can be mainly recommended for the roofs of buildings of an auxiliary or temporary nature. The beams are made with parallel chords (chiefly in bridge building) and also with a double or single pitch.

In order to provide the requisite strength, the full height of the beams with parallel chords (for double-pitched beams, the height at one-quarter of the span and for single-pitched, the height in the middle of the span) must not be less than one-ninth of the span.

For beams with sloping upper chords, the height at the supports is taken as not less than 0.4 of the height in the middle of the span, in order to prevent the development of excessive shearing forces and, as a consequence, the

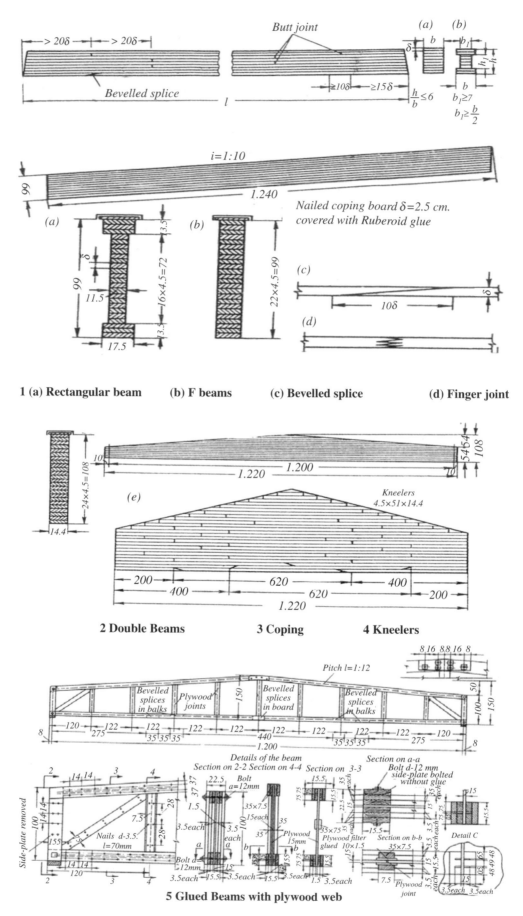

Figure 7.12 *Glued beams for roof (based on Russian practice).*

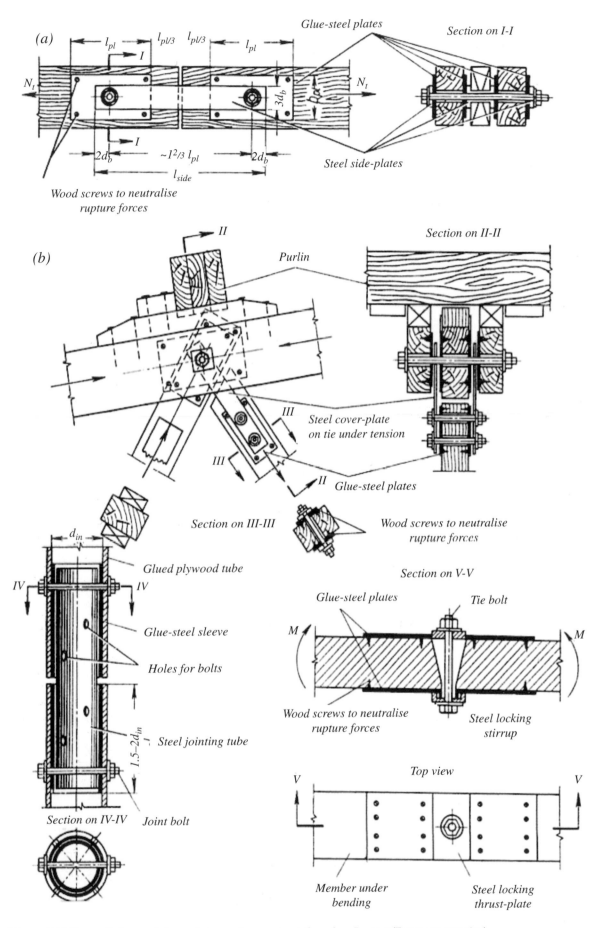

Figure 7.13 *Types of glue-steel (over plates and connectors) (based on Russian/European practice).*

difficulty of finding room for a large number of nails in the support panels. The beams consist of boards or squared timber chords, diagonally-crossed boarded web and strengthening rib. The chords and strengthening are nailed to the web.

In beams with boarded chords, the upper and lower chords consist of boards 4–6 cm thick. The diagonal web is made up of two layers of boards 1.9–3 cm thick, inclined at an angle of 45–30° to the chords. The total thickness of the web should, as a rule, be equal to or nearly equal to that of one board of the chords. The strengthening ribs are spaced 1/8–1/10 of the span apart; usually, they are arranged to coincide with the purlins of the roof covering or other concentrated loads in order to distribute them as uniformly as possible between the members of the beam.

The chords are fixed to the web with the calculated number of nails. The length of the nails should be not less than $4d_{nail}$, after deducting the three gaps of 2 mm each and the length of the nail point $1.15d_{nail}$. The strengthening ribs are fastened with the same nails as the chords. In the support ribs, the spacing of the nails is the same as that in the adjoining panels of the chord in order to provide a sound fixing for the ends of the boards forming the webs. In the intermediate ribs, the nails are arranged structurally, usually in two rows.

Both layers of the board in the web are fastened together with shorter and thinner nails, and the ends clenched over. The nails are spaced so that the free length of the boards does not exceed 30 times their thickness in order to prevent bulging.

Beams with chords of squared timbers (Fig. 7.14) are used for heavy loads, mainly in bridges. In these beams, each layer of the web is nailed separately to its semichord (the balk) with nails driven from the face of the web. The beams are thus prepared in the form of two half-beams, which are then put together and fastened with bolts and by nailing the two layers of the web together. In these beams, the thickness of the chords is not limited by the length of the nails, so that very stout sections of squared timber can be used. As a rule, the length of the chord nails is taken as three times the thickness of the boards of the web. The splices in these beams are also arranged differently, the side-plates being fitted above and below the chord timbers, and the tension splices fixed to the chord with blind dowels.

7.2.5 Triangular timber truss with collar beam and struts (based on Russian practice)

Analytical and design techniques are clearly established for trusses and collar beams. When designing trusses having an upper chord of plate-dowelled balks, it must be borne in mind that owing to the existence of a large stress in the lower chord, the point of application of this stress on the support joint is of considerable importance, as it influences the number of dowels in the joint lines of the compound members. The calculations

of the spindle in the centre joint of the lower chord must be carried out according to the resultant force in the struts.

The slenderness ratio of the plate must not be more than 200, while the effective length for this is taken to be the distance between the joint hinge and the first row of nails. The nails for fastening the plates to the struts (Fig. 7.15) are arranged opposite one another and are computed according to the usual rules. Owing to the likelihood of shrinkage cracks in the middle of the face sides of the boards and balks, the number of rows of nails should, as a rule, be even.

In some cases, owing to the narrowness of the lattice members, the nails cannot be driven opposite one another, and in order to accommodate them, the plates have to be made longer. A reduction in the length of the plate is obtained by using steel stub dowels or coach screws, screwed into a pre-drilled hole, for which the spacing standards are less than those for nails.

7.2.6 Arches, portals and frames (based on Russian practice)

7.2.6.1 Introduction

Arches can be three hinged types or two hinged types (Fig. 7.16), rigidly supported at bases and are designed with various boundary conditions. They can take the following shapes: circular, parabolic elliptical, gothic, Tudor, reverse curve and A-frame. Either graphical or mathematical procedures can be used in the design process. Similarly, portals and frames can be designed under various loading conditions and are shown in Figs 7.20 to 7.28. In Figs 7.16 and 7.17, the design of a tied arch, using a steel rod as a tie has been recommended by the Russian and European codified method. The design data provide flexibility for an arch width $b < 14$ and > 14 for varying height 'h'. The complete steel rod details can also be seen in Fig. 7.16.

Timber arches have one, two and three hinges. They can be rigid throughout with rigid support. They can be circular, elliptical, parabolic, conidial, Tudor, gothic, reverse curved and A-frame. A curved tapered member such as a tudor arch can be made up of two parts: a curved beam of constant section and a haunch that is attached mechanically. The three-hinged arch is probably the most commonly used glulam member that is both tapered and curved. The difficulty of making fixed end connections makes the three-hinged arch a much better choice. Figures 7.18 and 7.19 show several shapes that may be used as timber portals. Generally all loading can initially be treated as medium-term loading.

7.2.6.2 Design data (based on British practice)

Centres of frames	3.0 m to 4.5 m
Dead load (based on slope areas)	
Timber boarding	0.45 kN/m²

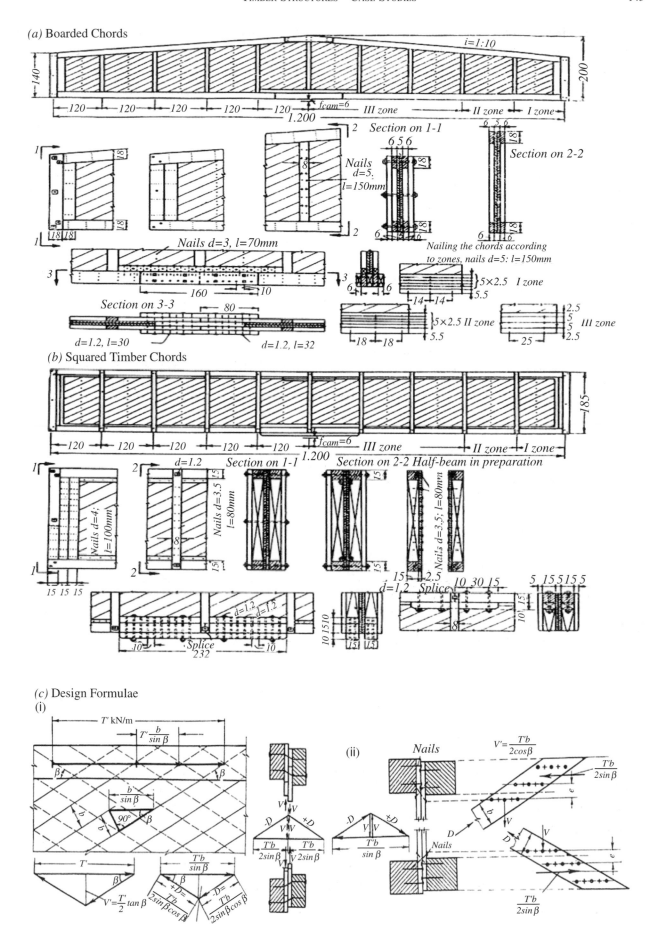

Figure 7.14 *Nailed I-beam with diagonal crossed web (based on Russian/European design practice).*

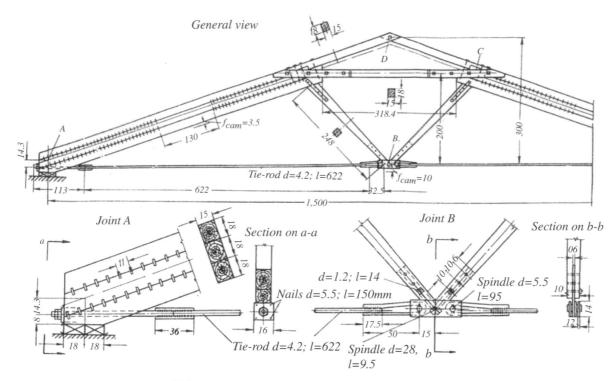

Triangular truss of 15m span with collar beam and struts

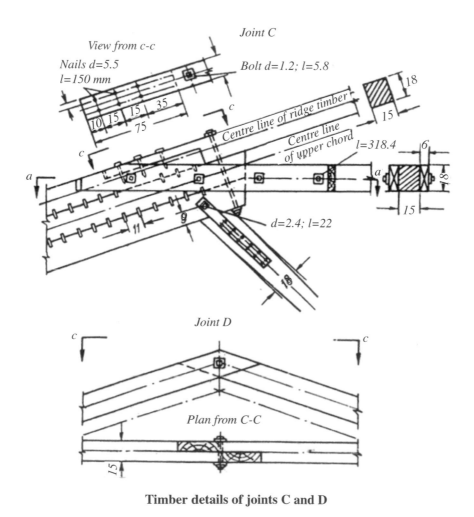

Timber details of joints C and D

Figure 7.15 *Triangular truss with collar beam and struts (based on Russian/European practice).*

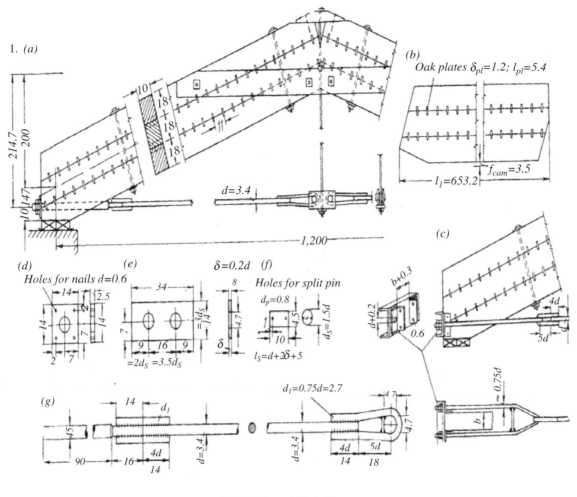

1. Stuctural Details

Coefficients k'_W for varying ratios b_1/b

b_1/b	1/2	1/3	1/4
k'_W	0.90	0.80	0.75

Least values of the ratio $\frac{l}{h}$ recommended for glued I-section beams

b_1/b	1/2	1/3	1/4
l/h	12	15	18

Figure 7.16 *Three-hinged arch (based on Russian/European codified methods).*

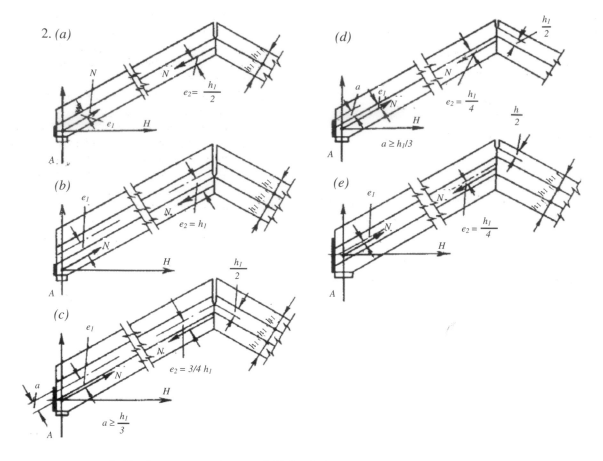

2. Design parameters and Calculations

Table 7.1 Coefficients k_W for rectangular glued beams of varying height h

Width of beam b in cm	Coefficient k_W for heights or beam h, cm					
	14–50	60	70	80	90	100 and over
$b < 14$	1.00	0.95	0.90	0.85	0.80	0.75
$b \geq 14$	1.15	1.05	0.95	0.90	0.85	0.80

The check on the section of the beams forming the upper chord of the arches is carried out as for a compressed-flexed strut:

$$\sigma = \frac{N}{F_{net}} + \frac{MR_c}{k_W W_{net} \xi R_{bend}} \leq R_c = 1 \tag{7.1}$$

When checking the glue line in laminated beams for shear according to the formula $\tau = \frac{QS}{Jb}$ the design resistance to shear is multiplied by the coefficient 0.5 to allow for possible faults in the spreading of the glue.

Figure 7.17 *Three hinged arch (based on Russian/European codified methods).*

Insulation, felt, etc.	0.5 kN/m²
Slates	0.25 kN/m²
Imposed loads (based on plan area)	0.75 kN/m²
Portal spans	12–15 m (laminated) 12–24 m (solid timber)
Eaves height	2.5 m to 7.5 m
Angle on top of eaves	>21.8–38.66°
Basic wind pressure (if not specified)	0.479 kN/m²

Materials and material properties

Laminated timber

European Redwood Grade SS. Strength Class C24
BS 5628 Part 2
$E = E_{mean} \times K_{20}$; lamination thickness, $t < 20$ mm
$\sigma_{c,ll} = \sigma_{c,g,ll} \times K_3$ K_3 = Load duration factor = 1.25

For curved glued laminated

$$\min \ \frac{r}{t} = \frac{E_{mean}}{70} ; \tag{7.2}$$

$$r_{mean} = r + 0.5h$$

Interaction Equation
Applied radial stress

$$\sigma_r = \frac{3M}{2bhr_{mean}} = \frac{\sigma_{m.a.II}}{\sigma_{m,adm,II}\left(1 - \frac{1.5\sigma_{c,a,II}}{\sigma_E} \times K_{12}\right)}$$

$$+ \frac{\sigma_{c.c.II}}{\sigma_{c,adm,II}} = \leq 1 \qquad (7.3)$$

where σ_e is the Euler critical stress $= \dfrac{\pi^2 E}{(L_e/i)^2}$

$$= \frac{\pi^2 E}{\lambda^2} \qquad (7.4)$$

The negative bending moment at the knee tends to decrease the radius of curvature and the radial stress will be in compression perpendicular to the grain, therefore:

$$\sigma_r \leq 1.33 \times \sigma_{c.g.\perp} \times K_3 \times K_{18} \qquad (7.5)$$

Important note: If Finnish birch faced plywood is used, flange section sizes must be increased from 50 × 125 nom. to 50 × 150 nom.

Nails to be sherardised or galvanised round wire.

7.2.6.3 Design data (Based on American practice)

In American practice, these design steps are followed:

1. Determine the applied loads. This will include snow and wind loads, as well as roof dead loads. If these loads are brought to the arch member by roof decking, they will be uniformly distributed loads; if roof purlins are used, the loads will reach the arch member as concentrated loads. The weight of the arch member itself must be estimated (and later confirmed or revised).

2. Sketch to scale the estimated outline of the arch members, so that the location of the member centre-lines can be estimated. At this step, be certain that the limits on allowable curvature are not exceeded.

3. For each combination of loads to be considered, compute the arch reactions. These will include both horizontal and vertical reactions at each end, and also the force transferred from one segment of the arch to the other at the central "hinge." (The central connection may be merely a connection that is incapable of transferring bending moment from one segment to another.)

4. Determine the area required at the base to resist shear. Setting the allowable longitudinal shear stress equal to the computed unit stress gives

$$\text{Req } A = bd = 1.5V/F \qquad (7.6)$$

5. For the first trial, use a crown depth dc = 1.5b and tangent-point depth dt = 1.5d(base).

Members of three-hinged arches are subject to both compression and bending moments. Therefore, an interaction equation must be used in design. For designing three-hinged arches, the American Institute of Timber Construction (AITC) recommends using:

$$f_c / F_c' + f_{bx} / F_{bx}' \leq 1.0 \qquad (7.7)$$

AITC also recommends that the volume factor, C_v, for the arch segments be solved by assuming that the ratio of member effective length to member depth equals 21. Carrying this assumption into the volume-factor equation presented earlier results in

$$C_v = K_L (5.125/b)^{1/x} (12/d)^{2/x} \qquad (7.8)$$

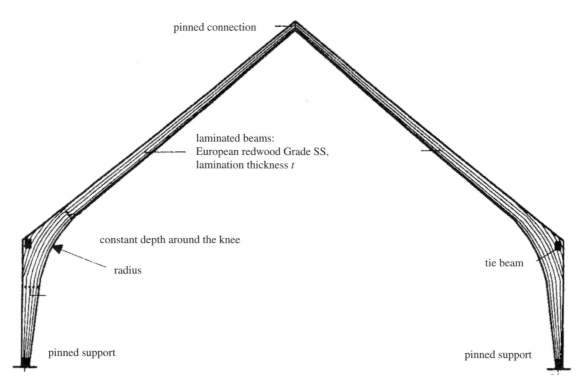

pinned connection

laminated beams:
European redwood Grade SS,
lamination thickness t

constant depth around the knee

radius

tie beam

pinned support

pinned support

Figure 7.18 *Arches/frames laminated portal.*

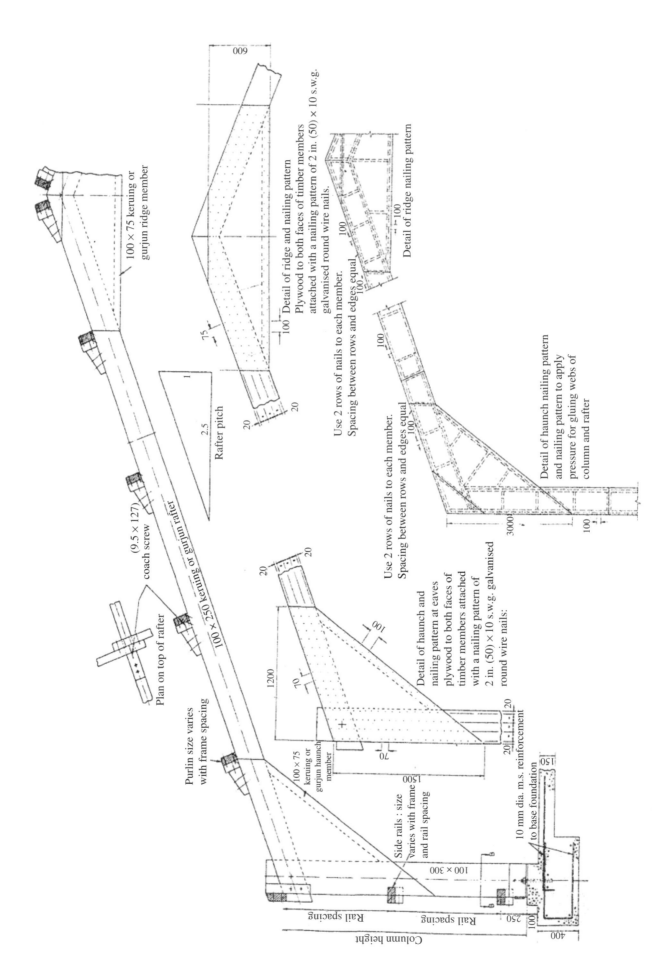

Figure 7.19 *Solid and box type portals (British practice) (adopted with permission from Mettem, C.J. Structural Timber Design and Technology, Longman, 1986).*

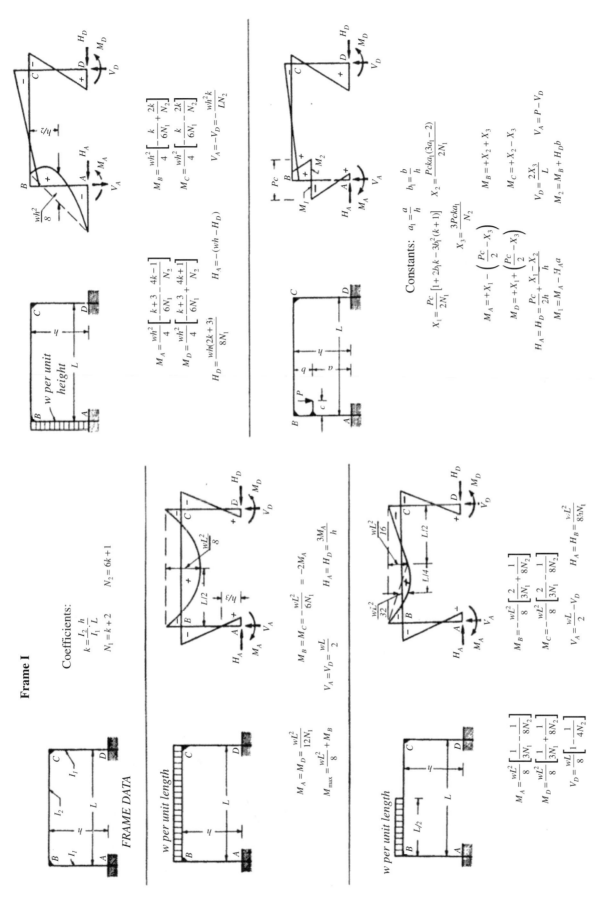

Figure 7.20 *Rectangular timber portals: formulae (uniformly distributed loads frame 1 and a concentrated bracket loads).*

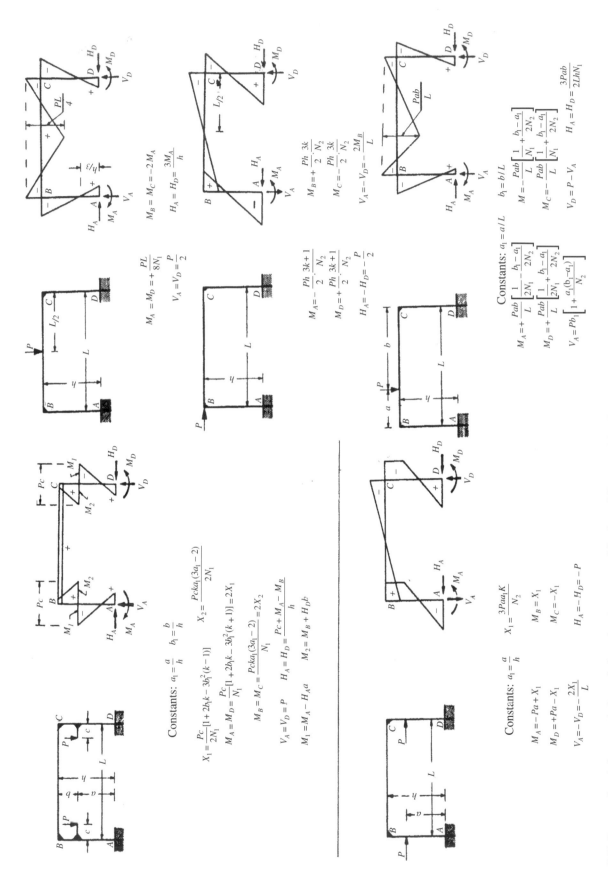

Figure 7.21 *Rectangular timber portals frame 1: formulae (concentrated loads at different positions).*

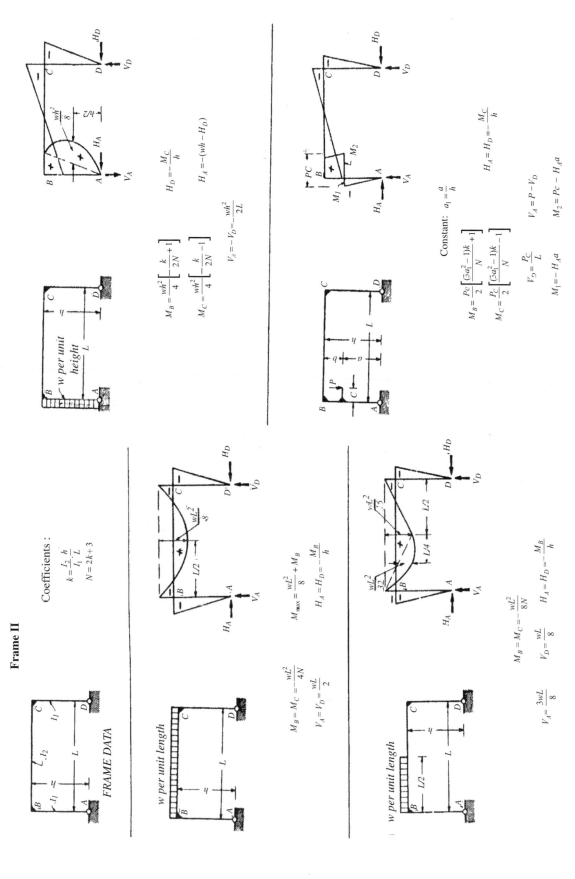

Figure 7.22 *Rectangular timber portals frame II: formulae (uniformly distributed loads at different positions).*

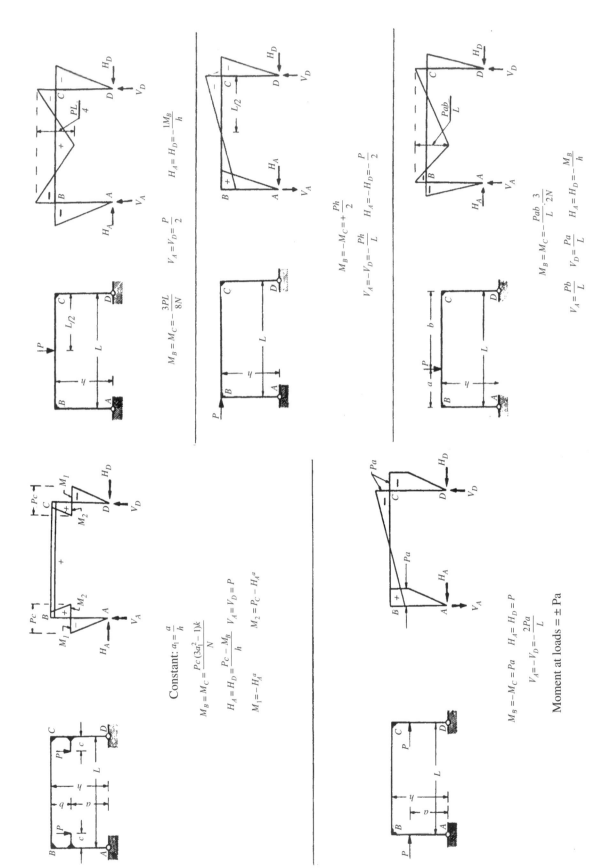

$$M_B = M_C = -\frac{3PL}{8N} \qquad V_A = V_D = \frac{P}{2} \qquad H_A = H_D = -\frac{1M_B}{h}$$

$$M_B = -M_C = +\frac{Ph}{2} \qquad H_A = -H_D = -\frac{P}{2}$$

$$V_A = -V_D = -\frac{Ph}{L}$$

$$M_B = M_C = -\frac{Pab}{L} \cdot \frac{3}{2N} \qquad H_A = H_D = -\frac{M_B}{h}$$

$$V_A = \frac{Pb}{L} \qquad V_D = \frac{Pa}{L}$$

Constant: $a_1 = \frac{a}{h}$

$$M_B = M_C = \frac{Pc(3a_1^2 - 1)k}{N}$$

$$H_A = H_D = \frac{Pc - M_B}{h} \qquad V_A = V_D = P$$

$$M_1 = -H_A^a \qquad M_2 = P_C - H_A^a$$

$$M_B = -M_C = Pa \qquad H_A = H_D = P$$

$$V_A = -V_D = -\frac{2Pa}{L}$$

Moment at loads = ±Pa

Figure 7.23 *Rectangular timber portals frame II: formulae (concentrated loads at different positions).*

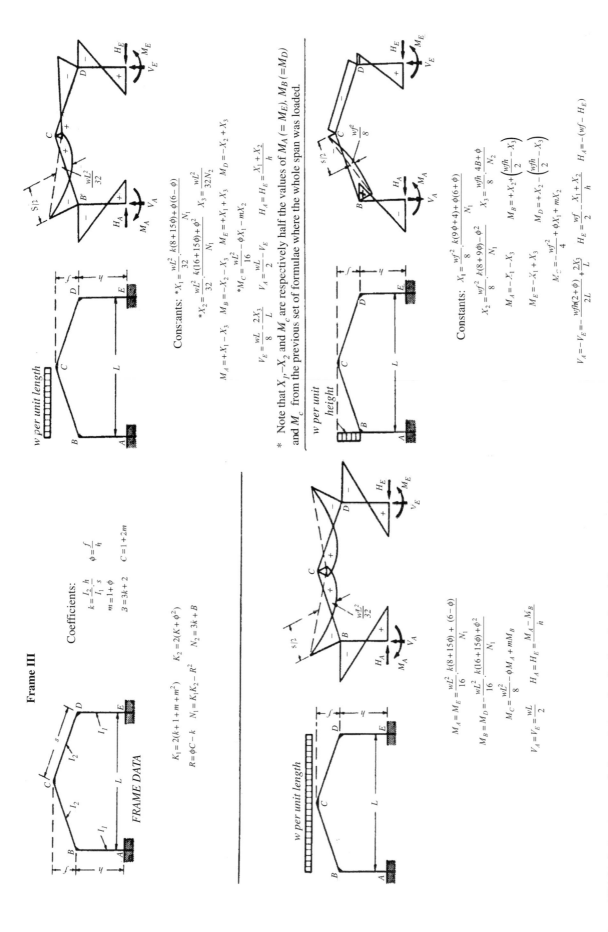

Figure 7.24 *Pitched timber portals frame I: formulae (uniform loads at different positions).*

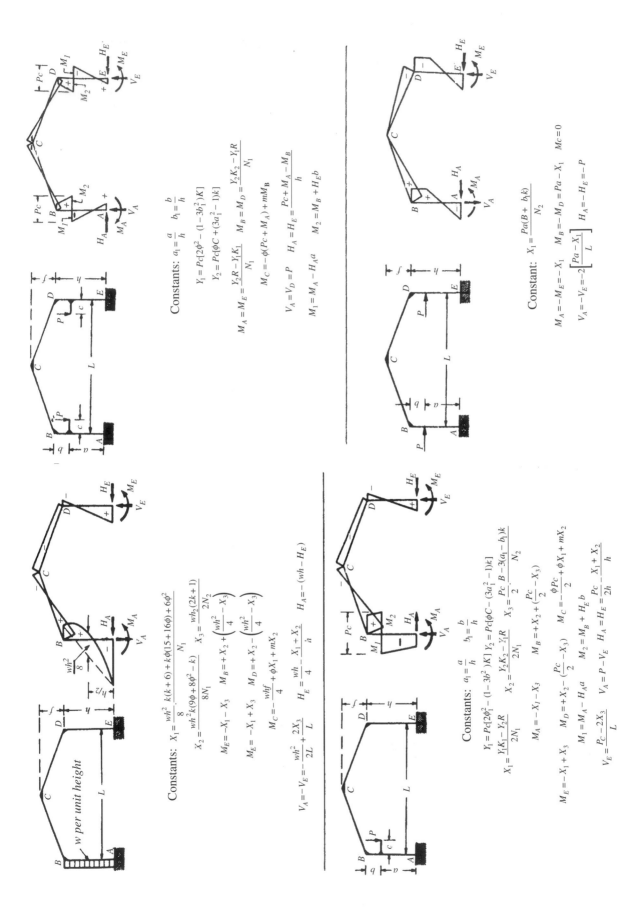

Figure 7.25 *Pitched timber portals frame II: formulae (uniform loads and concentrated loads at different positions).*

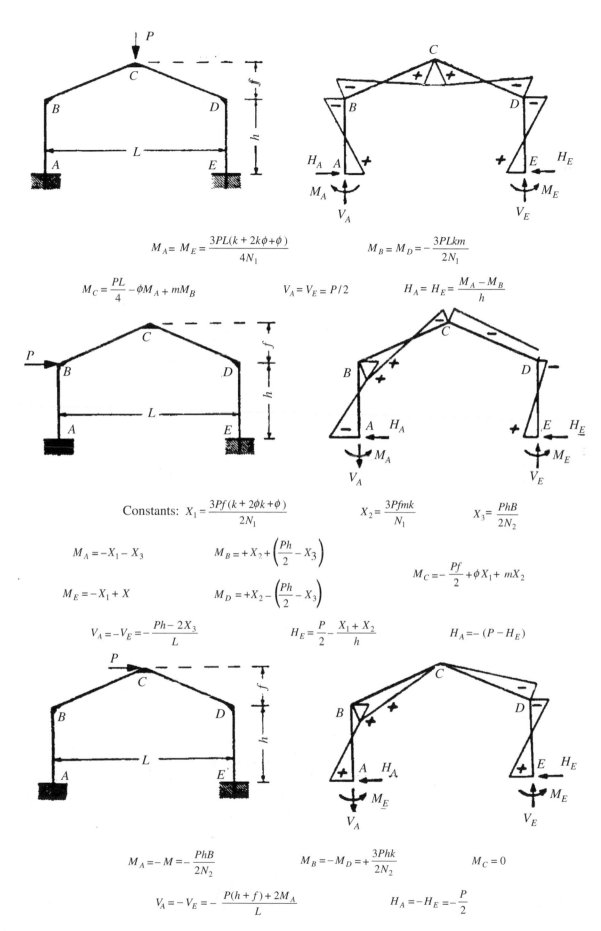

$$M_A = M_E = \frac{3PL(k + 2k\phi + \phi)}{4N_1} \qquad M_B = M_D = -\frac{3PLkm}{2N_1}$$

$$M_C = \frac{PL}{4} - \phi M_A + m M_B \qquad V_A = V_E = P/2 \qquad H_A = H_E = \frac{M_A - M_B}{h}$$

Constants: $X_1 = \dfrac{3Pf(k + 2\phi k + \phi)}{2N_1} \qquad X_2 = \dfrac{3Pfmk}{N_1} \qquad X_3 = \dfrac{PhB}{2N_2}$

$$M_A = -X_1 - X_3 \qquad M_B = +X_2 + \left(\frac{Ph}{2} - X_3\right)$$

$$M_E = -X_1 + X \qquad M_D = +X_2 - \left(\frac{Ph}{2} - X_3\right)$$

$$M_C = -\frac{Pf}{2} + \phi X_1 + m X_2$$

$$V_A = -V_E = -\frac{Ph - 2X_3}{L} \qquad H_E = \frac{P}{2} - \frac{X_1 + X_2}{h} \qquad H_A = -(P - H_E)$$

$$M_A = -M = -\frac{PhB}{2N_2} \qquad M_B = -M_D = +\frac{3Phk}{2N_2} \qquad M_C = 0$$

$$V_A = -V_E = -\frac{P(h + f) + 2M_A}{L} \qquad H_A = -H_E = -\frac{P}{2}$$

Figure 7.26 *Pitched timber portals frame III: formulae (concentrated loads).*

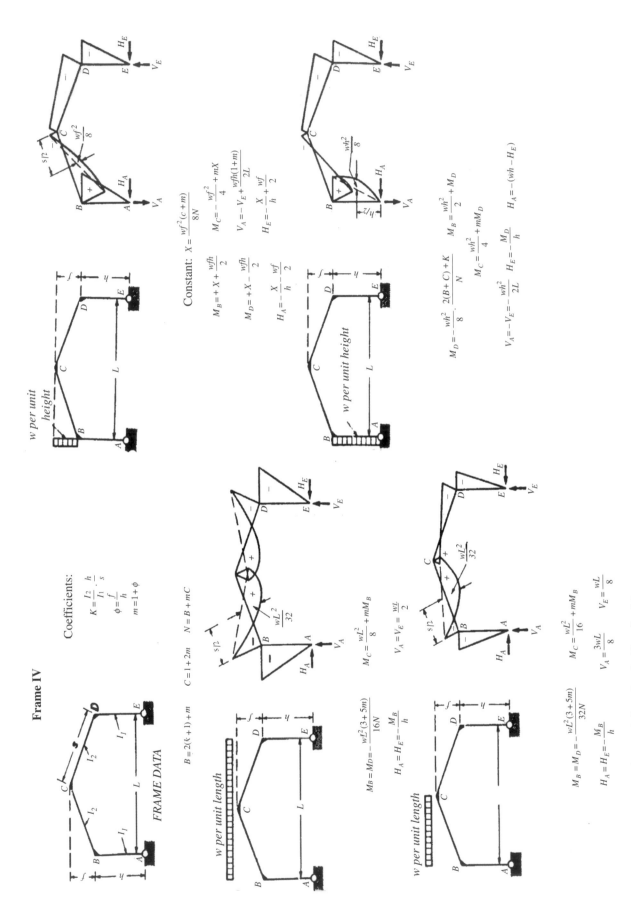

Figure 7.27 *Pitched timber portals frame IV: formulae (uniform loads).*

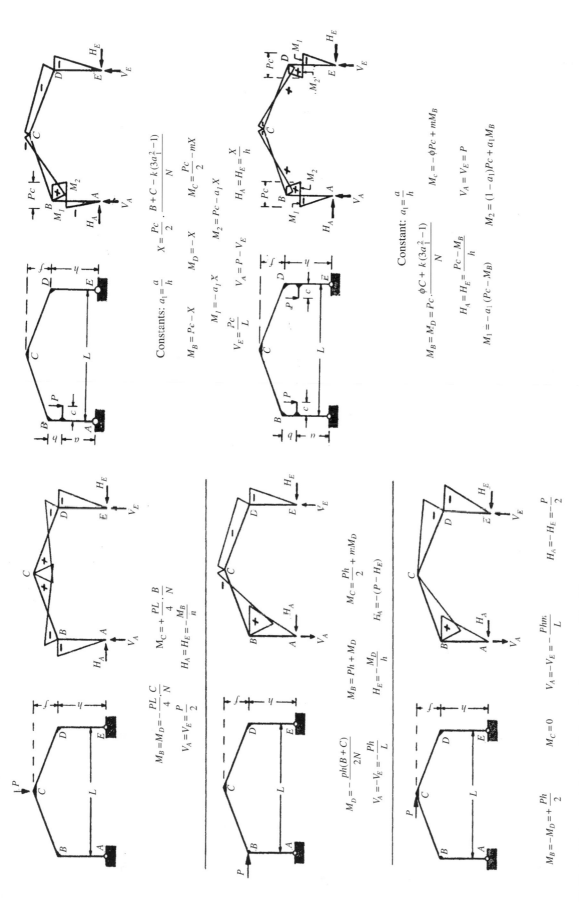

Figure 7.28 *Pitched timber portals frame V: formulae (concentrated loads).*

Since the depth, d, may vary from point to point along the arch member, so also does the volume factor.

Figures 7.20 to 7.28 indicate the analytical formulae for various timber frames/portals. They are used by designers for the quick evaluation of design quantities.

7.2.6.4 Knee-braced frames in timber (Russian codified approach)

The light-pitched knee-braced frame systems are the most widely used for timber frame work in single-storey buildings, where the braces bear at their top ends in the principal rafter (Fig. 7.29, upper diagram). The top ends of the knee-braces are usually fitted at one-third of the span of the rafter, and the bottom end at a height of not less than 2.25 m from the floor level. The angle of slope varies from 30 to 45°.

The braces are connected to the posts and rafters by stepped joints and timber-dogs, and as with the usual values of constant and temporary loads, they only take up compression stresses. Tensile stresses can arise in the end braces of double-span systems due to the action of wind loads, and also when a snow load on one side of the roof-slope is more than twice the dead load. In this case, in order to take up the tensile stresses, the brace is fastened to the post and rafter with a pair of bolted side-plates.

Structures of this type usually span openings of 4–9 m. The following main types of knee-braced frames are distinguished: single-braced, braced and straining-beam, and light-pitched knee-braced frames. A feature of these systems is the work not only of the beams but also of the posts on the one-sided vertical loading with lateral bending from the thrust transferred through braces to the posts. By fitting ties in the systems, this thrust can be taken up and an increase in the rigidity of the columns is obtained by using lattice-brace tower supports.

Under horizontal wind loads the columns of all systems work under lateral bending. These are used as load-carrying structures of a temporary or auxiliary nature (for shed, warehouses, stores, barns, scaffolding, staging and so on), in house building and also for trestles.

The load-carrying capacity of beams under heavy loads can be substantially increased by fitting braces and straining beams under them. In addition, when the posts have a hinged support on the foundations, the braces provide lateral rigidity to the system, connecting the posts and strainers in a single framed structure.

The main advantages of knee-braced frames are: (i) they are easy to make; (ii) local round timber can be used without requiring highly-qualified workers; (iii) the work of the logs under compression with bending is fully utilised.

When designing light-pitched knee-braced frames, the run-off (taper) of the log must be borne in mind, arranging the posts with the thin ends at the top, and in the outer bay of the rafter with the thin end on the end post.

The ends of the posts have a hinged support directly on the stone foundation or can be partly restrained by means of short reinforced concrete posts sent into the ground, and fastened above ground level to the posts with bolts. A typical design/detailing of a knee-braced frame is shown in Fig. 7.29 (lower diagram).

7.2.7 Timber grid work (Russian practice)

7.2.7.1 General introduction

A grid structure has a three-dimensional form with the assembly of elements inclined at any angle and resisting loads in any direction. The members can be rolled, extruded or fabricated sections supporting superstructures made in steel, glass, fabric, aluminium, wood, plastics or a mixture of these materials. Space frames and trusses with pinned or rigid points can be employed as supporting media to form spatial structures. Here timber as a material for grid work is under consideration.

In the past few centuries, thousands of such structures have been built in one form or another and numerous publications have been written concerning various aspects of their design and construction. The material is generally timber, steel, aluminium, concrete, plastics, fabric and glass.

Innumerable possible combinations have been discovered for unique structures, with regular and irregular forms having simplistic curvatures, such as domes, or anticlastic hyperbolic or elliptical structures with negative curvatures; design features have used single and multiple grids.

The classification of grid structures is extremely difficult owing to the great variety of possible forms. They can divide into three broad categories:

1. Skeleton and brace frame works;
2. Stressed skin systems
3. Suspended (cabled membrane) structures.

The behaviour of these three classes of grid structure differs, and the various kinds of analysis that exist also differ, depending upon the choice of materials and system. The skeleton or braced frame works are also known as latticed structures, braced frames or reticulated structures. For reasons of utility and economy, the majority of these follow regular geometric forms and can be further categorised as positive Gaussian synclastic and negative Gaussian (anticlastic) structures, as has already been introduced in the shell-type structures. Stressed skin domes and barrel vaults are typical examples of this category of structure. Cable roofs belong to the category of suspended structures.

7.2.7.2 Single-layered grid system

Here the grid system is built up of single layers of various patterns. They are plane grids as shown in Fig. 7.30. Two-way and three-way grids are popular for their strength and uniformity of stress under extreme loading conditions.

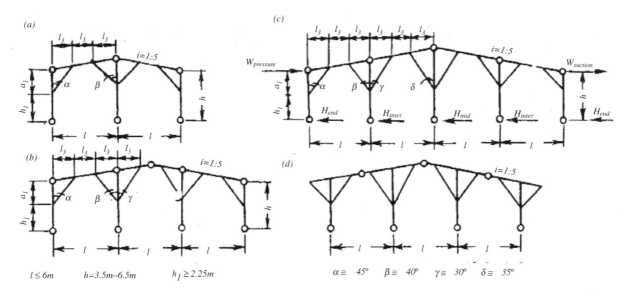

Principal Types of Skeletal Knee-braced frame system

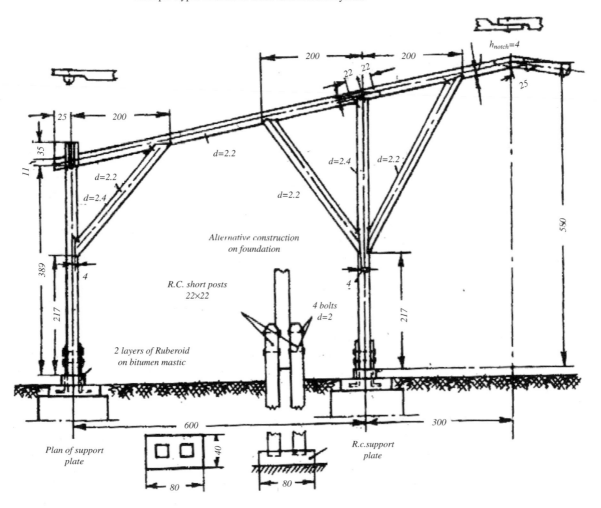

Design and Detailing of a Knee-braced Frame
Light-Pitched Knee-braced System

Note: All dimensions are in cm

Figure 7.29 *Timber knee-braced frames (based on Russian practice).*

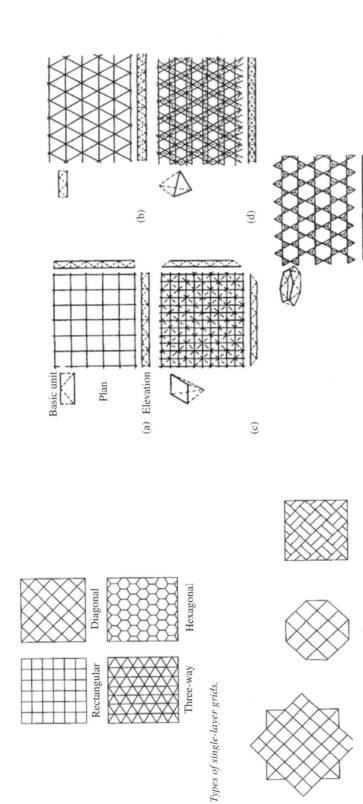

Basic unit

Plan

(a) Elevation

(b)

(c)

(d)

(e)

Types of double-layer grids. (a) Two-way lattice grid; (b) three-way lattice grid; (c) two-way space grid with square pyramids; (d) three-way space grid with triangular pyramids; (e) three-way space grid with tetrahedronal units.

Diagonal

Rectangular

Hexagonal

Three-way

Types of single-layer grids.

c

b

a

Types of diagonal grids.

Types of triangular grids.

Figure 7.30 *Types of grid-work.*

Table 7.3 Comparison of various types of pyramid

Type of pyramid	n	2α (°)	β (°)	γ (°)	2δ (°)	$f = n\alpha/\pi$	A_T/A_P
Triangular	3	101.5	45.0	26.6	104.5	0.85	1.41
		75.5	63.4	45.0	78.5	0.63	2.24
Square	4	70.5	45.0	35.3	120.0	0.78	1.41
		60.0	54.7	45.0	109.5	0.67	1.73
Hexagonal	6	44.4	45.0	40.9	138.6	0.74	1.41
		41.4	49.1	45.0	135.6	0.69	1.53
Cone	∞	–	45.0	45.0	180.0	0.71	1.41

For a flate plate $f = 1$ and for a hollow bar $f = 0$.

Table 7.4 Properties of regular polyhedra

Sl. No.	Regular polyhedra	Symbol	l/R	Dihedral angle	θ	R/r
1.	Tetrahedron	3^3	1.633	70°32′	109°28′	3.00
2.	Cube	4^3	1.155	90°	70°32′	1.732
3.	Octahedron	3^4	1.414	109°28′	90°	1.732
4.	Dodecahedron	5^3	0.714	116°34′	41°49′	1.258
5.	Icosahedron	3^5	1.051	138°11′	63°26′	1.258

In the above table l is the length of edge of regular polyhedra, θ is the angle subtended by edge of center, R is the radius of circumscribed sphere, and r is the radius of inscribed sphere.

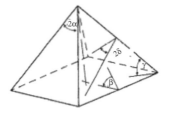

A_T = Total area of the pyramid wall
A_P = The plan area of the pyramid

Figure 7.30a *Principal elements of a pyramid.*

7.2.7.3 Double-layered grid system

These consist of two plane grids forming top and bottom layers parallel to each other and are connected by vertical and diagonal members. They may be lattice grids or true space grids consisting of prefabricated tetrahedral, octahedral and skeleton pyramids as shown in Fig. 7.30 and 7.30a. Tables 7.3 and 7.4 give various parameters for these types of shapes. It should be noted that the grid pattern of the top layer may or may not be identical to the bottom layer.

An improvement and extension of the skeleton type of double-layer grid is a stressed skin system which combines several advantages of the skeleton and sheet systems: great structural efficiency with many advantages of prefabrication.

Timber or aluminium, which are normally avoided for such structural elements because of their low Young's modulus, can be used in stressed skin space systems in which the stress distributed is membranal. In these systems, axial forces are of paramount importance, with virtual elimination of bending stresses. These systems consist of a large number of prefabricated three-dimensional units, made from thin sheets of steel, aluminium, plywood or plastic interconnected along the top edges by means of bolting, riveting, welding or gluing. The apices of all units are connected together by tie members, forming a

two-way or three-way bottom grid, and with corrugated roof sheeting fixed to the top edges or flanges to act as the integral load-carrying element of the whole structure.

7.2.7.4 Stressed skin system

The folded plate system is a typical example of a stressed skin structure in which the external loading is resolved into components acting in the planes. These components are called skin forces and can be resisted by the stiffness of the plates. The stressed skin system consists of a large number of prefabricated three-dimensional units, made from thin sheets of steel, aluminium, plywood or plastic interconnected along the top edges by means of bolting, riveting, welding or gluing. The apices of all these units are connected together by the members forming a two-way or three-way bottom grid. Flat or corrugated roof sheeting is secured to top players or edges.

Design Example 7.1 Slab analogies (Makowski formulae) double-layer grids in timber

Double-layer grids may be replaced by an equivalent plate to obtain indirect solutions to the stress distribution in real structures. Only 10–15% errors exist in solutions obtained by exact analysis.

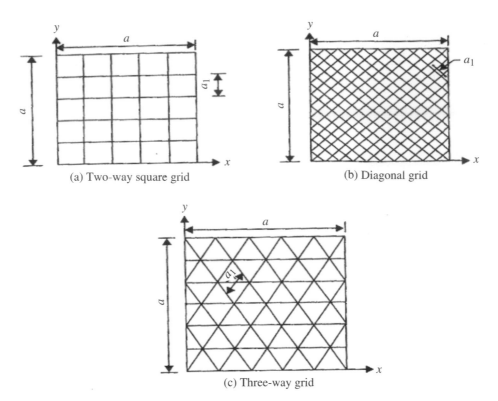

Figure 7.31 *Double layer grids.*

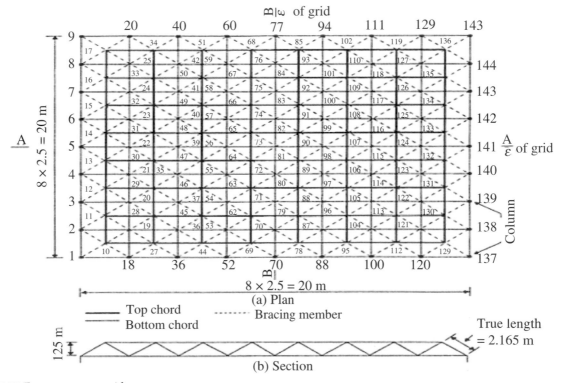

Figure 7.32 *Two-way square grid.*

Design Example 7.2 Square on square offset

Boundary conditions:
   P = applied load
   EI = flexural rigidity of an equivalent slab
   Maximum shear in the middle of the edge beam
   Maximum deflection at the centre $x = y = \dfrac{a}{b}$

Solution
(a) *Two-way square grids*

Double-layer   With boundary conditions

$$\omega = \text{deflection} = \frac{8p}{\pi^2 EI} \sin\frac{\pi x}{a} \sin\frac{\pi y}{a} a_1 a^4$$

$$= \omega_{\max} = \frac{8 p a_1 a^4}{\pi^6 EI}. \tag{7.9}$$

M_x = Bending moment

$$= M_y = \frac{8p}{\pi^4}\sin\frac{\pi x}{a}\sin\frac{\pi y}{a}a_1a^4 = M_x = M_y \quad (7.10)$$

max (per truss) $= \frac{8pa_1a^2}{\pi^4}$ $\quad (7.11)$

$$Q_x = Q_y = V = \frac{8p}{\pi^3 EI}\cos\frac{\pi x}{a}\sin\frac{\pi y}{a}a_1 = V_{max}\text{ (per truss)}$$

$$= Q_{max} = \frac{8pa_1a}{\pi^3} \quad (7.12)$$

(b) *Diagonal grids*

$$\omega_{max} = \frac{2\sqrt{2}pa_1a^4}{\pi^6 EI} \quad (7.13)$$

$$M_{max} = \frac{2\sqrt{2}pa_1a^2}{\pi^4} \quad (7.14)$$

$$V_{max} = Q_{max} = \frac{4pa_1a}{\pi^3} \quad (7.15)$$

(c) *Three-way grids*

$$\omega_{max} = \frac{8\sqrt{3}pa^4a_1(c_1)}{9\pi^5 EI} \quad (7.16)$$

P = 1.66 kN/m²; R=1.25 m
a_1 = 2.5 m; a = 20 m; E = 0.0072 N/m²
A_T = area of top chord members = 2030 mm²
A_B = area of bottom chord members = 855 mm²

Solution

$$M_{max} = \frac{8pa_1a^4}{\pi^4} = 135.50\,\text{kN}$$

$$I = \frac{A_T A_B}{A_T+A_B}h^2 = \frac{A_T}{2}h^2 = 0.00158\,\text{m}^4$$

$$F = \frac{135.50}{1.25} = 108.40\,\text{kN}$$

$$V = Q_{max} = 21.30\,\text{kN}$$

$$\omega_{max} = \frac{8qa_1a^4}{\pi EI} = \frac{8\times1.66\times2.5\times20^4}{\pi^2\times0.0072\times0.00158}$$

Reticulated barrel (curved space grid)
A barrel shell supported at the end on diaphragms is reticulated barrel with length
L = 30 m, width = 10 m and rise 5 m. The barrel is composed of wood member mesh of triangular type.
The barrel load p = 0.35 kN/m²
a_1 = mesh = l = 1.5m; f_t = 140 N/m²

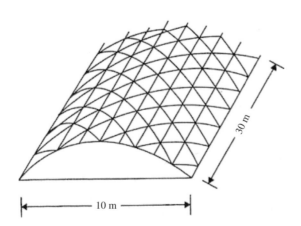

Figure 7.33 *Curved space grid.*

Data:
ρ_w = max.Weight of the member = 46.6 N/m²
$(N_x)_{max} = 0.47\frac{L}{R}pL$ Compressive

$$M_{max} = \frac{8\sqrt{3}paa_1(c_1)}{9\pi^3} \quad (7.17)$$

$$V_{max} = Q_{max} = \frac{8\sqrt{3}paa_1(c_1)}{9\pi^2} \quad (7.18)$$

where

$$c_1 = \frac{2\cosh\left(\frac{\pi}{2}\right)-\frac{\pi}{2}\tanh\left(\frac{\pi}{2}\right)-2}{\cosh\frac{\pi}{2}} \quad (7.19)$$

Makowski space grid
Square on square offset
Data:
$(N_x)_{min} = -0.83\frac{L}{R}pL$ Tensile $\quad (7.20)$

$(N_{x\phi})max = 0.55pL$

N_ϕ = unit meridional force = $pR\cos\phi - N_y$ $\quad (7.21)$

Solution
N_x = longitudinal stress
$N_{x\phi}$ = Shear stress = τ
$(N_x)_{max} = 29.60\,\text{kN/m}$ at midspan at top; S_{max} =
$0.75\times0.35\times30-7.875\,\text{kN/m}$
$(N_x)_{min} = -52.30\,\text{kN/m}$ at midspan at bottom
$(N_\phi)_{max} = 0.35\times5 = 1.750\,\text{kN/m};$
at top
$(N_\phi)_{min} = 0;$
at bottom
$(N_\phi) = 1.75\times\frac{2}{\pi} = 1.115\,\text{kN/m}$
at $\cos\phi = \frac{2}{\pi}$
at neutral axis
(i) *At midspan top of the reticulated barrel*
$$C_x = \frac{1.5}{2\sqrt{3}}[3(29.6-1.75)] = 36.2\,\text{kN}$$
$$C_r = \frac{1.5}{2\sqrt{3}}\times1.75 = C_s = 1.516\,\text{kN}$$

(ii) At midspan bottom
$$C_x = \frac{1.5}{2\sqrt{3}}(3)(-52.30) = -67.3\,\text{kN}$$
$$C_r = C_s = 0$$

(iii) At supports (neutral axis)
$$C_x = \frac{1.5}{2\sqrt{3}}(-1.115) = -0.483\,\text{kN}$$
$$C_r = \frac{1.5}{\sqrt{3}}(1.115-\sqrt{3}(7.875)) = -10.85\,\text{kN}$$
$$C_s = \frac{1.5}{\sqrt{3}}(1.115+\sqrt{3}(7.875)) = 12.78\,\text{kN}$$

A_{max}(bar)=C_s/140(mid-span) = 258.6 mm²
A_m= mesh area=$\frac{\sqrt{3}}{4}l^2 = 0.975$ m² $\quad (7.22)$
ρ_w = weight of bars of triangular
mesh =$\frac{1}{2}3lA\rho$ $\quad (7.23)$
$=1.5\times1.5\times1000\times258.5\times\frac{0.0078}{100} = 45\text{N} < 46.6\text{N}$

7.2.7.5 Timber grid roofs – structural detailing

Case Study No 1 (Based on European codes of practice)

Figure 7.34 gives the design detailing of a sports centre using timber grid work involving timber lattice beam construction spanning an area of 30 m × 45 m with a three-part sports hall. The lattice beams having two-part glulam chords which are connected by single glulam diagonals and which act as the main load-bearing system while covering the entire width of the building. The support reactions from each pair are carried back to concrete columns by four raking struts. The secondary load-bearing members of the glued laminated timber are continuous over the entire length of the building above the top chord and are then propped by knee braces from the bottom chords. The roof decking is supported by solid timber purlins of cantilever sections.

The architects were Wagner, Waner, Falterer and Dielershim. The structural engineers responsible for the structural design were Natterer and Dittrich of Munich. The building is a constructed facility since 1984.

Case Study No 2 (Based on European codes of practice)

The structural detailing given here is for the grid roof for the Weiherhof Sports Centre at Karlsruhe in Germany. Figure 7.35 gives major structural details of connections. A plan and typical structural elevations are given with various dimension in metres and millimetres.

A truss construction with raised glazed section is to allow daylight into the interior of this sports centre. The main Warren girders spanning 28.8 m are supported on reinforced concrete columns spaced 7.50 m apart. The chords are formed from twin glued laminated

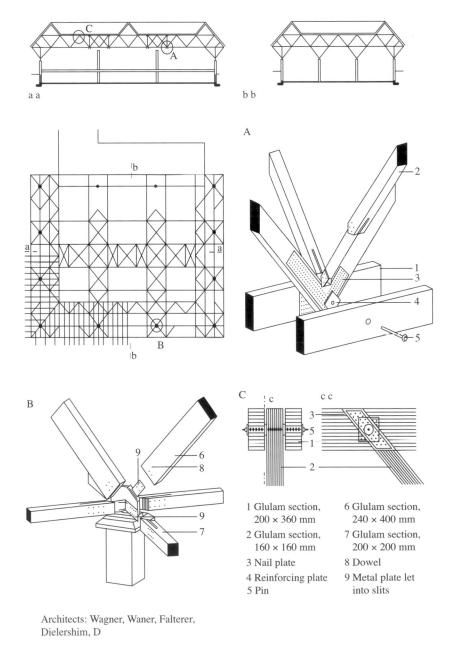

1 Glulam section, 200 × 360 mm	6 Glulam section, 240 × 400 mm
2 Glulam section, 160 × 160 mm	7 Glulam section, 200 × 200 mm
3 Nail plate	8 Dowel
4 Reinforcing plate	9 Metal plate let into slits
5 Pin	

Architects: Wagner, Waner, Falterer, Dielershim, D

Figure 7.34 *Grid roof for a sports centre (based on European practice) Eching. D 1984.*

timber sections dowelled together. The diagonals in compression are also glued laminated timber sections, but the (rising) di agonals in tension are steel rods. The compressive forces are transferred by way of direct contact with end plates, the tensile forces by steel parts in order to brace themselves against the compression members with end plates.

The various forces in the chords are transferred via nailed connections. Struts at 3.60 m centres perpendicular to the axis of the girder provide lateral restraints to the chords, support the raised roof sections and act as intermediate supports for the suspended purlins. The secondary load-bearing members are suspended continuous purlins made from glued laminated timber. At the quarter-points of the main trusses, these are made into a complete truss so as to connect the main girders and thus create torsion-resistant pairs.

7.2.8 Shell roofs and stressed skin surfaces

7.2.8.1 General introduction to shell roof structures

Shells are spatially curved surface structures which support externally applied loads. A great variety of shell roofs have been designed and constructed in many parts of the world. Large spans are generally covered by the shells. The most popular shells in timber are kite shaped hyperbolic paraboloid shells (HP) and folded plates.

The classification of shell surfaces can be made in a variety of ways. Figure 7.36 gives a detailed classification of stressed skin surfaces which are shells and folded plates. In this section, the emphasis is placed on HP shells and plates.

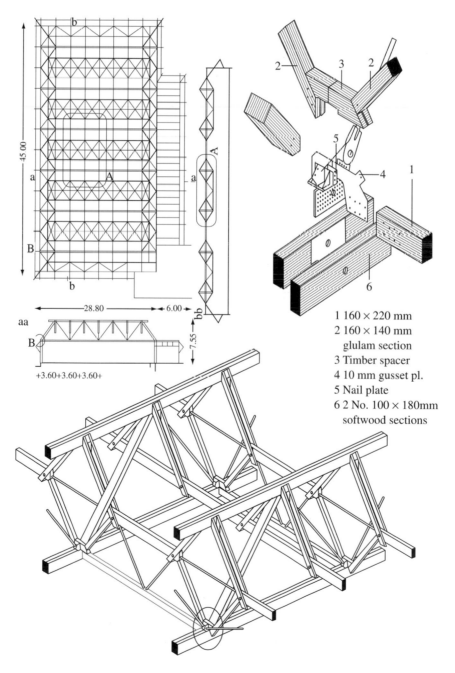

1 160 × 220 mm
2 160 × 140 mm
  glulam section
3 Timber spacer
4 10 mm gusset pl.
5 Nail plate
6 2 No. 100 × 180mm
  softwood sections

Figure 7.35 *Grid roof for Weiherhof centre in Germany (based on European practices).*

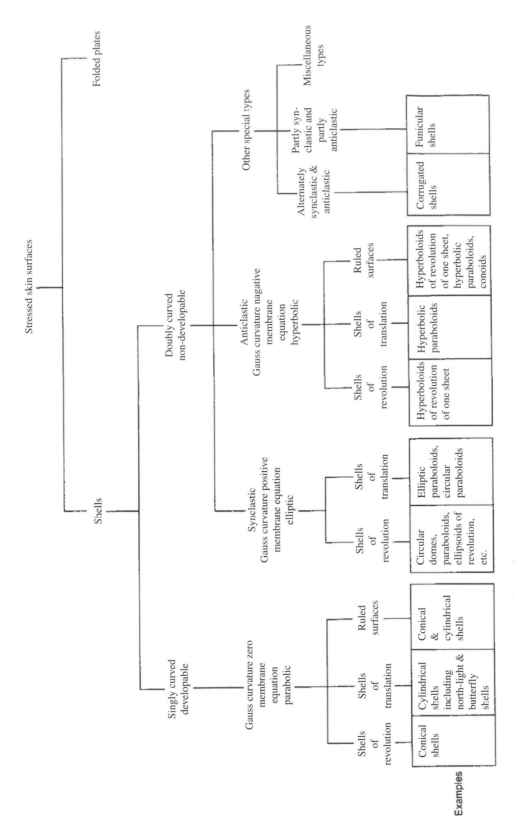

Figure 7.36 *Detailed classification of stressed skin surfaces.*

7.2.8.2 Hyperbolic paraboloid-shaped structures

Surface definition

The doubly curved surface of the hyperbolic paraboloid (HP) shell is defined as a surface of translation. A vertical parabola having an upward curvature is moved over another parabola with a downward curvature such that the parabola of translation is lying in a plane perpendicular to the first but moving parallel to it.

Figure 7.37 shows a variety of roof shapes made of HP timber shells. Figure 7.38 shows an HP shell with a corner dipped and having shape methodology of a saddle-shaped HP shell. Figure 7.38(g) then shows an analysis for a skewed HP timber shell.

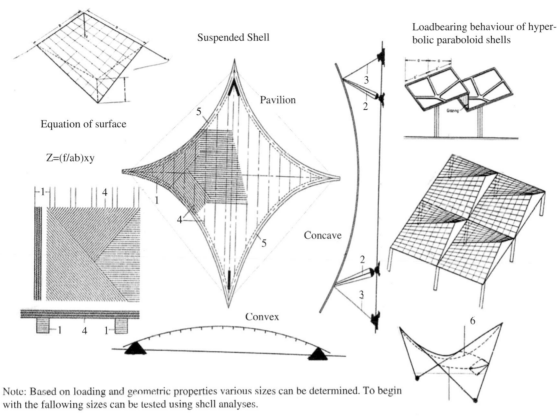

Equation of surface

$$Z=(f/ab)xy$$

Note: Based on loading and geometric properties various sizes can be determined. To begin with the fallowing sizes can be tested using shell analyses.

1 Suspended rib, 200 × 240 mm
2 Cruciform leg (with opposing tapers) in compression, 280 × 500–2500 mm + 280 × 500–1600mm

3 Tensioning cable, 91 or 217 individual wires, 7 mm dia., as parallel-lay wire rope
4 Layers of boards, 1 No. 24 mm and 2 No. 16 mm
5 Edge member, 360 × 1400 mm, twisted and in double curvature

Dortmund, D; 1969

Structural engineer: G. Scholz, Munich

Consultants: Natterer Bois Consult, Etoy, CH
A suspended shell similar to a diamond on plan. The edge members of glued laminated timber, 2 No. 180 mm deep × 1400 mm wide, twisted and in double curvature,

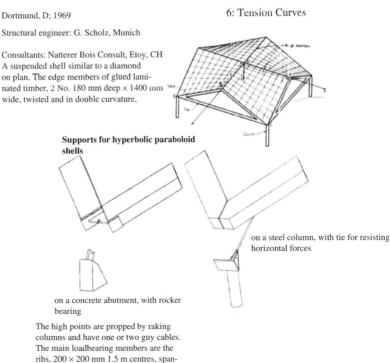

6: Tension Curves

Supports for hyperbolic paraboloid shells

on a steel column, with tie for resisting horizontal forces

on a concrete abutment, with rocker bearing

The high points are propped by raking columns and have one or two guy cables. The main loadbearing members are the ribs, 200 × 200 mm 1.5 m centres, spanning up to 65 m in concave from between the high points and the edge members.

Figure 7.37 *Timber hyperbolic paraboloid shell roof (based on EC-5 and European practices).*

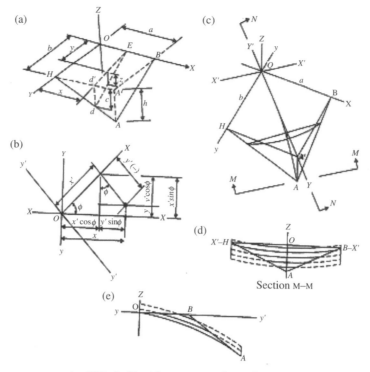

An HP shell with a corner dipped.

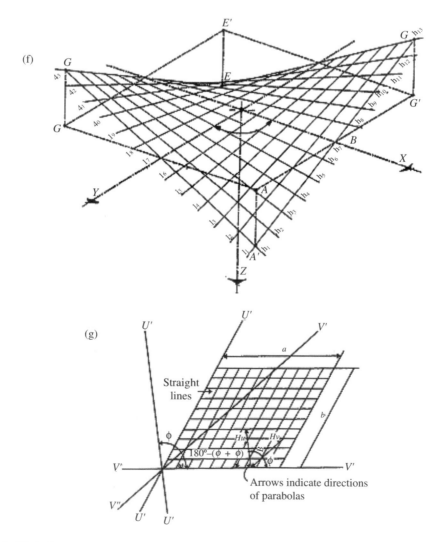

Figure 7.38 *Hyperbolic parabolid shells.*

7.2.8.3 Design equations

The HP may be generated by opposite generatrix and directrix lines. Some of them are shown in Fig. 7.38 and some appear saddle shaped. This type of shell is also known as a Hypar shell. A kite-shaped Hypar shell is shown in Fig. 7.38(f) and can be rail generated by moving along the y axis. A straight line that remains parallel to the x–z plane pivots while sliding along the straight line ABG'. The resulting surface is represented by grid lines h_n and i_n and every point on it may be considered to be the intersection of two such lines contained in the surface. The surface may be visualised by considering the horizontal plane $A'C'E'G'$ to be warped vertically, depressing corners A and E to new positions A' and E'. Straight lines h_n and i_n are now longer in the warped surface than in the projected horizontal surface in order that an intersection such as A may remain under A'.

Geometry

Figure 7.38(a), triangle $HA'A$

$$\frac{c}{h} = \frac{x}{a} \text{ or } c = \frac{xh}{a} \tag{7.24}$$

Figure 7.38(a), triangle $Ed'd$

$$\frac{z}{c} = \frac{y}{b} \text{ or } z = \frac{xyh}{ab} \text{ or } z = kxy \tag{7.25}$$

where OX and OY are rotated through an angle $\phi = 45$ such that OY' lies in a vertical plane with OA

$$x = x' \cos \phi - y' \sin \phi = 0.707(x' - y') \tag{7.26}$$
$$y = (-)(-)y' \cos \phi + x \sin \phi = 0.707(x' + y')$$

Substituting Eq. (7.24) into Eq. (7.25), the value of Z becomes

$$Z = 0.5[(x')^2 - (y')^2] \tag{7.27}$$

Figure 7.38(b) shows a rotated position. If a typical parabolic arch is cut parallel to the y–z plane indicating that x' is constant, then

$$Z' = -0.5k(y')^2 \tag{7.28}$$

If the surface comprises two opposite arches, the load intensity is shared equally, i.e. each has $w/2$:

$$\text{Internal moment} = \frac{w}{2}\left(\frac{L^2}{8}\right) \tag{7.29}$$

$$-H \times h_{xy} = \frac{w}{2}\left(\frac{L^2}{8}\right) \tag{7.30}$$

Hence the horizontal thrust is:

$$H = \frac{-w}{4L^2 \times 4h_{xy}} \tag{7.31}$$

where h_{xy} is the height at the cut end using Eq. (7.31), where $Z' = h_{xy}$, $y' = L/Z$ and

$$h_{xy} = \frac{-0.5kL^2}{4} \tag{7.32}$$

Substituting into Eq. (7.31), the value of the horizontal thrust H becomes

$$H = \frac{wab}{2h} \tag{7.33}$$

where w is the uniform load.

Skewed hyperbolic paraboloid

The same basic approach is applied to the more general case of roofs skewed in plan, as shown in Fig. 7.38. In this case, the surface is defined by

$$Z = \left(\frac{h}{ab}\right)u'v' \tag{7.34}$$

where u' and v' represent skewed coordinates.

On the surface, the two systems of straight lines parallel to the coordinates are u' and v'. Distances u and v are measured parallel to the u' and v' axes, respectively.

The skewed coordinate versus normal coordinate skew angle is w, and the angle of rotation is ϕ. The position of the parabolas is given by

$$\sin \phi = \frac{\sin \omega}{\sqrt{2}} \tag{7.35}$$

The thrust of the skewed shell can easily be calculated as

$$H_{v''} = \left(\frac{wab}{4h}\right)\sqrt{2}\,\frac{\sin \omega}{\sin(\omega - \phi)} \tag{7.36}$$

$$H_{v'''} = \left(\frac{wab}{4h}\right)\sqrt{2}\,\frac{\sin \omega}{\sin(\omega + \phi)} \tag{7.37}$$

Shear V at the boundaries will be

$$V = \frac{wab}{2h}\sin \omega \tag{7.38}$$

7.2.8.4 Structural detailing of hyperbolic paraboloid shells (HP): case studies

A sequence of HP shells with structural details for two cases of study is given. Section (A) shows a leisure pool at Freiburg, Germany supporting ten HP shells when placed on a hexagonal plan over columns, they make a roof for a multi-purpose hall. They have been designed according to European codes and practices.

(A) Leisure pool roof at Freiberg, Germany (Fig. 7.39A)

A sequence of 10 hyperbolic paraboloid shells forms the roof over this swimming pool. Each shell segment rests on four reinforced concrete columns placed in two rows 21 m apart, the segments cantilever out at the facade and the high points at mid-span are supported by trussed struts.

The three layers of 22 mm diagonal boarding on the shell segment enable each one to act as a shear-resistant secondary load-bearing system. The edge members are two-part, twisted, glued laminated sections. The joints between the segments are used to admit daylight and for ventilation. Overall stability is guaranteed by the shell segments acting as plates. Continuity results from the trussing and the fixed-base columns. The roof covering is of PVC coating on polyurethane foam insulation. The shell segments were built on the ground adjacent to the site and lifted into place with a crane.

This structure has been conceived by the architect H.D. Hecker and designed by M. Scherberger, structural engineer, Freiberg, Germany.

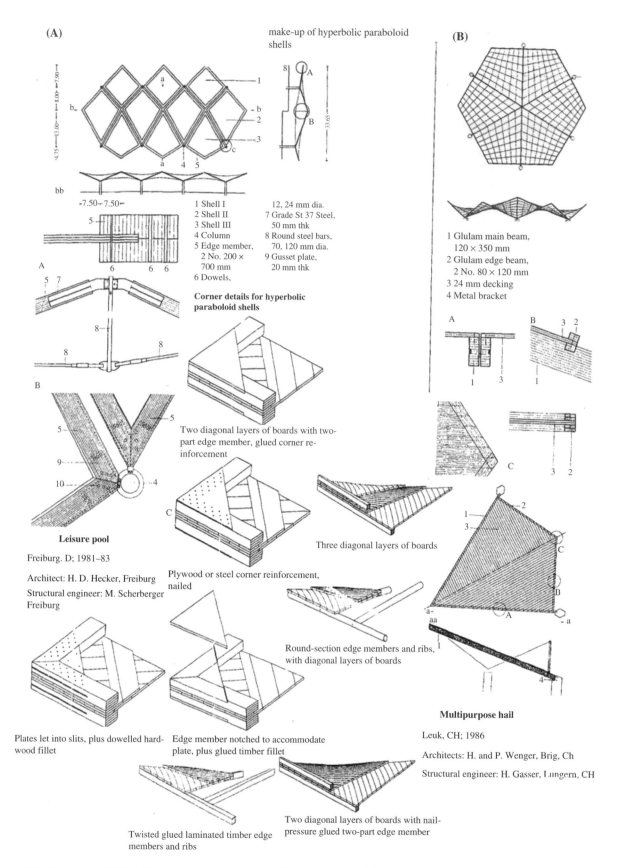

(A)

make-up of hyperbolic paraboloid shells

1 Shell I
2 Shell II
3 Shell III
4 Column
5 Edge member,
   2 No. 200 ×
   700 mm
6 Dowels,

12, 24 mm dia.
7 Grade St 37 Steel,
   50 mm thk
8 Round steel bars,
   70, 120 mm dia.
9 Gusset plate,
   20 mm thk

Corner details for hyperbolic paraboloid shells

Two diagonal layers of boards with two-part edge member, glued corner re-inforcement

Three diagonal layers of boards

Leisure pool

Freiburg. D; 1981–83

Architect: H. D. Hecker, Freiburg

Structural engineer: M. Scherberger Freiburg

Plywood or steel corner reinforcement, nailed

Round-section edge members and ribs, with diagonal layers of boards

Plates let into slits, plus dowelled hard-wood fillet

Edge member notched to accommodate plate, plus glued timber fillet

Twisted glued laminated timber edge members and ribs

Two diagonal layers of boards with nail-pressure glued two-part edge member

(B)

1 Glulam main beam,
   120 × 350 mm
2 Glulam edge beam,
   2 No. 80 × 120 mm
3 24 mm decking
4 Metal bracket

Multipurpose hall

Leuk, CH; 1986

Architects: H. and P. Wenger, Brig, Ch

Structural engineer: H. Gasser, Lungern, CH

Figure 7.39 *Sequence of HP shells supported on concrete columns – structural detailing.*

(B) Multi-purpose hall at Leuk, Switzerland (Fig. 7.39B)

The hexagonal multipurpose hall is covered by a timber shell of six identical hyperbolic paraboloid segments. The 260 m^2 roof is supported at the six low points. Each segment was produced on a jig using two diagonal layers of 24 mm boards and edge members of glued laminated timber. The parallel boards were curved and laid without open joints. The boards were glued together and to the edge members using resorcinol resin, with pressure applied by nails or screws.

The finished segments were lifted into place by crane and joined together. The underside has been left exposed, while the topside is insulated and covered with a synthetic roofing felt.

The architects for this hall were H. and P. Wenger of Brig, Switzerland and the structural engineer was H. Gasser of Lungern, Switzerland.

7.2.8.5 Timber dome (Fig. 7.40)

The next example is a timber dome for a warehouse with a point load at the apex due to a conveyor belt. The construction consisted of eight three-pin arches spanning 94.60 m made from 200 mm wide glued laminated timber with depth varying from 1400 to 2260 mm, with on-site splice. The apex ring was assembled on site with direct-contact joints filled with epoxy resin. Lateral restraint is provided by knee-braces on one side. Intermediate support is provided at mid-span to accommodate the loads at roof level, which are carried to the supports around the circumference by means of diagonals. There is solid timber bracing in four bays and around the apex. Overall stability is ensured by the three-dimensional arrangement of the arches as well as the roof bracing. The roof covering is of sheet aluminium.

7.2.9 Folded plates

Various geometrical descriptions of folded plate structures are known, and many analytical and numerical techniques are available to solve such structures. In reality, a prismatic folded plate is nothing but a cylindrical shell itself, except that the cross-section of the former system is the approximation of the given smooth directrix curve of the latter system as inscribed piecewise linear segments. Analysis using folded plate systems are free of most of the problems associated with the solutions of cylindrical shells.

Some of the advantages are:

(a) Variation in the thickness of the shells is possible. However, the plate thickness and plate width of a particular plate strip should only vary within the bounds of the plate zone.

(b) The boundary conditions, other than simply supported at two traverses of curvilinear boundaries, can be taken care of using suitable basic functions.

(c) The loads applied at the folded plates are transformed into equivalent fold-line loads in the form of line loads along the fold lines.

(d) It is not necessary to provide edge beams since they are simply treated as the first and last plates in the folded plate system.

(e) Analysis of the pre-stressed folded system can be carried out without additional modification to the method of calculation.

(f) The effect of the presence of stringers at the node lines can be easily accommodated compared with shells with edge beams.

(g) The folded plate analysis is primarily based on the behaviour of individual plates and plates meeting at the fold under different actions. The analysis is easier when they are continuously placed as computed with similar situations in the case of shells.

7.2.9.1 Some common shapes

Prismatic folded plate structures consist of a number of shapes joined smoothly along their common edges to span between stiff diaphragms. Several terms and conditions exist for the adoption of one shape or another – the most important ones on the analytical side are the resisting of compressive forces and the accommodation of tensile reinforcement for bending.

The net plate load in the plane of the kth plate is

$$P_k = P_{i,j} - P_{i-1,j} \qquad (7.16)$$

where, in P, the first suffix i stand for the node and the second suffix j stands for the plate. If the direction of P_k is from right to left, ie towards the decreasing order of the node numbers, it is positive. So also is the in-plane plate displacement δ.

7.2.9.2 Definitions

Reference is made to Fig. 7.41, in which the transverse section of a folded plate system is shown. The edges or ridges or nodes or folds are identified as $0, 1, 2, k-2, k-1, k+1, k+2, \ldots, n$.

The first plate is marked between folds 1 and 0, any other plate, say, the kth plate is identified between folds k and $k-1$. If the fold numbers progress from left to right, the kth plate will be marked whose right-hand side fold number is k.

The individual plates are, in general, inclined and attached to the neighbouring plates at angles other than $0°$ or $180°$. The inclined position of any plate is defined in two ways: its slope and its angle of inclination. These are defined as:

- *Slope*: Measured from the horizontal (anticlockwise positive), the angle of inclination of the kth plate is the slope ψ_k. It is observed at the fold $k-1$ of the plate k.
- *Inclination*: The angle ϕ_k the $(k+1)$th plate makes with the kth plate, and is considered to be positive if measured clockwise from the kth plate line. This angle is observed at the fold k of the $(k+1)$th plate. (If the situation demands, ϕ_k will be written alternatively as ϕ_k later.)

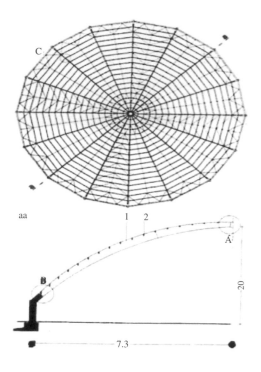

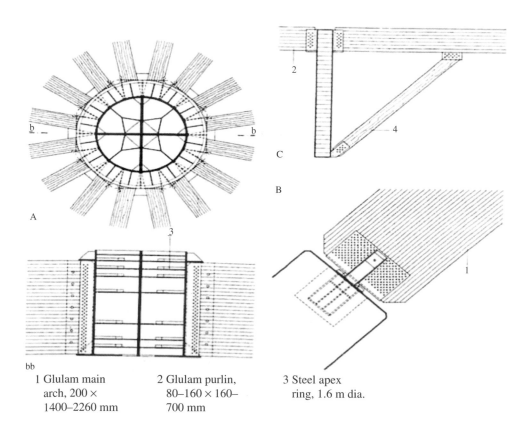

1 Glulam main
 arch, 200 ×
 1400–2260 mm

2 Glulam purlin,
 80–160 × 160–
 700 mm

3 Steel apex
 ring, 1.6 m dia.

Structural engineers: Brüninghoff Building
Department, Heiden, D

Figure 7.40 *Timber made dome over warehouse – Walsum Germany (based on European practices).*

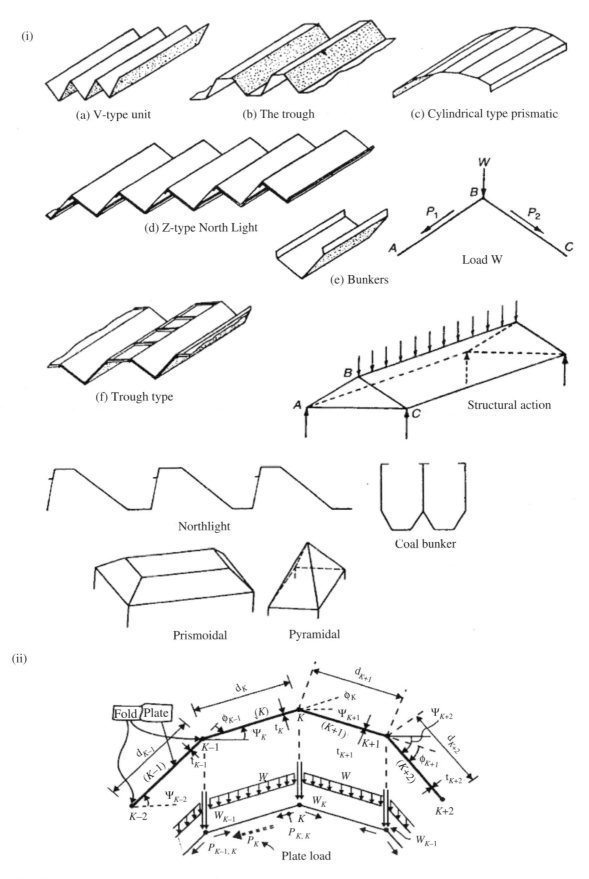

(i)

(a) V-type unit

(b) The trough

(c) Cylindrical type prismatic

(d) Z-type North Light

(e) Bunkers

Load W

(f) Trough type

Structural action

Northlight

Coal bunker

Prismoidal

Pyramidal

(ii)

Fold Plate

Plate load

Figure 7.41 *Folded plates in timber.*

If the kth plate is vertical, $\phi_k = +90°$; and if horizontal, $\phi_k = 0°$.

If the cross-section is symmetrical about the fold plates attached on either side, we have, $\phi_k = -\psi_{k+1}$, and $\phi_k = 2\psi_k$.

The width and thickness of any plate k are d_k and t_k, respectively, whereas the lengths of all of the plate strips are the same and equal to the longitudinal span L itself. In a folded plate system, the width d_k is much less than the length L. It is not more than 3 to 5 m in practice. However, since the individual plate strips are normally asked to behave as thin beams bending in their own planes, it is essential that L/d_k ratios are at least around 3 or more. This particular criterion not only caters for the requirement of one-dimensional beam behaviour of the individual plate strips, but also assures that the transverse strain $\acute{\varepsilon}_2$ across the width of the plates and the shearing strain γ_{12} in the plane of the plates remain within very negligible values.

The assumptions of the geometrical hypothesis as $\acute{\varepsilon}_2 = \gamma_{12} = 0$ along with the statically indeterminate hypothesis that $M_{12} = M_1 = Q_1 = 0$, the cylindrical shell solution methodology can be greatly simplified. Of course, the applicability of the method would remain restricted to medium and moderately long span shells.

Therefore, for long and medium span folded plate systems, the individual plate strips, when selected with L/d_k ratios > 3, justify their behaviour as simple beams in their planes with the consequence that the longitudinal stresses across the width d_k of plates thus developed would be linear.

The individual plate k is inclined. Any action on this plate (acting vertically on it, say) has two components: one is the in-plane and the other is the normal to the plate surface. The in-plane component causes the plate strip to behave as a beam and the corresponding action is known as plate action. The normal component of the load causes the plate to bend out of plane, and this behaviour is termed slab action. Both the plate action and slab action develop simultaneously in the individual plates of the folded plate system.

The given uniform load on the kth and $(k+1)$th plates can be converted into equivalent fold line loads. At the fold k, the sum of the vertical components of the loading from the contribution of the two adjacent plates is W_k (Fig. 7.41). Similarly, W_1 values at all other nodes are known.

7.2.9.3 Basic assumptions

The following are the basic assumptions made for the analysis of folded plates:

1. The material of the folded plate is homogeneous, isotropic and obeys Hook's law.
2. The out-of-plane bending behaviour of the plates (characterised by the slab action) is due to the transverse action caused by the normal component of the loading. This action is borne out by the one-way slab action of the plate strip of unit width ($dx = 1$) that is continuous over the fold lines.

3. The in-plane bending behaviour of the plates (characterised by the plate action) is due to the in-plane component of the loading, causing longitudinal bending.
4. The variation of loading in the longitudinal direction is the same throughout.
5. The plates offer no resistance to twisting moment.

7.2.9.4 Methods of analysis

The following methods are available in the literature:

1. Energy method
2. Force method
3. Displacement method
4. Mixed method
5. Transfer matrix method
6. Finite difference method

Note: The reader is advised to consult the author's following reference for the analysis and design of folded plate structures: M.Y.H. Bangash & T Bangash, *Elements of Spatial Structures: Analysis and Design*, Thomas Telford, London, 2003.

Until recently, folded-plate roofs, shell-vaults and shell-domes have been for the most part made in the form of nailed structures put together on site. This method of construction is very labour-consuming to make and erect and has enhanced deformability owing to the pliability of the nail connectors. At the present time, these structures can be made in the factory as glued or box-plywood units, using industrial methods of assembly. The use of glue for jointing the members of folded-plate roofs or shells considerably improves their rigidity.

The fabrication of glued and box-plywood folded-plate structures and shells calls for experienced designing and testing during manufacture, erection, and under service conditions.

Table 7.3 gives the principal types with characteristic of spatial timber roof structures. The Russian Federation, following European Practices has given decent summary with types dimensional and sphere of adoption of the spatial structures.

7.2.9.5 Folded plates: case studies

Ribbed folded plate roofs: structural design/ detailing (based on European/Russian Practices)

Ribbed folded-plate roofs consist of flat, sloping surfaces with solid transverse diaphragms. The stiffening ribs (Fig. 7.42) to the end walls act as supports for the folds. The span L of the folded-plate roof is the distance between the centre lines of the end walls; the width of the fold B is the distance between the lower edges of the surfaces in cross-section.

Folded-plate roofs can be used in the form of independent folds of symmetrical outline over buildings

Table 7.3 The principal types and technico-economic characteristic of spatial timber roofing structures

Notation: $k_{o.w.}$ = coefficient of its own weight;
k_m = consumption of metal as a percentage of the weight of the structure, excluding tie rods

Name and brief description	Overall dimensions and technico-economic features	Sphere of adoption
	Braced vaults	
1. Segmental-lattice barrel vaults with tie rods or with transfer of the thrust direct to the supports, having notched joints in the metalless vaults and bolted in the metalled ones. A prefabricated factory-made structure (type *1a*)	$l = 12$–$80 * m;$ $f/l \geq 1/7; h/l \geq 1/100;$ $k_{o.w.} = 13$–15 In metalless vaults $k_m = 1$–$2\%;$ In metalled vaults $k_m = 3$–5%	Roofs over industrial premises, garages, hangars, theatres, exhibition pavilions and so on
2. Segmental-lattice pointed (ogival) vaults without ties, the thrust being transferred directly to the supports (type *1b*). Jointed as in item 1. A prefabricated factory-made structure	$l = 12$–$80 * m;$ $f/l \geq 1/3; f_1/l_1 \geq 1/15;$ $h/l \geq 1/100$ $k_{o.w.}$ and k_m as in Item 1	Cold roofs over industrial premises, warehouses, etc., with iron, slate or similar roofing materials.
	Folded-plate structures	
3. Ribbed folded-plate single- and multibay roofs of triangular outline supported on the end walls. Glued or nailed joints (type *3a*)	$l = 12$–30 m; $f/l = 1/2$–$1/9; l/B = 1$–$3;$ $\alpha \geq 35$–$45°;$ $k_{o.w.} = 15$–$20;$ $k_m = 3$–5% (if nailed)	Roofs over industrial premises, warehouses, garages and so on
	Shell-vaults	
4. Thin-walled shell barrel vaults supported on end walls and with horizontal bracing of the edges by trusses. Joints in separate members glued or nailed (type *2*)	$l = 20$–40 m; $f/l = 1/5$–$1/6;$ $l/B = 2.5$–$3; 2\alpha \geq 120°$ $k_{o.w.} = 10$–$15;$ $k_m = 4$–5% (if nailed)	Roofs over industrial premises, warehouses, garages, where a smooth ceiling is required
5. Ribbed barrel vaults supported on end walls. Joints glued or nailed (type *3b*)	$l = 20$–100 m; $f/l = 1/6$–$1/8; l/B \geq 4;$ $2\alpha \geq 90°; k_{o.w.} = 6$–$12;$ $k_m = 5$–8% (when nailed)	Roofs over industrial premises, hangars, warehouses and so on
	Domes	
6. Dome-shaped roofs of three-hinged arches with the thrust transferred direct to the supports	$l = 20$–70 m; $f/l = 1/2$–$1/6;$ $l/l = 1/30$–$1/40;$ $k_{o.w.} = 3$–$5\%;$ $k_m = 4$–6% (nailed); $k_{o.w.} = 2$–$4; k_m = 2$–3% (glued structures)	Roofs over circuses, riding schools, auditoriums, having a plan of circular outline, etc
7. Thin-walled spherical shell-domes. Glued or nailed joints (type *9*)	$l = 12$–35 m; $f/l = 1/2$–$1/6;$ $h_d/l = 1/200$–$1/250;$ $k_{o.w.} = 10$–$15;$ $k_m = 3$–5% (nailed)	Roofs over industrial buildings circular in plan
8. Ribbed spherical shell-domes. Glued or nailed joints (type *16*)	$l = 35$–60 m; $f/l = 1/2$–$1/6;$ $h_{rlb}/l = 1/50$–$1/70;$ $k_{o.w.} = 10$–$15;$ $k_m = 4$–6% (when nailed)	As for Items 6 and 7

(continued)

Table 7.3 (Continued)

9. Dome in the form of a coupled shell-vault, with a square or regular polygonal plan. Glued or nailed (types 6 and 7)	$l = 20\text{--}40$ m; $f/l = 1/2\text{--}1/6$; $h_{rlb}/l = 1/40\text{--}1/60$ (for ribbed domes): $k_{o.w.} = 10\text{--}15$; $k_m = 4\text{--}6\%$ (when nailed)	As for Items 6 and 7 with a square or polygonal plan. In the form of a semi-dome (niche) as a roof over a stage, platform, bandstand and so on
10. Spherical segmental-lattice domes. Characterized by their non-standard segments, the shape and size of which vary according to their height. Jointed as in Item 1. A prefabricated factory-made structure (type 11)	$l = 15\text{--}35$ m; $f/l = 1/2\text{--}1/6$; $h/l = 1/150$; $k_{o.w.} = 8\text{--}10$; In metalless domes $k_m = 1.2\%$; In metalled domes $k_m = 3\text{--}5\%$	As for Items 6 and 7
11. Domes in the form of coupled segmental-lattice vaults with a square or regular polygonal plan. Characterised by their standard segments, jointed as in Item 1. A prefabricated factory-made structure (type 8)	$l = 15\text{--}45$ m; $f/l = 1/2\text{--}1/6$; $h/l \geq 1/150$; $k_{o.w.}$ and k_m—as in item 10	As for Items 6 and 7

Note: The folded-plate structures (Item 3), the shell-vaults (Items 4 and 5), the spherical shell-domes (Items 7 and 8) and the domes in the form of a coupled shell-vault (Item 9) are factory-made when formed of glued or box-plywood units and site-built when the members are nailed together.

rectangular in plan: garages, warehouses, sheds and so on, mainly in cases where the side walls are required to have large openings. They can also be used very effectively in the form of combinations of symmetrical folds, multibay in width and multispan in length, with intermediate column supports. In this case, however, the construction is made more complicated owing to heavy loads set up by snow pockets between the folds. The panel of the fold is in the form of a nailed I-beam with a diagonally crossed web; it can also be in the form of a glulam or box-plywood beam.

Roof-coverings of flat or corrugated asbestos-fibre sheeting can be recommended. The ribs, providing transverse stiffening of the panel, are fitted from 1.5 to 3 m apart. In order to minimise the rotation of the folds set up by one-sided loads, vertical columns are fitted along the edge of one panel of the fold, hinge-connected to the ribs and the foundations. The erection of folded-plate roofs is carried out with scaffolding supports. By utilising the stiff transverse ribs during erection, the scaffolding can be confined to supporting the lower ends of the ribs.

Folded plate-cum-triangular timber truss system (based on Eurocode-5 and European practices) (Fig. 7.43)

This example is of a multipurpose hall on a hexagonal plan measuring 32 m across and covered by a radial folded plate construction. Roof plates are resolved into triangular trusses, with 140×240 mm and 180×240 mm glulam sections. A glulam tension ring resists the

horizontal thrust due to the vertical loads. Forces from horizontal loads are transferred into concrete walls. Steel columns carrying the roof arc arranged in pairs in the centre of the façade.

For connections, plates are let into slits and steel brackets are fixed with concealed nailing or dowels. There is a glulam node at the apex. The exposed roof decking above the trusses carries the deck roof construction. Bracing at roof level is by way of the trusses.

The design was completed in 1977 by the architects A. Frank and W. Wicker of the Building Department of the City of Munich and structural engineers Natterer and Dittrich of Munich, Germany.

7.2.10 Vehicle and foot bridges in timber

7.2.10.1 General introduction to types of wooden bridges

A bridge is subdivided into (a) the superstructure, (b) the substructure and (c) the foundation. The bridge deck system is the part of the superstructure directly carrying the vehicular loads. It is furnished with balustrades or parapets, crash barriers, highway surfacing, footpaths, traffic islands, railway tracks on ties, expansion joints and drainage systems.

The substructure comprises piers, columns or abutments, capping beams and bearings. The foundations consist of reinforced concrete footings, spread foundations, rafts bearing directly on soil or rock and capping slabs supported on piles, wells and caissons. The superstructure

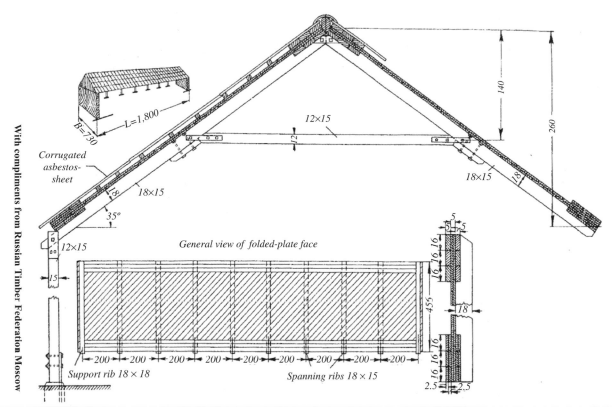

Figure 7.42 *Ribbed folded plates in timber (based on Russian design practices).*

of the bridge deck system can be any one or a combination of the following: planks, coffered planks, grids, beams, girders, cantilevers, frames, trusses and arches, cable-stayed and suspension types. The following classified system lists fully the types of bridges constructed in reinforced, pre-stressed and composite materials.

Wooden planks

1. Solid planks – Supported directly on piers, with or without intermediate supports
2. Void planks of haunches for the beams
3. Coffered planks – they act like a grid
4. Above with wooden beams of reinforced concrete and pre-stressed concrete wooden and steel beams.

Wooden beams

1. Longitudinal stringed beams with webs spaced apart and integral with the long deck,
2. Longitudinal and transverse beams forming a grid system integral with the deck,
3. Inverted longitudinal beams, trusses and girders, fully or partially integral with the deck,
4. A single central longitudinal spine beam, T-beam; truss and girder composite or monolithic with deck.

Wooden boxes

1. A single longitudinal box beam or several box beams with and without cantilevered top flanges comprised of a double webbed single unicellular box, and twin or multiple unicellular boxes with or without cross-beams or diaphragms.

Timber frames (with or without struts)

They may have members in one or more plane. They may be portal frames (single or multiple), vierendeel girders trestle piers, spill-through abutments and towers for cable-stayed or suspension bridges.

Short-span bridges over highways or rivers or flyovers over freeways.

Arches made in timber

They are classified as:

1. solid arches
2. open spandrel arches
3. solid spandrel arches
4. tied arches
5. funicular arches
6. strut-frame with inclined legs.

Suspension and cable-stayed bridges

Suspension bridges with draped cables and vertical or triangulated suspender hangers are adopted for spans exceeding 300 m. *Cable-stayed bridges* are economical over the span range of the order of 100 to 700 m with concrete deck, pylons and frames. For cable-stayed bridges, the elevational and transverse arrangements are:

1. elevational arrangement: single, radiating, harp, fan, star and combination

Bracing against critical deformation of outer edge

as a result of a uniformly distributed load on one side

Architect: A. Frank, Building Department, W. Wicker KG

Structural engineers: Natterer and Dittrich Planungsgesellschaft, Munich

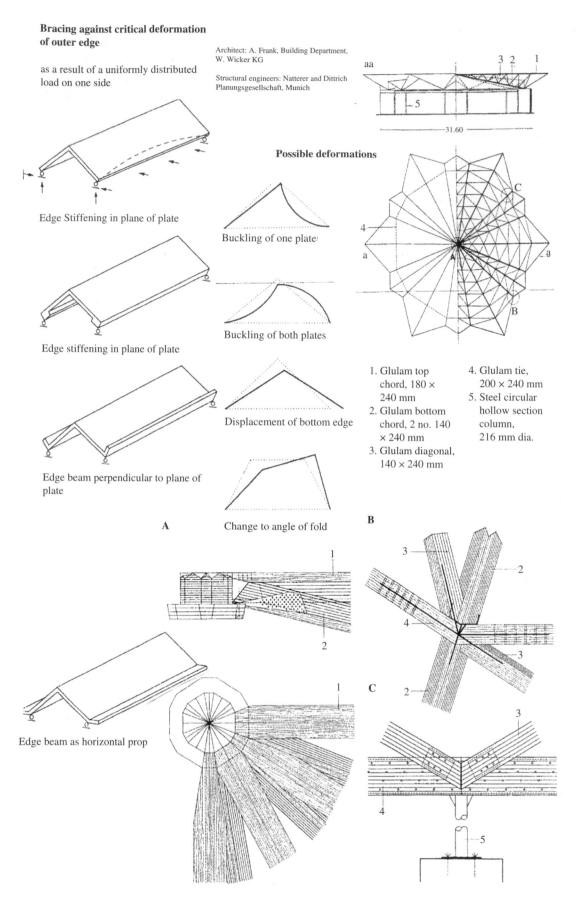

Edge Stiffening in plane of plate

Edge stiffening in plane of plate

Edge beam perpendicular to plane of plate

Edge beam as horizontal prop

Possible deformations

Buckling of one plate

Buckling of both plates

Displacement of bottom edge

Change to angle of fold

A

B

C

1. Glulam top chord, 180 × 240 mm
2. Glulam bottom chord, 2 no. 140 × 240 mm
3. Glulam diagonal, 140 × 240 mm
4. Glulam tie, 200 × 240 mm
5. Steel circular hollow section column, 216 mm dia.

Figure 7.43 *Fold plates roof/pavilion – Hartwald Clinic (based on European practices).*

2. transverse arrangement:

single plane
    (vertical – central or eccentric) | No cables single, double,
double plane | triple, multiple or combined.
    (vertical or sloping)

7.2.10.2 Types of loads acting on bridges

These are classified as follows.

1. *Permanent and long-term loads*: dead; superimposed; earth pressure and water pressure of excluded or retained water.
2. *Transient and variable loads (primary type)*: vehicular loading; railway loading; footway loading and cycle loading.
3. *Short-term load*: erection loads; dynamic and impact loads.
4. *Transient forces*: braking and traction forces; forces due to accidental skidding and vehicle collision with parapet or with bridge supports.
5. *Lurching and nosing by trains.*
6. *Transient forces due to natural causes*: wind action; flood action and seismic forces.
7. *Environmental effects*: Loads generated due to creep, shrinkage of concrete; prestress parasitic moments or reactions and prestrain and temperature range or gradient.

Relevant codes are consulted for the application of these loads on bridge structures.

7.2.10.3 Substructures supporting deck structures

The deck structures are supported directly on:

1. mass concrete or masonry gravity abutments
2. closed-end abutments with solid or void walls such as cantilevers, struts or diaphragms
3. counterforted or buttressed waits or combinations
4. open-end or spill-through abutments with trestle beams supported on columns.

The intermediate piers can be of the following types:

1. solid or void walls with or without capping beams
2. single solid or void columns with or without caps
3. trestles and bents
4. specially shaped columns, e.g. V-shaped or fork shaped, etc.

In most cases, bridge bearings are needed to transmit deck loads to substructures and to allow the deck to respond to environmental and vehicle loads.

7.2.10.4 Bridges – Case studies: Specifications

This section contains the relevant specifications of the British, European and American codes on the design/ detailing of steel bridges and their important components. In some cases, the practices are self-explanatory and need no additional text to clarify them. For thorough explanations of theory, design analysis and structural detailing, refer to *Prototype Bridge Structures: Analysis and Design* by the author, published by Thomas Telford, London, 1999.

General

Standard highway loading consists of HA and HB loading. HA loading is a formula loading representing normal traffic in Great Britain. HB loading is an abnormal vehicle unit loading. Both loadings include impact.

Loads to be considered

The structure and its elements shall be designed to resist the more severe effects of either:

(a) design HA loading
(b) design HA loading combined with design HB loading.

Notional lanes, hard shoulders, etc.

The width and number of notional lanes, and the presence of hard shoulders, hard strips, verges and central reserves are integral to the disposition of HA and HB loading. Requirements for deriving the width and number of notional lanes for design purposes are specified in the highway codes. Requirements for reducing HA loading for certain lane widths and loaded length are specified.

Distribution analysis of structure

The effects of the design standard loadings shall, where appropriate, be distributed in accordance with a rigorous distribution analysis or from data derived from suitable tests. In the latter case, the use of such data shall be subject to the approval of the appropriate authority.

Type HA loading (Fig. 7.44)

Type HA loading (Tables 7.4 and 7.5) consists of a uniformly distributed load (see Clause 8.1.2 of the code) and a knife edge load combined, or of a single wheel load.

Nominal uniformly distributed load (UDL)

For loaded lengths up to and including 50 m, the UDL, expressed in kN per linear metre of notional lane, shall be derived from the equation:

$$W = 336\left(\tfrac{1}{L}\right)^{0.67} \tag{7.39}$$

and for loaded lengths in excess of 50 m but less than 1600 m the UDL shall be derived from the equation:

$$W = 36\left(\tfrac{1}{L}\right)^{0.1} \tag{7.40}$$

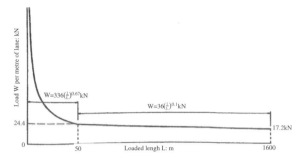

Figure 7.44 *HA loading.*

Table 7.5 Factors for limit state for combination of loads (HA type)

Design HA loading—for design HA load considered alone, VFL shall be taken as follows:

	For the ultimate limit state	For the serviceability limit state
For combination 1	1.50	1.20
For combinations 2 and 3	1.25	1.00

Table 7.6 Type HA uniformly distributed load

Loaded length: m	Load: kN/m	Loaded length: m	Load: kN/m	Loaded length: m	Load: kN/m
2	211.2	55	24.1	370	19.9
4	132.7	60	23.9	410	19.7
6	101.2	65	23.7	450	19.5
8	83.4	70	23.5	490	19.4
10	71.8	75	23.4	530	19.2
12	63.6	80	23.2	570	19.1
14	57.3	85	23.1	620	18.9
16	52.4	90	23.0	670	18.8
18	48.5	100	22.7	730	18.6
20	45.1	110	22.5	790	18.5
23	41.1	120	22.3	850	18.3
26	37.9	130	22.1	910	18.2
29	35.2	150	21.8	980	18.1
32	33.0	170	21.5	1050	18.0
35	31.0	190	21.3	1130	17.8
38	29.4	220	21.0	1210	17.7
41	27.9	250	20.7	1300	17.6
44	26.6	280	20.5	1400	17.4
47	25.5	310	20.3	1500	17.3
50	24.4	340	20.1	1600	17.2

where L is the loaded length (m) and W is the load per metre of notional lane (kN). For loaded lengths above 1600 m, the UDL shall be agreed with the appropriate authority.

Nominal knife edge load (KEL)

The KEL per notional lane shall be taken as 120 kN.

Distribution

The UDL and KEL shall be taken to occupy one notional lane, uniformly distributed over the full width of the lane and applied as specified in Clause 6.4.1 of the code.

Dispersal

No allowance for the dispersal of the UDL and KEL shall be made.

For all public highway bridges in Great Britain, the minimum number of units of type HB loading that shall normally be considered is 30, but this number may be increased up to 45 if so directed by the appropriate authority.

The overall length of the HB vehicle shall be taken as 10, 15, 20, 25 or 30 m for inner axle spacings of 6, 11, 16, 21 or 26 m, respectively (Fig. 7.45), and the effects of the most severe of these cases shall be adopted. The overall width shall be taken as 3.5 m. The longitudinal axis of the HB vehicle shall be taken as parallel to the lane markings.

Contact area

Nominal HB wheel loads shall be assumed to be uniformly distributed over a circular contact area, assuming an effective pressure of 1.1 N/mm^2.

Design HB loading

For design HB load, y_{fL} shall be taken as shown in Table 7.7.

7.2.10.5 Railway bridge live load

Standard railway loading consists of two types, RU and RL. RU loading allows for all combinations of vehicles currently running or projected to run on railways in Europe, including the UK, and is to be adopted for the design of bridges carrying main line railways of 1.4 m gauge and above.

RL loading is a reduced loading for use only on passenger rapid transit railway systems on lines where main line locomotives and rolling stock do not operate.

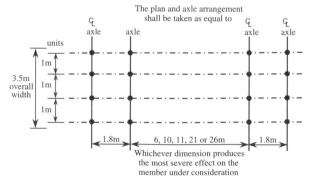

Figure 7.45 *Dimensions of HB vehicle for 1 unit of nominal loading (1 unit = 10 kN per axle, i.e. 2.5 kN per wheel).*

Table 7.7 Factors for limit state for combination of loads (HB type)

	For the ultimate limit state	For the serviceability limit state
For combination 1	1.30	1.10
For combinations 2 and 3	1.10	1.00

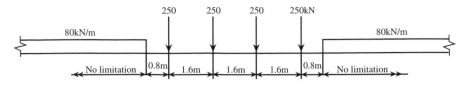

Note: see dynamic effects, below, for effect of additions to this loading

Figure 7.46 *Type RU loading.*

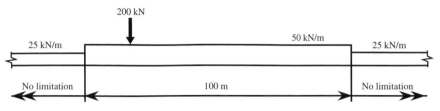

Note: see dynamic effects, below, for effect of additions to this loading

Figure 7.47 *Type RL loading.*

Type RU loading

Nominal type RU loading consists of four 250 kN concentrated loads preceded, and followed, by a uniformly distributed load of 80 kN/m. The arrangement of this loading is as shown in Fig. 7.46.

Type RL Loading (Fig. 7.47)

Nominal type RL loading consists of a single 200 kN concentrated load coupled with a uniformly distributed load of 50 kN/m for loaded lengths up to 100 m. For loaded lengths in excess of 100 m, the distributed nominal load shall be 50 kN/m for the first 100 m and shall be reduced to 25 kN/m for lengths in excess of 100 m.

Alternatively, two concentrated nominal loads, one of 300 kN and the other of 150 kN, spaced at 2.4 m intervals along the track, shall be used on deck elements where this gives a more severe condition. These two concentrated loads shall be deemed to include dynamic effects.

The standard railway loadings specified above (except the 300 kN and 150 kN concentrated alternative RL loading) are equivalent static loadings and shall be multiplied by appropriate dynamic factors to allow for impact, oscillation and other dynamic effects including those caused by track and wheel irregularities.

Type RU loading

The dynamic factor for RU loading applied to all types of track shall be as given in Table 7.7.

In deriving the dynamic factor, L is taken as the length (m) of the influence line for deflection of the element under consideration. For unsymmetrical

Table 7.8 Dynamic Factor for type RU loading

Dimension L: m	Dynamic factor for evaluating bending moment	Dynamic factor for evaluating shear
Up to 3.6	2.00	1.67
From 3.6 to 6.7	$0.73 + \dfrac{2.16}{\sqrt{(L - 0.2)}}$	$0.82 + \dfrac{1.44}{\sqrt{(L - 0.2)}}$
Over 6.7	1.00	1.00

influence lines, L is twice the distance between the point at which the greatest ordinate occurs and the nearest end point of the influence line. In the case of floor members, 3 m should be added to the length of the influence line as an allowance for load distribution through track.

Type RL loading

The dynamic factor for RL loading, when evaluating moments and shears, shall be taken as 1.20, except for unballasted tracks where, for rail bearers and single-track cross girders, the dynamic factor shall be increased to 1.40.

7.2.10.6 Road traffic actions and other actions specifically for road bridges

Loads due to road traffic, consisting of cars, lorries and special vehicles (e.g. for industrial transport), give rise to vertical and horizontal, static and dynamic forces. The load models defined in this section do not describe actual loads. They have been selected so that their effects (with dynamic amplification included unless otherwise specified) represent the effects of the actual traffic. Where traffic outside the scope of the load models specified in this section needs to be considered, then complementary load models, with associated combination rules, should be defined or agreed by the client.

Separate models are defined below for vertical, horizontal, accidental and fatigue loads.

Loading classes

The actual loads on road bridges result from various categories of vehicles and from pedestrians. Vehicle traffic may differ between bridges depending on traffic composition (e.g. percentages of lorries), density (e.g. average number of vehicles per year), conditions (e.g. jam frequency), the extreme likely weights of vehicles and their axle loads, and, if relevant, the influence of road signs restricting carrying capacity.

These differences justify the use of load models suited to the location of a bridge. Some classifications are defined in this section (e.g. classes of special vehicles).

Table 7.9 Number and width of lanes

Carriageway width, W	Number of notional lanes	Width of a notional lane	Width of the remaining area
$w < 5.4$ m	$n_1 = 1$	3 m	$w - 3$ m
5.4 m $\leq w < 6$ m	$n_1 = 2$	$\dfrac{W}{2}$	0
6 m $\leq w$	$n_1 = \text{Int}\left(\dfrac{W}{3}\right)$	3 m	$W - 3 \times n_1$

Note: for example, for a carriageway width of 11 m, $n_1 \text{Int}\left(\dfrac{w}{3}\right) = 3$, and the width of the remaining area is $11 - 3 \times 3 = 2$ m.

Others are only suggested for further consideration (e.g. choice of adjustment factors α and β defined in Clause 4.3.2(7) of the code for the main model and in Clause 4.3.3 for the single axle model) and may be presented as loading classes (or traffic classes).

Divisions of the carriageway into notional lanes

The widths w_1 of notional lanes on a carriageway and the greatest possible whole (integer) number n_1 of such lanes on this carriageway are shown in Table 7.8

7.2.10.7 Highway loads based on EC3 loading

Location and numbering of the lanes for design (EC3) (ENV 1995)

The location and numbering of the lanes should be determined in accordance with the following rules:

1. the locations of notional lanes are not necessarily related to their numbering
2. for each individual verification (e.g. for a verification of the ultimate limit states of resistance of a cross-section to bending), the number of lanes to be taken into account as loaded, their location on the carriageway and their numbering should be so chosen that the effects from the load models are the most adverse.

Vertical loads – characteristic values

General and associated design situations

Characteristic loads are intended for the determination of road traffic effects associated with ultimate limit-state verifications and with particular serviceability verifications (see ENV 1991-1, 9.4.2 and 9.5.2, and ENV 1992 to 1995). The load models for vertical loads represent the following traffic effects.

1. Load model 1: concentrated and uniformly distributed loads, which cover most of the effects of the traffic of lorries and cars. This model is intended for general and local verifications.
2. Load model 2: a single axle load applied on specific tyre contact areas which covers the dynamic effects of normal traffic on very short structural elements. This model should be separately considered and is only intended for local verifications.

3. Load model 3: a set of assemblies of axle loads representing special vehicles (e.g. for industrial transport) which may travel on routes permitted for abnormal loads. This model is intended to be used only when, and as far as required by the client, for general and local verifications.
4. Load model 4: a crowd loading. This model should be considered only when required by the client. It is intended only for general verifications. However, crowd loading may be usefully specified by the relevant authority for bridges located in or near towns if its effects are not obviously covered by load model 1.

Load models 1 and 2 are defined numerically for persistent situations and are to be considered for any type of design situation (e.g. for transient situations during repair works). Load models 3 and 4 are defined only for some transient design situations. Design situations are specified as far as necessary in design Eurocodes and/or in particular projects, in accordance with definitions and principles given in ENV 1999-1. Combination for persistent and transient situations may be numerically different.

Main loading system (load model 1)

The main loading system consists of two partial systems as detailed below.

Double-axle concentrated loads (tandem system: TS), each axle having a weight:

$$\alpha_Q Q_k \tag{7.40}$$

where α_Q are adjustment factors.

No more than one tandem system should be considered per lane; only complete tandem systems shall be considered. Each tandem system should be located in the most adverse position in its lane (see, however, below and Fig. 7.48). Each axle of the tandem model has two identical wheels, the load per wheel being therefore equal to $0.5\alpha_Q Q_k$. The contact surface of each wheel is to be taken as square and of size 0.40 m.

Uniformly distributed loads (UDL system), having a weight density per square metre:

$$\alpha_q q_k \tag{7.41}$$

where α_q are adjustment factors.

These loads should be applied only in the unfavourable parts of the influence surface, longitudinally and transversally.

Load model 1 should be applied on each notional lane and on the remaining areas. On notional lane number 1, the load magnitudes are referred to as $\alpha_{Qi} Q_{jk}$ and $\alpha_{qi} q_{jk}$ (Table 7.10). On the remaining areas, the load magnitude is referred to as $\alpha_{qr} q_{rk}$.

Table 7.10 Basic values

Location	Tandem system axle loads, Q_{1k}: kN	UDL system q_{ik} Q_{1k}(or q_{rk}): kN/m^2
Lane number 1	300	9
Lane number 2	200	2.5
Lane number 3	100	2.5
Other lanes	0	2.5
Remaining area (q_{rk})	0	2.5

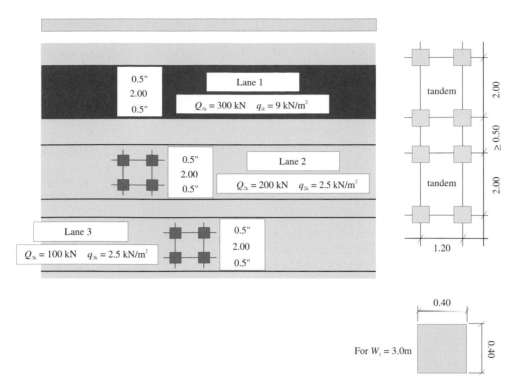

Figure 7.48 *Example of lane numbering in the most general case and load model 1.*

Unless otherwise specified, the dynamic amplification is included in the values for Q_{ik} and q_{ik}, the values of which are given in Table 7.9.

For the assessment of general effects, the tandem systems may be assumed to travel along the axes of the notional lanes.

Where general and local effects can be calculated separately, and unless otherwise specified by the client, the general effects may be calculated:

1) by replacing the second and third tandem systems by a second tandem system with axle weight equal to:
   $(200\alpha_{Q2} + 100\alpha_{Q3})$ kN (although relevant authorities may restrict the application of this simplification), or
2) for span lengths greater than 10 m, by replacing each tandem system in each lane by a one-axle concentrated load of weight equal to the total weight of the two axles. However, the relevant authorities may restrict the application of this simplification. The single axle weight is:

   $600\alpha_{Q1}$ kN on lane number 1
   $400\alpha_{Q2}$ kN on lane number 2
   $200\alpha_{Q3}$ kN on lane number 3.

The values of the factors α_{Qi}, α_{qi} and α_{qr} (adjustment factors) may be different for different classes of route or of expected traffic. In the absence of specification, these factors are taken as equal to 1. In all classes, for bridges without road signs restricting vehicle weights:

$\alpha_{Q1} \geq 0.8$ and

for $i \geq 2$, $\alpha_{qi} \geq 1$; this restriction is not applicable to α_{qr}. Note that α_{Qi}, α_{qi} and α_{qr} factors other than 1 should be used only if they are chosen or agreed by the relevant authority.

Single axle model (load model 2)

This model consists of a single axle load $\beta_Q Q_{ak}$ with Q_{ak} equal to 400 kN, dynamic amplification included, which should be applied at any location on the carriageway. However, when relevant, only one wheel of $200\,\beta_Q$ (kN) may be considered. Unless otherwise specified, β_Q is equal to α_{Q1}.

Unless it is specified that the same contact surface as for load model 1 should be adopted, the contact surface of each wheel is a rectangle of sides 0.35 m and 0.60 m as shown in Fig. 7.49.

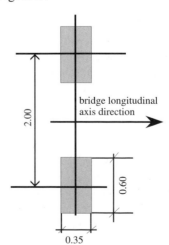

Figure 7.49 *Load model 2.*

Set of models of special vehicles (load model 3)

When one or more of the standardized models of this set is required by the client to be taken into account, the load values and dimension should be as described in annex A of the code concerned.

The characteristic loads associated with the special vehicles should be taken as nominal values and should be considered as associated solely with transient design situations.

Unless otherwise specified the following should be assumed.

1. Each standardized model is applicable on one notional traffic lane (considered as lane number 1) for the models composed of 150 or 200 kN axles, or on two adjacent notional lanes (considered as lane numbers 1 and 2; see Fig. 7.51) for models composed of heavier axles. The lanes are located as unfavourably as possible in the carriageway. For this case, the carriageway width may be defined as excluding hard shoulders, hard strips and marker strips.
2. Special vehicles simulated by the models are assumed to move at low speed (not more than 5 km/h); only vertical loads without dynamic amplification have therefore to be considered.
3. Each notional lane and the remaining area of the bridge deck are loaded by the main loading system. On the lane(s) occupied by the standardized vehicle, this system should not be applied at less than 25 m from the outer axles of the vehicle under consideration.

Crowd loading (load model 4)

Crowd loading, if relevant, is represented by a nominal load (which includes dynamic amplification). Unless otherwise specified, it should be applied on the relevant parts of the length and width of the road bridge deck, the central reservation being included where relevant. This loading system, intended for general verifications, is associated solely with a transient situation.

Dispersal of concentrated loads

The various concentrated loads to be considered for local verifications, associated with load models 1, 2 and

3, are assumed to be uniformly distributed across their whole contact area. The dispersal through the pavement and concrete slabs is taken at a spread-to-depth ratio of 1 horizontally to 1 vertically down to the level of the centroid of the structural flange below (see Figs 7.50 and 7.51).

7.2.11 Girders and trusses

See Figs 7.52 to 7.57.

7.2.11.1 Strutted-girder bridges

Efforts to utilize the qualities of strutted-girder (or braced-girder) systems, using local, round timber with a minimum amount of processing, converting them into easy-to-erect, site-built structures, resulted in the wide adoption during wartime of what is known as the strutted-girder truss with spans from 10 to 24 m, a ratio of height to span $h/l = 1/5-1/7$ and $K_{o.w} = 5-7$. These trusses are free from the main disadvantages of strutted systems, as they can be fabricated near the site of erection and set up in finished form, often consisting of complete or partly assembled structures similar to boarded and nail beams. Figure 7.57 shows the spanning structure of a bridge with strutted-girder trusses and the detail of the principal joints. All the joints and splices are connected with steel dowels and bolts. The trusses have no connectors working under shear stress, and this makes them very reliable structures, not requiring complicated forging or an excessive use of metal.

7.2.11.2 Howe-Zhuravsky trusses

Howe-Zhuravsky trusses (Fig. 7.58) are strutted systems of cords and crossed diagonal struts made of round or squared timbers, and uprights of steel ties. When squared timbers are used, the trusses can be made in the bridge-building workshops, with all of the wooden members treated with preservative solutions. The length of the panels depends on the loading conditions: with a local load on the panel $a = 1.4-2.5$ m; with a load on the joint $a = 3-5$ m. In order to reduce the length of the panels in bridges of long spans, trussed-beams can be included. Local transfer of the load is possible with a deck-truss and light loading; with a through-truss,

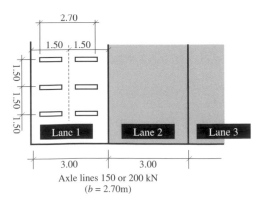

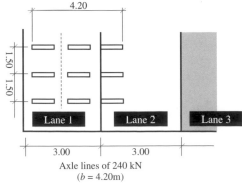

Figure 7.50 *Location of special vehicles.*

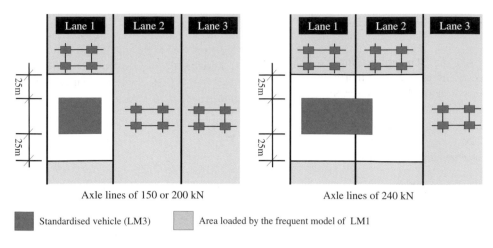

Axle lines of 150 or 200 kN   Axle lines of 240 kN

■ Standardised vehicle (LM3)   ▨ Area loaded by the frequent model of LM1

Figure 7.51 *Simultaneity of load models 1 and 3.*

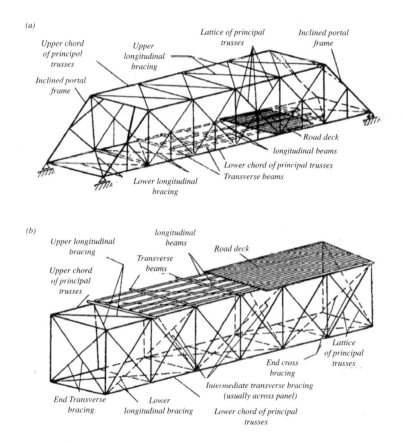

Figure 7.52 *Diagrammatic layout of spanning structure; (a), through-truss type; (b), deck-truss type.*

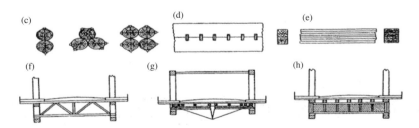

Figure 7.53 *Types of transverse beam; c, clustered logs; d, Derevyagin beam; e, glulam beam; f, lattice beam; g, trussed beam; h, beam with diagonal web (Russian practice).*

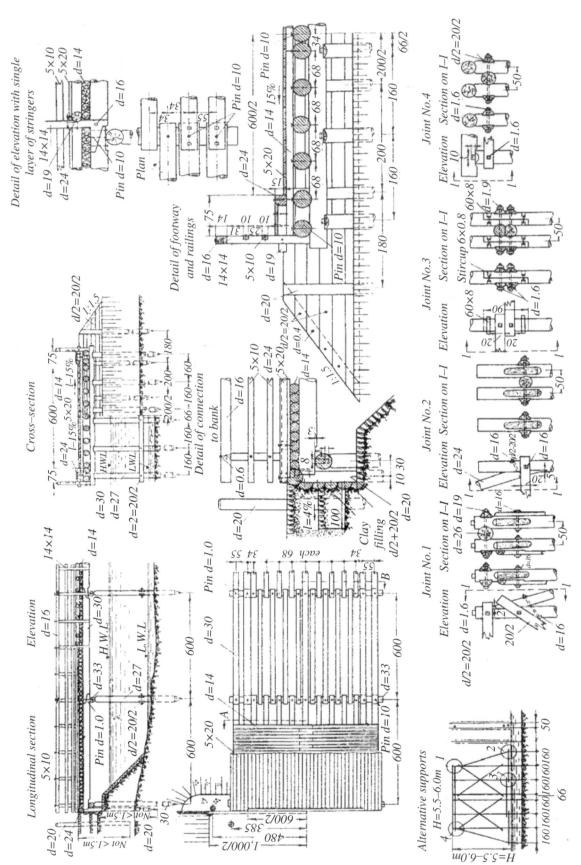

Figure 7.54 Details of railroad girder bridge with single-layer; closely-spaced stringers.

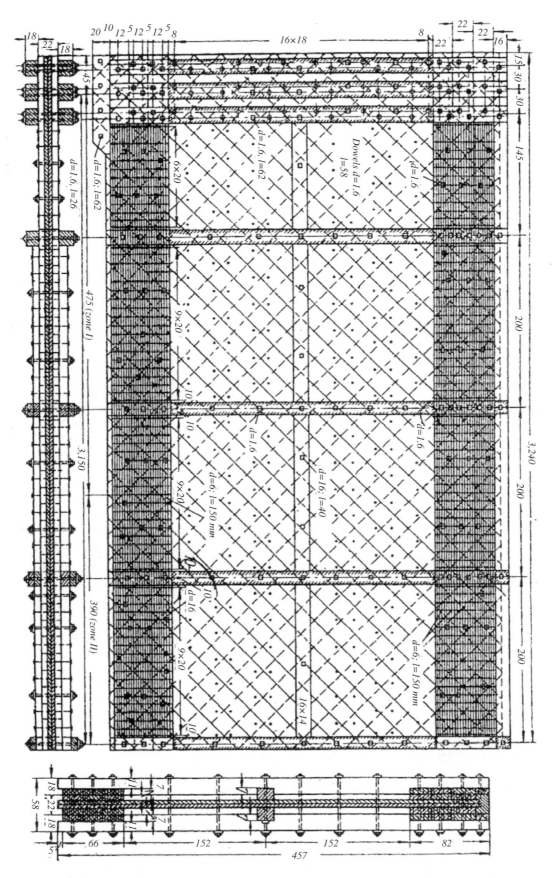

Figure 7.55 *Details of boarded and nailed girders with a span of 31.5 m.*

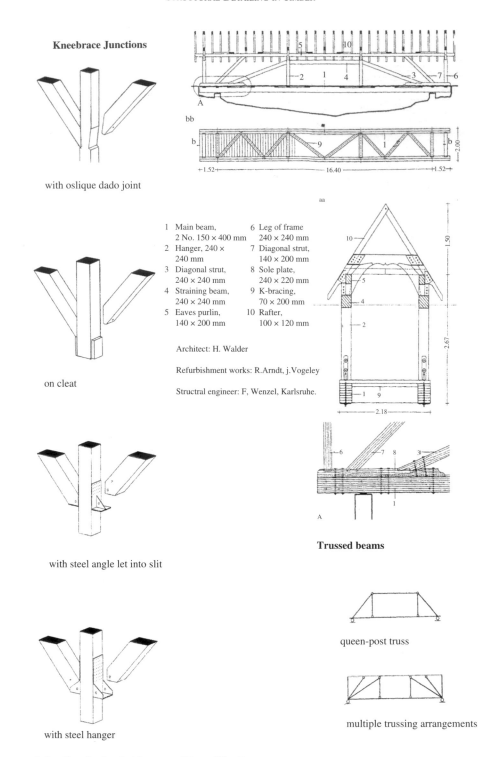

Kneebrace Junctions

with oslique dado joint

on cleat

with steel angle let into slit

with steel hanger

1 Main beam,   6 Leg of frame
   2 No. 150 × 400 mm   240 × 240 mm
2 Hanger, 240 ×   7 Diagonal strut,
   240 mm   140 × 200 mm
3 Diagonal strut,   8 Sole plate,
   240 × 240 mm   240 × 220 mm
4 Straining beam,   9 K-bracing,
   240 × 240 mm   70 × 200 mm
5 Eaves purlin,   10 Rafter,
   140 × 200 mm   100 × 120 mm

Architect: H. Walder

Refurbishment works: R.Arndt, j.Vogeley

Structral engineer: F, Wenzel, Karlsruhe.

Trussed beams

queen-post truss

multiple trussing arrangements

Figure 7.56 *Structural details of a footbridge over River Alb, Germany.*

there is, for the most part, transfer of the load through the joints.

The construction of the joints designed to take the plain thrust of the struts on oak blocks or metal shoes, only allows them to resist compressive stresses. In order to ensure that, during the passing of temporary loads, all of the struts remain under compression and that their ends will not come away from their bearings, the tie rods are pre-stressed during assembly and adjusted regularly during the service life of the bridge. The calculation of the forces can be carried out with sufficient accuracy without taking this pre-stressing into account, on the assumption

that one diagonal is stressed (under compression) in each panel. The timber chords are made with two, three, or four squared timbers throughout their length, with gaps between them for the ties and for ventilation. Since building practice and special research have shown the unreliability of connecting the tension splices in the chords with metal prismatic keys, the side-plates should be sheet steel (known as "combs") connected with thin stub dowels of steel. The metal chords are a steel section.

The rising struts are made with two or three timbers, and the descending struts correspondingly of one or two

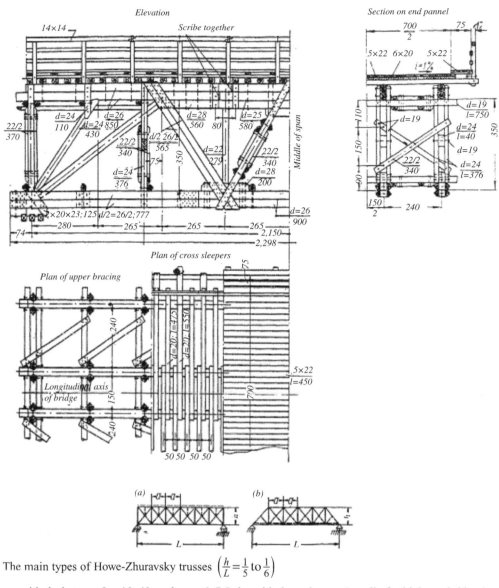

The main types of Howe-Zhuravsky trusses $\left(\frac{h}{L}=\frac{1}{5}\text{ to }\frac{1}{6}\right)$

a—with deck-truss, L = 10–40 m; $k_{o.w.}$ = 6–7.5; *b*—with through-truss (usually for highway bridges), L = 20–50 m; $k_{o.w.}$ = 5–7

Camber must be applied to Howe-Zhuravsky trusses, amount in highway bridges to $f_{eam} = \frac{1}{300}\,l$, in railway bridges $f_{eam} = \frac{1}{500}\,l$. Trusses of this design are extremely reliable and sufficiently rigid if the rods are tightened up periodically during service life.

Figure 7.57 *Strutted-girder truss bridge – 21.5 m span (based on Russian practice).*

timbers; the angle of slope varies from 35° to 60°. The oak joint blocks have their grain across the plane of the truss. Tie rods of diameters from 20 to 100 mm are used, the ends fitting into sockets and provided with locking nuts; the construction of the joints must allow for easy re-adjustment during service life.

7.2.11.3 Complex combined bridge system

For highway bridges with wide spans, trusses of combined systems (Fig. 7.58) provide the most efficient solution. With high banks, it is advisable to adopt flexible thrust arches with a bracing truss. A more common design consists of flexible arches with a bracing truss (girder), the chords of which, working under tensile stresses, take

up the thrust of the arch; under bending, the bracing truss mainly resists the temporary one-sided loads.

The effectiveness of adopting combined systems for timber is governed by the following considerations. When the span is increased to 40–50 m, the stresses in the lattice of timber bridge trusses with parallel chords rise considerably, which sometimes create difficulties in the construction of the joints. The transfer to a girder truss with a curved or polygonal outline for the top chord and substantially reducing the amount of stress in the lattice due to dead loads makes the construction of the joints very much easier. These trusses have definite advantages over flat timber girder trusses with parallel chords.

Among their disadvantages are the difficulty of protecting them against decay; the lack of standardisation

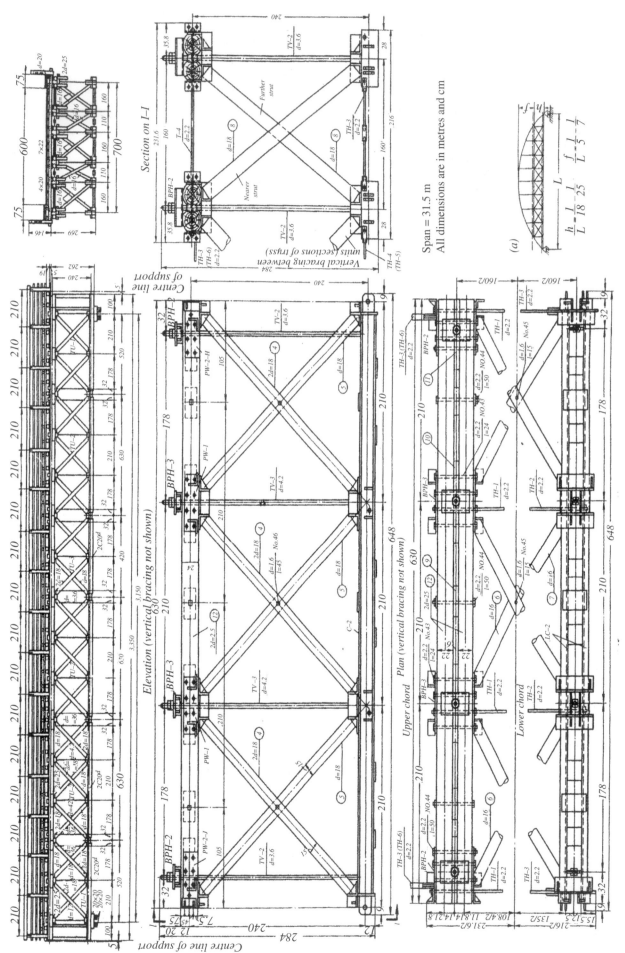

Figure 7.58 *Timber highway bridge (Howe-Zhuravsky system) checked with European Highway Loading (Russian practice).*

in the members and joints; the difficulty of obtaining panels of practicable lengths at the optimum angle of slope of the diagonals; and the excessive length of the lattice members, greater than that for the available timber.

Combined systems give a practical solution for timber spans of about 50 m and over. With all of the advantages of the trusses with a curved upper chord, they have none of the drawbacks indicated above. At the same time, the trusses can be protected effectively against damp and decay by building an impermeable road deck along the upper chord of the bracing truss and with the application of sheathing, with gaps for ventilation, to the arches and trusses.

7.2.12 Timber foot bridges

7.2.12.1 Introduction

Based on BD 29/87, pedestrian foot bridges may be constructed of timber. They are generally used both by pedestrians and cyclists. Timber bridges are designed by both permissible and limit-state methods for timber bridges and, where precise information on vibration is not available due to structural damping, it shall be taken as 0.04. Where the footway is curved in elevation, the slope, either of the bridge structure or adjoining loadings, shall not be steeper than 1 in 20. Where such limitations are unachievable, it should be limited to 1 in 12. Ramps shall not be steeper than 1 in 12. For hand rails and parapets, reference is made to EN 1317 Part 6. For a loading design, standards such as BD 37 (DMRB.1.3.14) are used.

Figure 7.59 gives structural detailing of a cable supported footbridge. Figure 7.60 shows structural detailing of the Aare Bridge in Switzerland.

7.2.12.2 Footbridge loading

Bridges supporting footways or cycle tracks only

Nominal live load

The nominal live load on elements supporting footways and cycle tracks only shall be taken as:

1. for loaded lengths of 30 m and under, a uniformly distributed live load of 5.0 kN/m^2
2. for loaded lengths in excess of 30 m, $k \times 5.0$ kN/m^2, where k is

$$\frac{\text{nominal HA UDL for appropriate loaded length (in kN/m)}}{30 \text{ kN/m}}$$

Special consideration shall be given to the intensity of the live load to be adopted on loaded lengths in excess of 30 m where exceptional crowds may be expected (as, for example, where a footbridge serves a sport stadium).

7.2.12.3 Nominal load on pedestrian parapets

The nominal load shall be taken as 1.4 kN/m length, applied at a height of 1 m above the footway or cycle track and acting horizontally.

7.2.12.4 Design load

For the live load on footways and cycle tracks and for the load on pedestrian parapets, γ_{fL} shall be taken as follows

	For the ultimate limit state	For the serviceability limit state
For combination 1	1.50	1.00
For combination 2 and 3	1.25	1.00

7.2.12.5 Collision load on supports of foot/cycle track bridges

The load on the supports of foot/cycle track bridges shall be a single load of 50 kN applied horizontally in any direction up to a height of 3 m above the adjacent carriageway. This is the only secondary live load to be considered on the bridges.

7.2.12.6 Associated nominal primary live load

Design load

For combination 4, γ_{fL} shall be taken as follows.

For the ultimate limit state	For the serviceability limit state
1.25	1.00

7.2.12.7 Vibration serviceability

Consideration shall be given to vibration that can be induced in foot/cycle track bridges by resonance with the movement of users. The structure shall be deemed to be satisfactory where its response, as calculated in accordance with a specific code such as HA Loading, complies with the limitations specified therein.

7.2.12.8 Elements supporting footways or cycle tracks and a highway or railways

Nominal live load

On footways and cycle tracks carried by elements that also support highway and railway loading, the nominal live load shall be taken as 0.8 of the value specified in any code such as HA Loading or Railway Loading as appropriate, except where crowd loading is expected, in which case, special consideration shall be given to the intensity of live loading to be adopted. Where the footways (or footway and cycle track together) is wider than 2 m, these intensities may be further reduced by 15% on the first metre in excess of 2 m and by 30% on the second metre in excess of 2 m. No further reduction shall be made for widths exceeding 4 m. These intensities may be averaged and applied as a uniform intensity over the full width of the footway.

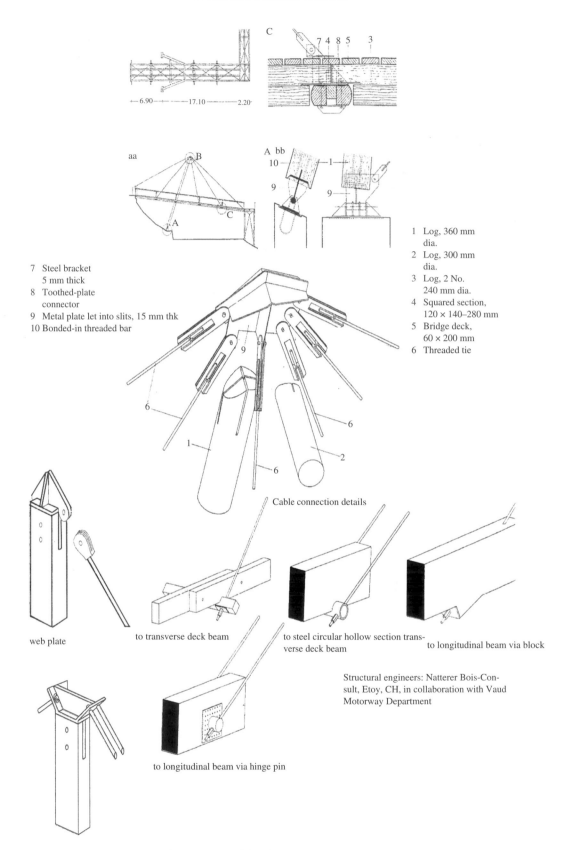

7   Steel bracket
    5 mm thick
8   Toothed-plate
    connector
9   Metal plate let into slits, 15 mm thk
10 Bonded-in threaded bar

1   Log, 360 mm
    dia.
2   Log, 300 mm
    dia.
3   Log, 2 No.
    240 mm dia.
4   Squared section,
    120 × 140–280 mm
5   Bridge deck,
    60 × 200 mm
6   Threaded tie

Cable connection details

web plate

to transverse deck beam

to steel circular hollow section transverse deck beam

to longitudinal beam via block

to longitudinal beam via hinge pin

Structural engineers: Natterer Bois-Con-
sult, Etoy, CH, in collaboration with Vaud
Motorway Department

Figure 7.59 *Cable supported timber footbridge – structural details.*

Kneebrace junctions

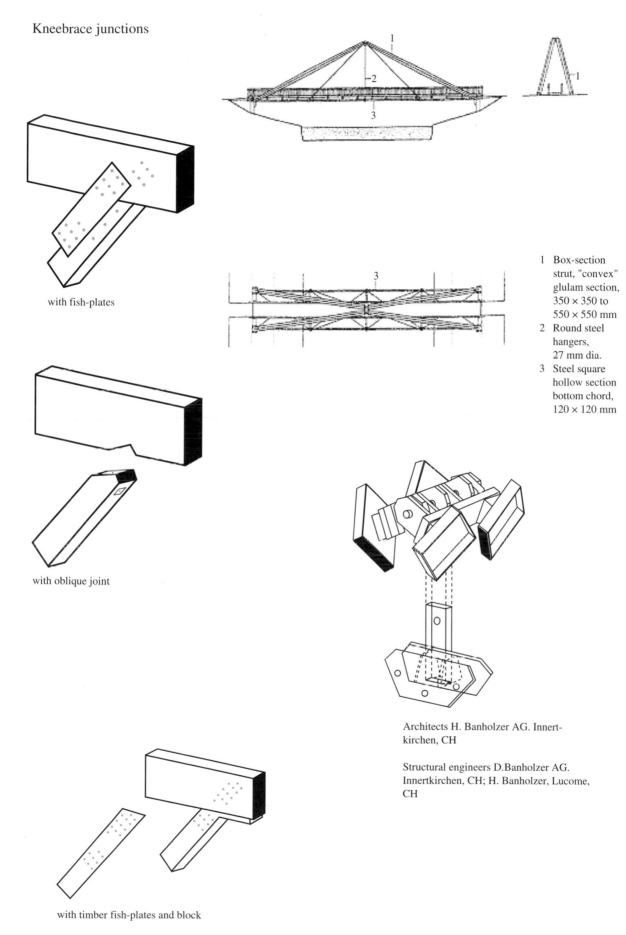

with fish-plates

with oblique joint

1 Box-section
  strut, "convex"
  glulam section,
  350 × 350 to
  550 × 550 mm
2 Round steel
  hangers,
  27 mm dia.
3 Steel square
  hollow section
  bottom chord,
  120 × 120 mm

Architects H. Banholzer AG. Innertkirchen, CH

Structural engineers D.Banholzer AG.
Innertkirchen, CH; H. Banholzer, Lucome,
CH

with timber fish-plates and block

Figure 7.60 *Cable supported timber footbridge – structural detailing, Aare Bridge, Switzerland.*

Where a main structural member supports two or more highways, traffic lanes or railway tracks, the footpath and cycle track loading to be carried by the main member may be reduced to:

On footway: 0.5 of the value given in 7.1.1 (a) and (b) as appropriate

On cycle tracks: 0.2 of the value given in 7.1.1 (a) and (b), as appropriate.

Special consideration shall, however, be given to structures where there is a possibility of crowds using cycle tracks, which could coincide with exceptionally heavy highway loading.

Nominal wheel load

Where the footway or cycle track is not protected from highway traffic by an effective barrier, any four wheels of 25 units of HB loading acting alone in any position shall be considered.

Associated nominal primary live load

Associated nominal primary live load on the carriageway or rail track shall be derived and applied in accordance with 6.4 or clause 8, as appropriate of the HB Loading.

Load due to vehicle collision with parapets

Where the footway or cycle track is not protected from the highway traffic by an effective barrier, the nominal loads on elements of the structure supporting parapets shall be as specified in 6.8 of the HB Loading.

Design load

γ_{fL} to be applied to the nominal loads shall be as follows:

1. For live loading on footways and cycle tracks, as specified in 7.1.4 of the code
2. For highway live loading, as specified in 6.2.7 and 6.3.4 of the code
3. For railway live loading, as specified in 8.4 of the code
4. For loading derived from vehicle collision with parapets, as specified in 6.8 of the code HB loading.

The load-bearing structure is assembled from pressure-impregnated fir logs. The longitudinal beams for the bridge deck were made from edge-sawn logs dowelled together. Larch wood planks were used for the bridge deck. Threaded steel reinforcing bars were used for the cable stays and all wind bracing members.

Selected bibliography

Aicher, S. and Reinhardt, H.W., Joints in Timber Structures, University of Stuttgart S. A. R. L. Publications, 12–14 September 2001.

Aicher, S., Bornschlegl, V. and Herr, J., Numerical and full scale experimental investigations on glulam frame corners with large finger joints, International Conference of IUFRO S5.02, Copenhagen, Timber Engineering, 239–258, 1997.

Alcock N.W. et. al., Recording timber-framed buildings: an illustrated glossary, Practical Handbook in Archaeology 5, Council for British Archaeology, 1996.

Allen, H.G. and Bulson, P.S., Background to Buckling, Maidenhead: McGraw-Hill, 1980.

American Forest & Paper Association, Inc., Technical Report 10: Calculating the Fire Resistance of Exposed Wood Members, Washington, 1999.

American Plywood Association, Plywood Rigid Frames, Tacoma, Washington, 1969.

American Society for Testing and Materials (ASTM), Standard Test Methods for Structural Panels in Tension. D 3500-90. Annual Book of ASTM Standards, Section 4, Vol. 04.10. Wood, West Conshocken, PA: ASTM, 1996.

American Society for Testing and Materials, Annual Book of ASTM Standards, Part 22 Wood, Adhesives. ANSI/ASTM 1, 143–52, reapproved 1978.

Arthur, E. and Witney, D., The barn. A Vanishing landmark in North America, New York: Arrow Wood Press, 1972.

Ashton, L.A., Fire and Timber in Modern Building Design, Timber Research and Development Association, 1970.

Association of Consulting Engineers, Conditions of Engagement. Agreement D: Report and Advisory Services, Second Edition, London: ACE, 1998.

Australian / New Zealand Standard, AS/NZS 1328.2:1998, Glue laminated structural timber, Part2, Guidelines for AS/NZS 1328:Part 1 for the selection, production and installation of glue laminated structural timber, AS/NZS.

Australian / New Zealand Standard, AS/NZS 4634:1996, Adhesives, phenolic and aminoplastic for load bearing timber structures: classification and performance requirements, AS/NZS.

Australian New Zealand Standard, AS/NZS 1328:1998, 1 Glue laminated timber- Part1: Performance requirements and minimum production requirements, AS/NZS.

Australian Standard, AS 1720.1:1997, SAA Timber Structures Code, Part 1-Design Methods.

Avent, R.R., Epoxy Repair of Wood Trusses, Mississippi State University, Engineering and Industrial Research Station, MSU, 1977.

Baird J., Timber Specifies Guide, Oxford: BSP Professional Books, 1990.

Baird, J.A. and Ozelton, E.C., Timber Designers Manual, Second Edition, Blackwell Science (UK), 1984.

Baird, J.A. and Ozelton, E.C., Timber Destkners Manual, London: Granada, 1984.

Barnswell, P.S. and Adams, A.T., The House within. Interpreting medieval houses in Kent, Swindon: RCHM, 1994.

Beadell, S., Windmills, London: Bracken Books, 1972.

Beckell, D. and Marsh, P., An introduction to structural design timber, Surrey University Press, date.

Beckett, D., Bridges, (Great Buildings of the World), Hamlym, 1969.

Beckmann, P., Structural aspects of building conservation, McGraw-Hill International, 1995.

Beech, J.C., The Gluing of Timber Components with Elastomeric Glues, Information Sheet IS 21/76, Princes Risborough: BRE, 1976.

Benny, R.W., Remedial treatment of wood rot and insect attack in buildings, Watford: BRE, 1994.

Bergfelder, J., Näherungsverfahren zur Berechnung hölzerner Biegeträger, Bauingenieur, 49, S. 350–357, 1974.

Berni, C., Bolza, E. and Christensen, F.J., Characteristics, Uses and Properties of 190 South American Timbers, Melbourne, Australia: CSIRO Division of Building Research, 1980.

Blaß, H.-J., Ehlbeck, J., Linden, M.L.R. and Schlager, M., Trag-und Verformungsverhalten von Holz-Beton-Verbundkonstruktionen, Bauen mit Holz, 5, 392–399, 1996.

Blass, H.J., et al., Timber Engineering STEP 1 and STEP 2. Basis of design, material properties structural components and points, The Netherlands: Centrum Hout, 1995.

Bod g, J. and Jayne, B., Mechanics of Wood and Wood Composites, New York: Van Nostrand Reinhold Company, 1982.

Bodig, J. and Benjamin A.J., Mechanics of Wood and Wood Composites, NY: Van Nostrand Reinhold Company, 1982.

Bohannan, W., Affect of Size on Bending strength of Wood Members, US ForestProducts Laboratory, Research Paper FPL 56, 1966.

Bolton, A.J., Humphrey, P.E. and Kavvouras, P.K., The hot pressing of dry-formed wood-based composites, Part VI, The importance of stresses in the pressed mattress and their relevance to the minimization of pressing time and the variability of board properties, Holzforschung, 43, (6), 406–410, 1989.

Bolza, E. and Keating, W.G., African Timbers - The Properties, Uses and Characteristics of 700 Species, Melbourne, Australia: CSIRO Division of Building Research, 1972.

Bolza, E. and Keating, W.G., TM Characteristics, Properties and Uses of 362 Species and Species Groups from South-East Asia, Northern Australia and the Pacific Region, Melbourne, Australia: CSIRO Division of Building Research, 1981.

Booth, L.G. and Reece, P.O., A commentary on the British Standard Code of Practice CP112, E. & F.N. Spon Ltd., 1967.

Boström, L., Isf international RILEM Symposium on Timber Engineering. 5 NTRI/SARI Publications. Sopf. 13–14 1999. Stockholm Sweden.

British Standard Institution, BS EN 301:1992, Adhesives, phenolic and aminoplastic for load bearing timber structures: Classification and performance requirements, London: BSI.

British Standard Institution, BS EN 302:1992, Adhesives for load bearing timber structures: Test methods, London: BSI.

British Standard Institution, BS EN 386:1995, Glue laminated timber – Performance requirements and minimum production requirements, London: BSI.

British Standard Institution, BS EN 391:1995, Glue laminated timber- Delamination test of glue lines, London: BSI.

British Standard Institution, BS EN 392:1995, Glue laminated timber – Shear test of glue lines, London: BSI.

British Standards Institution, The structural use of timber Part 2. Metric units, CP 112, BSI, 1971.

Brock, G.R., The Strength of Nailed Joints, Forest Products Research Bulletin No. 41, London: HMSO, 1957.

Bromwell, M. (Editor), The international book of wood, Mitchells Building Series, Longman Scientific and Technical, 1976.

Brown, R.J., Timber-framed buildings of England, London: Robin Hale, 1986.

Brunskill, R.W., Illustrated handbook of vernacular architecture, London: Faber and Faber, 1971.

de Bruyne, N.A., et al., The theory and practice of gluing with synthetic resins, Large, Maxwell and Springer, date.

BSI, Extracts from British Standards for students of structural design, Fourth Edition, BSI, 1998.

Buchanan, A.H. and Lai, J.C., Glulam rivets in pine, Canadian Journal of Civil Engineering, 21, (2), 340–350, 1994.

Buchanan, A.H. and Moss, P., Design of epoxied steel rods in glulam timber, Proceedings of Pacific Timber Engineering Conference, Rotorua, New Zealand, March 1999, Forest Research Bulletin, 212, (3), 286–293, 1999.

Buchanan, A.H., Burning issues in timber engineering, Proceedings of Pacific Timber Engineering Conference, Rotorua, New Zealand, March, Forest Research Bulletin, 212, 1–11, 1999.

Building Research Establishment, A Handbook of Softwoods, London: HMSO, 1977.

Building Research Establishment, Choice of Glues for Wood, Digest No. 175, Princes Risborough:BRE, 1975.

Building Research Establishment, Technical Note No 25, Preservative Treatment of External Joinery Timber, Princes Risborough: Building Research Establishment, date.

Building Research Station, The collapse of a precast concrete building under construction. Technical statement by the Building Research Station, London: HMSO, 1963.

Burgess, H.J. and Masters, M.A., Span Charts for Solid Timber Beams (revised), Timber Research and Development Association, Publication No. TBL 34, 1976.

Burgess, H.J. and Peek, J.D., Span charts for solid timber beams, Timber Research and Development Association, 1968. *Now called the Princes Risborough Laboratory of the Building Research Establishment (see Acknowledgements).

Burgess, H.J., Collins, J.E. and Masters, M.A., Use of the TRADA Universal SpanChart for a Range of Load Cases, Timber Research and Development Association, Publication No TBL 47, 1972.

Burgess, H.J., Further Applications of TRADA Span Charts. Timber Research and Development Association, Publication No. TBL 42, 1971.

Burgess, H.J., Introduction to the design of ply-web beams, TRADA information bulletin E/1B/24, date.

Burgess, H.J., Johnson, V.C. and Mettem, C.J., Span Tables for Ridged Farm Portals in Solid Timber, Timber Research and Development Association, Publication No. E/IB/17, 1970.

Burgess, H.J., Joist Span Tables for Domestic Floors and Roofs to BS 5268, Timber Research and Development Association, Publication No. DA 6, 1985.

Burgess, H.J., Limit state design applied to mechanical fasteners. Timber Research and Development Association (unpublished), 1977. Burgess, H.J., Ties with lateral load, Proceedings, Meeting of International Union of Forestry Research Organizations, Oxford: Wood Engineering Group, 1980.

Burgess, H.J., Ridged Portals in Solid Timber, Timber Research and Developing Association Publication No. E#B/18, 1970.

Burgess, H.J., Span Tables for Floor Joists to BS 5268, Timber Research and Development Association, Publication No. DA 3.84, 1984.

Burgess, H.J., Span Tables for Floor Joists to BS 5268: Part 2 1984, Processed Timber Sizes, Strength Classes SC3 and SC4, Timber Research and Development Association Publication No. DA 3.84, 1976.

Burgess, H.J., Strength grouping of Malayan timbers, Malayan Forester, 33–36, 1956.

Cagle, C.V. (Editor), Handbook of Adhesive Bonding, Chapter 17, New York: McGraw-Hill, 1973.

Canadian Standards Association, Code for Engineering Design in Wood, CAN-086-M80, Ontario, Canada, 1980.

Canadian Standards Association, Douglas Fir Plywood, CSA 0121-M, Ontario, Canada, 1978.

Canadian Wood Council, Introduction to wood design, Ottawa, Ontario, 1999.

Canadian Wood Council, Wood Design Manual, Ottawa, Ontario, 1995.

Casilla, R.C. and Chow, S.Z., Press-Time Reduction by Preheating and Strength Improvement by Finger-Jointing Laminated Veneer Lumber, Forest Prod. J., 29, (11), 30–34, 1979.

Ceccotti, A. and Vignoli, A., The effects of seismic events on the behaviour of semi-rigid joint timber structures, Proceedings of the 1988 International Conference on Timber Engineering, 1, Madison, USA: Forest Products Research Society, 823–831, 1988.

Ceccotti, A., Holz-Beton-Verbundkonstruktion, Holzbauwerke nach EC 5, STEP 2, Bauteile, Konstruktion, Details, 1995.

CFI, Fir plywood web beam design, Templar House, 81–87 High Holborn, London WC1X 6L5, date.

Champman, S.J., MATLAB Programming for Engineers, Thomsons Learning, 1999.

Charles, F.W.B. with Charles, M., Conservation of timber buildings, Cheltenham:Stanley Thomas (Publishers) Ltd, 1984.

Chugg, W.A. and James, P.E., The Gluability of Hardwoods for Structural Purposes, Timber Research and Development Association, Research Report 2/R11/22, 1965.

Chugg, W.A., Glulam: the Theory and Practice of the Manufacture of Glued Laminated Timber Structures, London: Ernest Benn, 1964.

Chu Yue Pun, The relation of strength properties of wood to its specific gravity, Malaysian Forester, 38,(4), 284–93, 1975.

CIB-WI8 Timber Structures, Structural Timber Design Code, CIB Publication 66, Rotterdam: International Council for Building Research Studies and Documentation, 1983.

Clifton-Taylor, A., The pattern of English building, London: Faber and Faber, 1972.

Coates, R.C., Coutie, M.G. and Kong, F.K., Structural Analysis, 3rd Ed., UK: Van Nostrand Reinhold, 1988.

Cockcroft, R., Timber Preservatives and Methods of Treatment, Timberlab Paper No 46, Princes Risborough: Building Research Establishment, 1971.

Cocke, T. et al., Recording a church: an illustrated glossary. Practical Handbook in Archaeology 7, Council for British Archaeology, 1996.

Condit, C.W., American Building Art – The Nineteenth Century, New York: Oxford University Press, 1960.

Cook, R.D., Malkus, D.S. and Plesha, M.E., Concepts and Applications of Finite Element Analysis, Third Edition, John Wiley & Sons, Inc., 1989.

Cooper, B.M., Writing technical reports, Middlesex: Penguin Books, 1964.

Core, H.A., Costs, W.A. and Day, A.C., Wood Structure and Identification, Syracuse, USA: Syracuse University Press, 1979.

Corkhill, T., A glossary of wood, London: Stobart Davis, 1979.

Council of Forest Industries of British Columbia (CFI), 44 mm Hem-Fir, Canadian Metric Timber for the British Builder, date.

Council of Forest Industries of British Columbia, 44 mm Hem-Fir, Templar House, 81–87 High Holborn, London WC1X 6L5, date.

Council of Forest Industries of British Columbia, Canadian COFI Exterior Plywood, Vancouver, BC, Canada, 1979.

Council of Forest Industries of British Columbia, Canadian Fir Plywood Manual, date.

CSA, Engineering design in wood (limited states design), CSA 086.1-1994, Rexdale (Toronto), Ont.: Canadian Standards Association, 1994.

Curry, W.T., Interim stress values for machine stress-graded European Redwood and Whitewood, Timberlab paper No. 21, Princes Risborough, Bucks: Forest Products Research Laboratory, 1969.

Curry, W.T., Permissible spans for canvassing timbers, The Builder, Sept. 7–Oct. 19, 1962.

Curry, W.T., Working Stresses for Structural Laminated Timber, Forest Products Research, Special Report 15, London: HMSO, 1961.

De Vos, P., Draft Firecode—Performance-based, Boutek, CSIR, Pretoria, 1999.

Department of the Environment, Transport and the Regions (2000). Building Regulations 2000, London: HMSO, 2000.

Desch, H.E. (revised Dinwoodie, J.M.) Timber, its structure, properties and utilisation, Sixth Edition, Hamphire/London: Macmillan Education, 1981.

Desch, H.E., rev. Dinwoodie, Timber: its Structure, Properties and Utilization, Sixth Edition, London: Macmillan, 1981.

Desh, H.E. and Dinwoodie J.M., Timber: structure, properties, conversion and use, Seventh Edition, Macmillan Education, 1996.

Dhima, D., Vérification expérimentale de la résistance au feu des assemblages déléments en bois, CTICM, Ref. INC – 99/399 – DD/NB, 1999.

Edlin, H.L., The Forestry Commission, A Review of Progress to 1971, HMSO, 1971.

Edlin, H.L., Timber! Your Growing Investment, Forestry Commission, HMSO, 1969.

Ehlbeck, J.-Görlacher, Determination perpendicular-to-grain tensile stresses in joints with dowel-type fasteners, East Berlin: International Council for Building Research, W18-Timber Structures, 1988.

EN 10025, Warmgewalzte Erzeugnisse aus unlegierten Baustahlen – Technische Lieferbedinggungen, 1993.

EN 1363-1, Feuerwiderstandprufungen – Allgemeine Anforderungen, 1999.

EN ISO 4016, Sechskantschrauben mit Schaft – Produktklasse C, 2000.

Enger, K. and Kolb, H., Investigations on the ageing of glues for supporting structural wooden elements, Otto-Graf-Institute, Technical University, Stuttgart, date.

ENV 1995-1-1, Eurocode 5 Design of timber structures, Part 1-1 General rules and rules for buildings, Brussels: CEN, 1993.

ENV 1995-1-1.

ENV 1995-1-2, Eurocode 5 Design of timber structures, Part 1-1 General rules and rules for

ENV 1995-1-2.

Erfurth, T. and Rusche, H., The Marketing of Tropical Wood: A – Wood species from African tropical moist forests, Rome: Food and Agriculture Organization of the United Nations, 1976a.

Erfurth, T. and Rusche, H., The Marketing of Tropical Wood: B – Wood species from South American tropical moist forests, Rome: Food and Agriculture Organization of the United Nations, 1976b.

Erki, M.A., Modelling the load-slip behaviour of timber joints with mechanical fasteners, Canadian Journal of Civil Engineering, 18(4), 607–616, 1991.

Eurocode 5, First Draft EN 1995-1-1 – Design of Timber Structures – Part 1.1. General rules for buildings, date.

Everett, A., Mitchells building series: materials, Fifth Edition, Harlow: Longman Scientific and Technical, 1970.

Fantozzi, J. and Humphrey, P.E., Effects of bending moments on the tensile performance of multiple-bolted connectors, Part I, A technique to model such joints, Wood and Fiber Science, 27, (1), 55–67, 1995.

FAO, Appendices from The Marketing of Tropical Wood: C – Wood species from South-East Asian tropical moist forests (unpublished), Rome: Food and Agriculture Organization of the United Nations, 1976.

Farmer, R.H. (Editor), Handbook of Hardwoods, 2nd Edition (revised), London: HMSO, 1972.

Fielden, B., Conservation of historic buildings, London: Butterworth Scientific, 1982.

Fishlock, M., The great fire at Hampton Court, London:The Herbert Press Ltd, 1993.

Foliente, G., Hysteresis Modelling of Wood Joints and Structural System, ASCE Journal of Structural Engineering, 121, 1013–1022, 1995.

Fonather, J. et al., Versuchsbericht Kleinbrandversuche 1 Teil 1 (KBV 1/1), Wien: Institut für Konstructiven Ingenieurbau-Universität für Bodenkultur, 2000.

Forest Products Laboratory, Feasibility of producing high yield laminated structural products, Res. Pap. FPL-175, Madison, WI: USDA Forest Service, Forest Products Laboratory, 1972.

Forest Products Research Laboratory, Gluing Preservative-treated Wood, Technical Note No. 31, 1968.

Forman, W., Oxfordshire mills, Sussex: Phillimore, 1983.

Fornather, J. and Bergmeister, K., Versuchsbericht Kleinbrandversuche 1 Teil 2 (KBV ½), Wien: Institut für Konstructiven Ingenieurbau – Universität für Bodenkultur, 2001.

Fornether, J. et al., Versuchsbericht Kleinbrandversuche 2 Teil 1 (KBV 2/1), Wien: Institut fur Konstructiven Ingenieurbau – Universität fur Bodenkultur, 2001.

Foschi, R.O. and Longworth, J., Analysis and design of Griplam nailed connections, ASCE Journal of the Structural Division, 101(ST12), 2537–2555, 1975.

Foschi, R.O., Load-slip characteristics of nails, Wood Science, (71), 69–76, 1974.

Foster, R., Discovering English churches, London: British Broadcasting Corporation, 1981.

Fox, S.P. and Lincoln, R.G., Effect of plate and hole size on Griplam-nail load capacity, Canadian Journal of Civil Engineering, 6, (3), 390–393, 1979.

Freas, A.D. and Selbo, N.L., Fabrication and Design of Glued Laminated Wood Structural Member, US Department of Agriculture, Technical Bulletin No. 1069, 1954.

Freudenthal, A.M., The Safety of Structures, Proceedings of the American Society of Civil Engineers, October, 1945.

Gehri, E., Zur Berechnung und Bemessung von Fachwerkträgern mit Konterplatte in eingeschlitzten Hölzern, Schweizer Ingenieur und Architekt, H.25, 1983.

Ghugg, W.A., Glulam, Ernest Benn Ltd., 1964.

Giles, C. and Goodall, I.H., Yorkshire textile mills, London: HMSO, 1992.

Goetz, K.-H., Hoor, D., Moehler, K. and Natterer, J., Timber Design & Construction Sourcebook, McGraw-Hill, 1989.

Gordon, J.E., Structures, or why things don't fall down, Middlesex: Penguin Books, 1978.

Gordon, J.E., The new science of strong materials, or why you dont fall through the floor, Second Edition, Middlesex: Penguin Books, 1976.

Gower, Sir E., The complete plain words. Revised Edition by Sir Bruce Fraser, London: HMSO, 1973.

Grainger, G.D., Lateral distribution of a concentrated load on timber floor panels, Civil Engineering, Apr., 91–4, 1969.

Hall, F.A., Glued Laminated Timber Structures, Rainham Timber Engineering Co Ltd., date.

Harlt, H., Behaviour of timber and wood-based materials in fire (Lecture A13). STEP Timber Engineering, 1st Ed., The Netherlands: Centrum Hout, 1995.

Harlt, H., Fire resistance of timber members (Lecture B17). STEP Timber Engineering, 1st Ed., The Netherlands: Centurum Hout, 1995.

Harris, R., Discovering timber-framed buildings, Buckinghamshire: Shire Publications Ltd., 1978.

Harrison, H.W., Roofs and roofing-performance, diagnosis maintenance, repair and the avoidance of defects, Watford: BRE, 1996.

Haslam, J.M., Writing engineering specifications, London: Spon, 1988.

Heimeshoff, B., Berechnung von Rahmenecken mit Keilzinkenverbindungen Holzbaustatik aktuel, 7–8, 1976.

Hewett, C.A., Church carpentry. A study based on Essex examples, London: Phillimore, 1982.

Hewett, C.A., English cathedral and monastic carpentry, London: Phillimore, 1985.

Hewett, C.A., English historic carpentry, London: Phillimore, 1980.

Hibbit, Karlsson and Sorensen, Inc., ABACUS Finite Element Computer Program, 1998.

Hoadley, R.B., Understanding wood. A craftsmans guide to wood technology, The Taunton Press, 1980.

Howard, E.T. and Manwiller, F.G., Anatomical characteristics of southern pine stemwood, Wood Science, (22), 77–86, 1969.

Howard, F.E. and Crossley, F.H., English church woodwork. A study in craftsmanship during the medieval period, AD 1250–1550, London: Batsford, 1917.

Hsu, W.E., Laminated Veneer Lumber From Aspen, Proc. 22nd Inter. Particleboard/Composite Symp. Washington State University, Pullman, WA, 257–269, 1988.

Hughes, D.H. and Rajakaruna, M., Polyurethane adhesives for glulam timber in Australia, Proceedings of the International Conference on Wood and Wood Fiber Composites, Otto-Graf-Institute, University of Stuttgart, Stuttgart, Germany, 387–398 2000.

Hume, I., Office floor loading in historic buildings, English Heritage, 1994.

Humphery, P.E., A device to test adhesive bonds. US Patent number 5,170,028 (and other patents), Washington, DC. USA: US Patent Office, 1993.

Humphery, P.E., Adhesive bond-strength development in pressing operations: some considerations. Proc. First European Panel Products Symposium Llandudno, Wales, UWB, Bangor, LL57 2UW: UK BioComposites Centre, 145–155, 1997.

Humphery, P.E., Engineering composites from oriented natural fibres: a strategy, Proc. Second Pacific Rim BioBased Composites Conference. Vancouver, BC. Canada, 1994.

Humphery, P.E., Thermoplastic characteristics of partially cured thermosetting adhesive-to-wood bonds: the significance for wood-based composite manufacture. Proc. Third Pacific Rim BioBased Composites conference, Kyoto, Japan, 1960.

International Conference of Building Officials, Uniform Building Code, 5360 South Workman Mill Road, Whittier, California, 1979.

International Truss Plate Association, Roof Bracing for Fink and Fan Trusses, PO Box 44, Halesowen, West Midlands, UK, 1974.

ISO/WD 16670, Timber Structures-Joints made with mechanical fasteners-Quasi static reversed cyclic test method, ISO, 2000.

Ivester, F.D., Visual stress grading of timber, Timber Research and Development Association, 1969.

Jablonkay, P., Schrauben unter Ausziehbeanspruchung, Master Thesis, Zurich: Professuz für Holztechnologie, ETH Zürich, 1999.

James, J.G., The evolution of wooden bridge trusses to 1850, J1 IWSC 9, (3,4), 116–35, 168–93, 1982.

Jansson, I., Nailed joints with plywood, Modern Timber Joints, Timber Research and Development Association, 1968.

Jensen, J.L., et al., Axially loaded glued-in hardwood dowels, Wood Science and Technology, 35, (1/2), 73–83, 2001.

Jensen, J.L., Koizumi, A. and Sasaki, T., Timber joints with glued-in hard-wood dowels, Proceedings of Pacific Timber Engineering Conference, Rotorua, 100–107, 1999.

Jensen, J.L., Sasaki, T. and Koizumi, A., Plywood frame corner joints with glued-in hardwood dowels, Journal of Wood Science (to be published).

Johansen, K.W., Tests with timber joints (in Danish), Report No. 19, Lyngby: Laboratory for Building Statics, Technical University of Denmark, 1941.

Johansson, C.J., et al., Delamination test of glue lines according to EN391, Nordtest Project 1089-93, Swedish Testing and Research Institute, SP Report 67, 1994.

Johnson, V.C., Mettem, C.J. and Withers, J.F., Span Tables for Industrial Purlins, Information Bulletin N0. E/1B/27, Timber Research and Development Association, 1977.

Joint Committee RILEM/CIB-3TT, Testing methods for plywood in structural grades for use in load-bearing structures, Materials and Structures, 11, (62), 45–64, 1978a.

Joint Committee RILEMXIB-I|IT, Testing methods for timber in structural sizes, Materials and Structures, 11, (66), 445–52 a, 1978b.

Jung, J., Investigation of the Fracture of Butt Joints in Parallel-Laminated Veneer, Wood and Fiber Science, 15, (2), 116–134, 1984??1983??.

Jung, J., Investigation of various end joints in parallel-laminated veneer, Forest Prod. J., 34, (5), 51–55, 1984.

Kaminetzky, D., Design and Construction Failures: Lessons from forensic investigations, McGraw-Hill, 1991.

Kanócz, J.and Ducko, P., Experimentálne vyšetrovanie klincových spojkových spojov drevených konštrukcií, Inžinierske stavby, c˘.9–10, roc˘ník 43, 1995.

Kanócz.[initial] Únosnost spojkových spojov drevených konštrukcií a komplexná analýza priec˘ ne zataženého klinka, Košice: Habilitac˘ ná práca, 1998.

Karacabeyli, E. and Foschi, R.O., Glulam rivet connections under eccentric loading, Canadian Journal of Civil Engineering, 14, (5), 621–630, 1987.

Kasal, B. and Xu, H., A nonlinear nonconservative finite element with hysteretic behaviour and memory, 8th International ANSYS Conference, 3.235–3.242, date.

Kerr, J.S., The Conservation Plan, Fourth Edition, Sydney: National Trust of Australia, 1996.

Kidson, P., Murray, P. and Thompson, P., A history of English architecture, Middlesex: Penguin Books, 1965.

Kloot, N.H. and Schuster, R.B., Load Distribution of Wooden Floors Subjected to Concentrated Loads, Technological paper No. 29, Melbourne, Australia: CSIRO Division of Forest Products, 1963.

Kloot, N.H., The strength group and stress grade systems, Forest Products Newsletter, CSIRO, Sept.–Oct.,1–12, Melbourne, Australia: Forest Products Laboratory, 1973.

Kneidl, R. and Hartmann, H., Träger mit nachgiebigen Verbund-Eine Berechnung mit Stabwerksprogrammen, Bauen mit Holz, 5, 285–290, 1995.

Koizumi, A., et al., Moment-resisting properties of post-to-sill joints connected with hardwood dowels, Mokuzai Gakkaishi, 47, (1) 14–21, 2001.

Koizumi, A., et al., Withdrawal properties of hardwood dowels glued perpendicular to the grain, Mokazat Gakkaishi, 45, (3), 230–236, 1999.

Koizumi, A., et al., Withdrawal properties of hardwood dowels in end joints (1), Mokuzal Gakkaishi, 44, (1), 41–48, 1998.

Koizumi, A., et al., Withdrawal properties of hardwood dowels in end joints (2), Mokuzai Gakkdtsht, 44, (2), 109–115, 1998.

Kolb, H., Festigkeitsuntersuchungen an gestoßenen und gekrümmten Bauteilen aus Brettschichtholz, Holz als Roh-und Werkstoff, Bd.26, 243–253, 1968.

Kolb, H., Versuche an geleimten Rahmenecken und Montagestößen 2.Tekil, Reprint from Otto-Graf-Institute Publication, 1–10, 1968.

Kolllhann, F.F.P., Kuenzi, E.W. and Stamm, A.J., Principles afford Science and Technology: 11. Wood-based materials, Berlin: Springer-Verlag, 1975.

Kollmann, F.F.P. and Costs, W.A., Principles of Wood Science and Technology: 1. Solid wood, London: Allen and Unwin, 1968.

Komatsu, K., et al., Flexural behaviour of GLT beams end-jointed by glued-in hardwood dowels, Proceedings of CIB-W18 30th Meeting, Vancouver, 1–8, 1997.

König, J. and Walleij, L., One-demensional charring of timber exposed to standard and parametric fires in initially unprotected and postprotection situations, Trätec – Swed. Inst. For Wood Techn. Res., Report 19908029, 1999.

Kretschmann, D.E., Moody, R.C., Pellerin, R.F., Bendtsen, B.A., Cahill, J.M., McAlister, R.H. and Sharp, D.W., Effect of Various Proportions of Juvenile Wood on Laminated Lumber, FPL. Research Paper FPL-RP-521., U.S. Department of Agriculture Forest Service, 1993.

Kreuzinger, H. and Scholz, A., Wirtschaftliche Ausführungs-und Bemessungsmethoden von ebenen Holzelementen (Brücken, Decken, Wände), Zwischenbericht zu Forschungsprojekt, Technische Universität München, 10/98.

Kreuzinger, H., Träger und Stützen aus nachgiebig verbundenez Querschnittsteilen; Arbeitsgemeinschaft Holz e. V. (Hrsg.): Step 1: Holzbauwerke nach EC 5, Bemessung und Baustoffe, 1. Aufl. Fachverlag Holz, Dusseldorf, S, B11/1-B11/9, 1995.

Kruger, T.S. and Van Rengsburg, B.W.J., The matrix stiffness method adapted to incorporate semi-rigid connections, Journal of the S. A. Institution of Civil Engineers, 35 (4), 1993.

Kruger,T.S., Van Rensburg, B.W.J. and Du Plessis, G.M., Non-linear analysis of structural steel frames, Journal of Construction Steel Research, 34, 285–306, 1995.

Kruppa, J., Lamadon, T. and Rachet, P., Fire resistance tests of timber connections, CTICM, Ref. INC – 00/187 – JK/NB, 2000.

Laidlaw, W.B.R., Guide to British hardwoods, London: Leanard Hill (Books) Limited, 1960.

Laminated Timber Institute of Canada, Timber Design Manual (metric amyl), Ottawa, Canada, 1980.

Lancaster, O., Pillar to post and Home sweet homes, London: John Murray, 1938, reprinted 1963.

Langlands, I., The mechanical properties of South Australian plantation-grown Pinus Radiata Pamph, Coun. Scient. Ind. Res. Aust., No. 87, 1938.

Larson, D.S., Sandberg, L.B., Laufenberg, T.L., Krueger, G.P. and Rowlands, R.E., Butt Joint Reinforcement in Parallel-Laminated Veneer (PLV) Lumber, Wood and Fiber Science, 19, (4), 414–429, 1987.

Laufenberg, T.L., Exposure Effect upon Performance of Laminated Veneer Lumber and Glulam Materials, Forest Prod. J., 32, (5), 42–48, 1982.

Laufenberg, T.L., Parallel-laminated Veneer: Processing and Performance Research Review, Forest Prod. J., 33, (9), 21–28, 1983.

Lavers Gvendoline, M., The strength properties of timber, Ministry of Technology Forest Products Research Bulletin No. 50, London: HMSO, 1969.

Lavers, G., The strength properties of timber, Third Edition, Revised by G. L. Moore, Also 1997 supplement, London: BRE, 1983.

Leacroft, R., The development of the English playhouse, London: Eyre Methuen Ltd, 1973.

Lee Yew Hon, et al., The strength Properties of Some Malaysian Timbers, Trade Leaflet No. 34, Kuala Lumpur, Malaysia: Malaysian Forest Service, 1965.

Lee, J.N., Structural Behavior of Yellow-Poplar Laminated Veneer Lumber. Effect of Veneer-Joint Designs and Environmental Conditions, Ph.D. Dissertation, Auburn University, Auburn, Alabama, 1999.

Lee, Jong N., Tang, R.C. and Kaiserlik, J.H., Edgewise Static Bending Properties of Yellow-Poplar Laminated Veneer Lumber: Effect of Veneer-Joint Design, Forest Prod., J., 49, (7/8), 64–70, 1999.

Leicester, R.H. and Keating, W.G., Strength Classifications for Timber Engineering Codes. International Council for Building Research Studies and Documentation, Working Commission W18-Timber Structures, Oct. 1979, Bordeaux, France, 1979.

Leijten, A.J.M., An effective timber joint developed at TU-Delft, Proceedings of Pacific Timber Engineering Conference, Rotorua, New Zealand, March 1999, Forest Research Bulletin, 212, (3), 253–258, 1999.

Leijten, A.J.M., Steel reinforced joints with dowels and bolts, Proceedings of the 1988 International Conference on Timber Engineering, 2, Madison, USA: Forest Products Research Society, 474–488, 1988.

Libby, E.F., Radiocarbon dating, Second Edition, Chicago: University of Chicago Press, 1955.

Macandrews and Forbes Ltd., Design manual for timber connector construction, date.

Mack, J.J., The load-displacement curve for nailed joints, Journal of the Institute of Wood Science, 7, (6), 34–6, 1977.

Madsen, B., Behaviour of timber connections, North Vancouver, B.C.: Timber Engineering Ltd, 2000.

Madsen, B., Behaviour of Timber Connections, Vancouver: Timber Engineering Ltd, 2000.

Malaysian Timber Industry Board, The Malaysian Grading Rules for Sawn Hardwood Timber, Kuala Lumpur, Malaysia, 1968.

Margquis-Kyle, P. and Walker, M., The illustrated Burra charter: making good decisions about the care of important places, Australia ICOMOS 1992, UK: ICOMOS, 1992.

Mark, R.E., Adams, S.F. and Tang, R.C., Moduli of rigidity of Virginia pine and Tulip poplar related to moisture content, Wood Science, 2, (4), 203–211, 1970.

Masafumi INOUE et al., Experimental Study on Rigidity and strength of Joints Composed by Metal connector and Adhesive in Timber Structures, Proceedings of Pacific Timber Engineering Conference, Vol. 2, Gold Coast, Australia, 802–812, 1994.

Masafumi INOUE et al., Introduction of Timber Structures Constructed by New Connecting System, Proceedings of International Conference on Effective Utilization of Plantation Timber 99, Chi-Tou, R.O.C, 555–560, 1999.

Masafumi INOUE et al., Strength of Heavy Timber Frame Jointed by Metal Connector and Adhesive, Proceedings of 5th World Conference on Timber Engineering, Vol. 2, Montreux, Switzerland, 830–831, 1998.

Masafumi INOUE et al., Strength of Wooden Frames Jointed by Metal Connector and Adhesive, Proceedings of International Wood Engineering Conference, 6, Vol. 4, New Orleans, U.S.A., 65–72, 1996.

Masafumi INOUE et al., Strengthening Method for Wooden Frames with Brace, Proceedings of International Conference on Effective Utilization of Plantation Timber 99, Chi-Tou, R.O.C, 524–529, 1999.

Masafumi INOUE et al., Structural Behavior of Box Beam Jointed by New Connecting System, Proceedings of Pacific Timber Engineering Conference 99, Vol. 3, Rotorua, New Zealand, 123–130, 1999.

Masafumi INOUE et al., Structural behaviour of the new type of beam-column connection in timber structure, Proceedings of World Conference on Timber Engineering 2000, Whistler, Canada, P02-1-8, 2000.

McGowan, W.M., A nailed plate connector for glue-laminated timbers, Journal of Materials, 1(3): 509–535, 1966.

McKay, W.B., Building construction, London: Longmans, 1994 and later editions.

Measurement Group, Inc., Technical note: Errors due to transverse sensitivity in strain gages, TN-509, Raleigh, NC: Vishay Measurement Group, Inc., 1993.

Melville, I.A. and Gordon, I.A., The repair and maintenance of houses, London: Estates Gazette, 1973.

Mettem, C.J., The Influence of Specific Gravity on the Strength and Stiffness of Nailed Joints, Timber Research and Development Association, Research Report; E/RR/38, 1977.

Ministry of Technology, Symposium No 3, Fire and Structural use of Timber in Building, 1967.

Mirov, N.T., The Genus Pinus, New York: Ronald Press, 1967.

Moroney, M.J., Facts from figures, Penguin Books, date.

Morris, E.N., Calculation of Species Factor for Nailed Joints with Plywood Members under Short-term Loading, Department of Architecture and Building Science, University of Strathclyde, 1972.

National Building Agency – NBA Metric Component File, NBA, 1971.

National Building Studies – Special Report No 36, Prefabrication – A History of its Develpoment in Great Britain, HMSO, 1965.

National Design Specification for Wood Construction, 1997 edition, Washington, DC: American Forest and Paper Association, 1997.

National Lumber Grades Authority, Standard Grading Rules for Canadian Lumber, British Columbia, Canada: National Lumber Grades Authority, 1979.

Natrella, M.G., Experimental Statistics, US Department of Commerce, National Bureau of Standards Handbook, 1963.

Natterer, J., Concepts and details of mixed timber-concrete structures, Composite Construction - conventional and innovative, Conference Report, Innsbruck, 175–180, 1997

Newlands, J., The carpenter and joiners assistant, Glasgow: Blackie, 1857, Reprinted, London: Studio Editions, 1990.

Nicolson, A., Restoration – The rebuilding of Windsor Castle, London: Michael Joseph, 1997.

Niskanen, F., On the Strength and Elasticity Characteristics of Finnish Structural Birch Plywood, Finnish State Institute for Technical Research, Publication No. 81, 1963.

Norén, J., Load-bearing capacity of nailed joints exposed to fire, Fire and Materials, 20, 133–143, 1996.

Orica Chemicals, Material Safety Data Sheet, 1998.

Parsons, W.B., Engineers and Engineering in the Renaissance, MIT Press, 1968.

Pearce, D., Conservation today, London: Routledge, 1989.

Pearson, R.G., Kloot, N.H. and Boyd, J.L., Timber Engineering Design Handbook, Brisbane, Australia: Jacaranda Press, 1962.

Pearson, R.G., The Establishment of Working Stresses for Groups of Species, Technological Paper No. 35, Melbourne, Australia: CSIRO Division of Forest Products, 1965.

Perkins, N.S., Landsem, P. and Trayer, G.W., Modern Connectors for Timber Construction, US Department of Commerce and US Department of Agriculture, Forest Service (unnumbered publication), 1933.

Perkins, N.S., Plywood Properties, Design and Construction, Tacoma, Washington: Douglas Fir Plywood Association, 1962.

Peters, J.E.C., Discovering traditional farm buildings, Shire Publications Ltd., 1981.

Pevsner, N., An outline of European architecture, Middlesex: Penguin Books, 1943.

Phillips, R., Tress in Britain, Europe and North America, London: Pan Books Ltd., 1978.

Pierce, C. B., The Weibull Distribution and the Determination of its Parameters for Application to Timber Strength Data, Building Research Establishment Current Paper, CP 26/76, 1976.

PMBC, Fir plywood design fundamentals and physical properties, PMBC, United Kingdom Office, Templar House, 81–87 High Holborn, London WC1X 6L5, date.

PMBC, Fir plywood stressed skin panels, PMBC, Templar House, 81 High Holborn, London WC1X 6L5, date.

PMBC, Plywood folded plate design, PMBC, Templar House, 81–87 High Holborn, London WC1X 6L5, date.

Powell-Smith, V. and Billington, M.J., The Building Regulations explained and illustrated, Tenth Edition, Blackwell Science, 1995.

Pratt, G.H., Timber-drying Manual. Department of the Environment, Building Research Establishment, London: HMSO, 1974.

prEN 1995-1-2, Second Draft Eurocode 5 Design of timber structures, Part 1-1 General rules and rules for buildings, CEN, 2001.

Princes Risborough Laboratory, Gluing Preservative-treated Wood. Technical Note No. 31, 1968.

Princes Risborough Laboratory, The Movement of Timbers, Technical Note No.38, 1976.

Princes Risborough Laboratory, The Strength Properties of Timber, MTP |7 ..) Construction, Lancaster by 1974)

Prokon, Structural Analysis & Design – Users Guide, Putney, London: Prokon Software Consultants, 2000.

Pu, J.H. and Tang, R.C., Non-destructive Evaluations of Modulus of Elasticity of Southern Pine LVL: Effect of Veneer Grade and Relative Humidity, Wood and Fiber Science, 29, (3), 249–263, 1997.

Quiney, A., The traditional buildings of England, Thames and Hudson, 1995.

Racher, P., Moment resisting connections (Lecture C16). STEP Timber Engineering, 1st Ed., The Netherlands: Centrum Hout, 1995.

Rackham, O., The history of the countryside, London: J.M. Dent & Sons Ltd., 1986.

Rackham, O., Trees and woodland in the British landscape, London: J.M. Dent & Sons Ltd., 1976.

Reece, P.O., An Introduction to the Design of Timber Structures, London: Spon, date.

Report of the inquiry into the collapse of flats at Ronan Point, Canning Town, London: HMSO, 1968.

Report on the R&D research activities on the on-site glue joint for glulam beam joints, Japan Housing and Wood Technology Center (HOWTEC), 1992 (in Japanese).

RIBA, National Building Specification, London: RIBA Publication, 1973 and later.

Richardson, B., Remedial treatment of buildings, Second Edition, Construction Press, 1995.

Ridout, B., Timber decay in buildings-the conservation approach to treatment, London and New York: E. & F.N. Spon, 1999.

Roark, R.J., Formulas for Stress and Strain, Fourth Edition, New York: McGraw-Hill, 1956; BSI, 1994.

Robbins, Weier and Stockway[initials], Botany – An Introduction to Plant Science, New York: John Wiley and Sons Inc., date.

Salzman, L.F., Building in England down to 1540 - a documentary history, Oxford: OUP 1952, reissued 1997.

SAP 2000 PLUS, Integrated Finite Element Analysis and Design of Structures, Berkeley, California, USA: Computers and Structures, Inc., 1996.

Sasaki, T., et al., End joint with glued-in hardwood dowels in timber construction (1), Mokuzai Gakkaisht, 45, (1) 17–24, 1999.

Sasaki, T., et al., End joint with glued-in hardwood dowels in timber construction (2), Mokuzai Gakkaishi (to be published).

Schaffer, E.L., Jokerst, R.W., Moody, R.C., Peters, C.C., Tschernitz, J.L. and Zahn, J.J., Feasibility of Producing A High-Yield Laminated Structural Products – General Summary, U.S D.A. Forest Service. Research Paper FPL 175, Madison, WI, 1972.

Schneider, P., Schraubenauszugsversuche parallel zur Faser – Dichteabhängigkender maximalen Schubspannung, Zurich: Interner Versuchbericht Prof. Für Holztechnologie ETH Zürich, 1999.

Scholten, J.A., Timber Connector Joints: their strength and design, US Department of Agriculture, Technical Bulletin No. 865, 1944.

Schreilechner, G. et al., Versuchsbericht Kleinbrandversuche 2 Teil 2(KBV 2/2), Wien: Institut fur Konstruktiven Ingenieurbau – Universität für Bodenkultur, 2001.

Smith, I., Load sharing in floors: including the effect of random variations in .t beam stiffnesses and strengthens, Proceedings, Meeting of International Union of BE Forestry Research Organizations, Oxford: Wood Engineering Group, 1980.

Spécification technique belge, STS 2 eme partie : Matériaux : 04-Bois et Panneaux 1 Base de bois ; 06.8-Matériaux dassemblage pour charpenterie, édition décembre 1990.

Stability of Buildings, Institution of Structural Engineers, 1998.

Standards Association of Australia, Report on Strength Grouping of Timbers, Miscellaneous publication No. 45, Sydney: SAA, date.

Stanhope Properties, An assessment of the imposed loading needs for current commercial office buildings in Great Britain, OAP and Stanhope Properties PLC, 1992.

Step 1: Structural Timber Education Programme, Volume 1, First Edition, The Netherlands: Centrum Hout, 1995.

Step 2: Structural Timber Education ProgrammeVolume 2, First Edition, The Netherlands: Centrum Hout, 1995.

Stern, E.G., Research on jointing of timber framing in the United States of America Modern Timber Joints, Timber Research and Development Association, 1968.

Strength test on structural timber joints made with 20-gauge Gang-Nails, Ministry of Technology, Forest Products Research Laboratory, April 1966.

Structures en bois aux états limites – Introduction à 1Eurocode-5 – STEP 1: Matériaux et bases de calcul, Ed.Sedibois/Union Nationale française de Charpente, Menuiserie, Parquets, Paris (F), 1996.

Structures en bois aux états limites – Introduction à 1Eurocode-5 – STEP 2: Calcul de structures, Ed. Sedibois/Union Nationale française de Charpente, Menuiserie Parquets, Paris (F), 1997.

Suddars, R.W., Listed buildings, Second Edition, London: Sweet & Maxwell, 1988.

Sugiyama, H., On the racking stiffness of bearing w1979) alls of platform framing construction due to full-scale house test, and ASTM racking test (paper 12), Proceedings, Meeting of International Union of Forestry Research Organizations. Vancouver, BC, Canada: Wood Engineering Group, 1978.

Sunley, J. and Bedding, B., Timber in construction, London: Batsford Ltd., 1985.

Sunley, J.G., Grade Stresses for Structural Timbers (3rd Edition; metric units), Forest Products Research, Bulletin No. 47, London: HMSO, 1968.

Sunley, J.G., The Strength of Timber Struts, Forest Products Research, Special Report No. 9, London: HMSO, 1955.

Sunley, T.G. and Brock, G.R., Joints in timber, methods of test and the determination of working loads, Modern Timber Joints, Timber Research and Development Association, 1968.

Sunly, J.G., Timber, Building and Technological Advance, Timberlab Paper No 9, Princes Risborough Research Establishment, 1969.

Swedish Timber Council, The properties of Swedish Redwood and Whitewood, Swedish House, Trinity Square, London EC3, Swedish Timber Council, date.

Tang, R.C. and Pu, J.II., Edgewise bending properties of LVL: Effect of veneer grade and relative humidity, Forest Prod. J., 47, (5), 60–70, 1997.

Tang, R.C., Adams, S.F. and Mark, R.E., Moduli of rigidity and torsional strength of Scarlet oak related to moisture content, Wood Science, 3, (4), 238–244, 1971.

Taylor, G.R., The Doomsday Book, Thames and Hudson, 1970.

The Royal Institute of British Architects, The Industrialisation of Building, RIBA, 1965.

The structural use of hardwoods, Timber Research and Development Association, Hughenden Valley, High Wycombe, Bucks.

The Swedish Timber Council, Swedish Redwood and Whitewood, The Sawn Timber and Products, July 1971.

The Swedish Timber Council, The properties of Swedish Redwood and Whitewood, 1971.

Timber Drying. Timber Research and Development Association in conjunction with the Timber Drying Association, 1969.

Timber Research and Development Association, Grouping of Lao Timbers fora Community Building System for Laos, United Nations Industrial Development Organization, 1976.

Timber Research and Development Association, Load Tables for Glued Ply-box Beams to BS 5268, TRADA Publication No. DA 5.84, 1984b.

Timber Research and Development Association, Load Tables for Nailed Ply-box Beams to BS 5268, TRADA Publication No. DA 4.84, 1984a.

Timber Research and Development Association, Specification and Treatment ofExterior Plywood Wood Information Section 2/3, Sheet 11, 1981.

Timber Research and Development Association, Timber Frame Housing Design Guide,1966.

Timber structures – Test Methods – Determination of embedment strength value and foundation values for dowel type fasteners, NBN- EN 383, date.

Timber Trade Federation of the UK, Plywood for Building and Construction, date.

Timoshenko, S.P. and Gere, J.M., Mechanics of materials, Van Nostrand Reinhold Co., 1972.

Tottenham, H., The effective width of plywood flanges in stressed skin construction, Timber Research and Development Association E/RR/3, March 1958.

TRADA Technology Ltd., Timber frame construction, Second Edition, TRADA Technology, 1994.

TRADA Technology Ltd., Timber frame construction, Third Edition, TRADA Technology, 2001.

TRADA, Timber and Fire Protection, Timber Research and Development Association, 1953.

Trayer, G.W., The Bearing Strength of Wood Under Bolts. US Department of Agriculture, Technical Bulletin No. 332, 1932.

Tredgold, T., Elementary principles of carpentry, Third Edition, P. Barlow (Editor), London: John Weale, 1840.

Truss Plate Institute, Design Specification for Metal-connected Wood Trusses, Glenview, Illinois,1978.

Tschernitz, J.L., Schaffer, E.L., Jokerst, R.W., Gromala, D.S., Peters, C.C. and Henry, W.T., Hardwood Press-Lam Crossties: Processing and Performance, U.S.D.A. Forest Service. Research Paper FPL 313, Madison, WI, 1979.

Tuomi, R.L. and McCutcheon, W.J., Racking strength of light-frame nailed walls, Proceedings of the American Society of Civil Engineers, July 1978, 104 ST 7 Proc, Paper 13876, 1131–40, 1978.

United Kingdom, The Building Regulations, London: HMSO, 1976.

United Nations Economic Commission for Europe, ECE recommended Standard for Stress Grading of Coniferous Sawn Timber, ECE/FAO Timber Committee, Timber Bulletin for Europe 30, Supplement 2, Geneva, 1977.

Untersuchungsbericht: Auszugsversuche zum Tragverhalten von SFS Verbinder WT-T8,2x300, MPA BAU vom 14.02.2000. TU Munche.

US Department of Commerce, American Softwood Lumber Standard, National Bureau of Standards, Voluntary Product Standard PS 20-70, Washington, 1970,

US Department of Commerce, Construction and Industrial Plywood, National Bureau of Standards, Voluntary Product Standard PS 1-83, Washington, 1983.

US Forest Products Laboratory, Longitudinal Shrinkage of Wood. US Department of Agriculture, Publication No. 1093, 1960.

US Forest Products Laboratory, Nail holding power of American woods. US Forest Products Laboratory, Technical Note No. 236, 1931.

US Forest Products Laboratory, Wood Handbook: wood as an engineering material, US Department of Agriculture, Forest Products Laboratory, Agriculture, Handbook No. 72, 1974.

USDA. Forest Prod. Laboratory, Wood Handbook: Wood as an Engineering Material. Agric. Handbook 72, Washington, DC: U.S. Department of Agriculture, 1999.

Van Rensburg, B.W.J. and Karemaker, J.C., The design of fire resistant steel structures. Proceedings of the Conference on Computers in Civil Engineering, Johannesburg: S. A. Institution of Civil Engineers, 1989.

Van Rensburg, B.W.J. and Kruger, T.S., Analysis of steel gable frame structures, Proceedings of the 3rd International Kerensky Conference on Global Trends in Structural Engineering, Singapore, 1994.

Van Rensburg, B.W.J., The non-linear analysis of framed structures with the aid of the matrix displacement method, Proceedings of the Conference on Computers in Civil Engineering, Pretoria: S. A. Institution of Civil Engineers, 1982.

Vick, C.B. and Okkonen, A.E., Strength and Durability of One-part Polyurethane Adhesive Bonds to Wood, Forest Products Journal, 48, (11/12), 71–76, 1998.

Vining, S., An overview of Engineering Wood Products, Proceedings of Engineering Wood Products, Processing and Design, F.T. Kurpiel and T.D. Faust (Editors), 27–34, 1991.

Wainwright, R.B. and Keyworth, B., Timber frame construction, High Wycombe:Timber Research and Development Association, 1988.

Wardle, T.M., Glued scarf and finger joints for structural timber, Timber Research and Develpoment Association, 1967.

Watkin, D., A history of western architecture, Laurence King, 1992.

Watt, D. and Swallow, P., Surveying historic buildings, Donhead Publishing, 1996.

Wheatcroft, A (Editor), The making of Britain: The Norman Heritage; The High Middle Ages; The Age of Exuberance; The Georgian Triumph; The Transformation of Britain, London: Paladin Books, 1983.

White, M.G., The Inspection and Treatment of Houses for Damage by Wood-boring Insects, Timber lab Paper No 33, Princes Risborough: Building Research Establishment, 1970.

White, N.V., The Durability of Glues for Plywood Manufacture, Building Research Establishment, Paper UP 84/75 HMSO, London, 1975.

Wilkinson, T.L., Strength of Bolted Wood Joints with Various Ratios of Member Thicknesses, US Forest Products Laboratory, 1978.

Wilson, T.R.C., Strength-moisture Relations for Wood US Department of Agriculture Technical bulletin No. 282, 1932.

Wong Choong Ngok, Permissible Loads for Timber Connectors and Bolts Used in Malayan Timbers, Forest Research Institute, Research Pamphlet No.57, Kepong, Malaysia, 1969.

Wood Handbook: Wood as an Engineering Material. Forest Products Laboratory Agric. Handb. 72, U. S. Department of Agriculture, 1987.

Wood, L.W., Relation of Strength of Wood to Duration of Load, US Forest Products Laboratory, Report No. 1916, 1960.

Wood, M., The English medieval house, London: Ferndale Editions, 1981.

XP ENV 1995-1-1, Eurocode 5 Design of timber structures, Part 1-1 General rules and rules for buildings and French National Application Document, 1995.

Yeomans, D., The trussed: its history and development, Scholar Press, 1992.

Zavala, D. and Humphery, P.E., Hot pressing of veneer-based products: The interaction of physical processes, Forest Products Journal, 46, (1), 69–77, 1996.

Codes of Practice

1) British Standards Institution, BS 648:1964, *Schedule of weights of building materials,* London: BSI.
2) British Standards Institution, BS 4978:1996, *Specification for softwood grades for structural use,* London: BSI.
3) British Standards Institution, BS 5268:1996, 1998, 1978, *Structural use of timber:*
   Part 2: Code of practice for permissible stress design, materials and workmanship
   Part 3: Code of practice for trussed rafter roofs
   Part 4: Fire resistance of timber structures
   London: BSI.
4) British Standards Institution, BS 6399:1996, 1997, 1988, *Loading for buildings:*
   Part 1: Code of practice for dead and imposed loads
   Part 2: Code of practice for wind loads
   Part 3: Code of practice for imposed roof loads
   London: BSI.
5) British Standards Institution, BS 6446:1984, *Specification for manufacture of glued structural components of timber and wood based panel products,* London: BSI.
6) British Standards Institution, BS EN 385:1995, *Finger jointed structural timber – Performance requirements and minimum production requirements,* London: BSI.
7) British Standards Institution, BS EN 518:1995, *Structural timber – Grading – Requirements for visual strength grading standards,* London: BSI.
8) British Standards Institution, BS EN 519:1995 *Structural timber –Grading – Requirements for machine strength graded timber and grading machines,* London: BSI.
9) *Fir Plywood Web Beam Design,* COFI, Council of Forest Industries of British Columbia, date.
10) *Eurocode 1: Basis of design and actions on structures, DD ENV 1991 – 1 – 1Part 1: Basis of design (together with United Kingdom National Application Document),* BSI, 1996.
11) *Eurocode 5: Design of timber structures DD ENV 1995 – 1 – 1 Part 1. 1: General rules and rules for buildings (together with United Kingdom National Application Document),* BSI, 1994.
12) British Standards Institution, CP 112:1952, *Code of practice for the structural use of timber (withdrawn),* London: BSI.
13) British Standards Institution, BS 373:1957 1986, *Methods of testing small clear specimens of timber,* London: BSI.
14) British Standards Institution, CP 3: Chapter V. Part 1: 1967, *Dead and imposed loads,* London: BSI.
15) British Standards Institution, BS 4471:1987, *Specification for sizes of sawn and processed softwood,* London: BSI.
16) British Standards Institution, BS 4169:1988, *Specification for manufacture of glued-laminated timber structural members,* London: BSI.
17) British Standards Institution, BS 6399: Part 3, 1988, *Code of practice for imposed roof loads,* London: BSI.
18) British Standards Institution, BS 5589:1989, 1997, *Code of practice for preservation of timber,* London: BSI.
19) British Standards Institution, BS 1203:1979 1991, *Specification for synthetic resin adhesives (phenolic and aminoplastic) for Plywood,* London: BSI.
20) British Standards Institution, BS EN 335-1:1992, *Classification of hazard classes,* London: BSI.
21) British Standards Institution, BS 6742:1992, *Guide to evaluation of human exposure to vibration in buildings (1 Hz to 80 Hz),* London: BSI.
22) British Standards Institution, BS 1204:1993, *Specification for type MR phenolic and aminoplastic synthetic resin adhesives for wood,* London: BSI.
23) British Standards Institution, DD ENV 1995-1-1, Eurocode 5:1994, *Design of timber structures. Part 1. 1. General rules and rules for buildings,* London: BSI.
24) British Standards Institution, BS EN 336:1995, *Structural timber. Coniferous and poplar,* London: BSI.
25) British Standards Institution, BS EN 338:1995, *Structural timber. Strength classes,* London: BSI.
26) British Standards Institution, BS 4978:1996, *Specification for visual strength grading of softwood,* London: BSI.
27) British Standards Institution, BS 6399: Part 1:1996, *Code of practice for dead and imposed loads,* London: BSI.
28) British Standards Institution, BS 5756:1997, *Specification for visual strength grading of hardwood,* London: BSI.
29) British Standards Institution, BS 6399: Part 2:1997, *Code of practice for wind loads,* London: BSI.
30) British Standards Institution, BS 476: Part 7:1997, *Method of test to determine the classification of the surface spread of flame of products,* London: BSI.

Research Papers

PRO 8:   *International RILEM Symposium on Timber Engineering* (ISBN: 2-912143-10-1), Edited by L. Boström.[published in or by??]

PRO 9:   *2nd International RILEM Symposium on Adhesion between Polymers and Concrete ISAP 99* (ISBN: 2-912143-11-X), Edited by Y. Ohama and [initial] Puterman. [published in or by??]

PRO 10:   *3 rd International RILEM Symposium on Durability of Building and Construction Sealants* (ISBN: 2-912143-13-6), Edited by A.T. Wolf. [published in or by??]

PRO 14:   *Integrated Life-Cycle Design of Materials and Structures (ILCDES 2000) (ISBN: 951-758-408-3)*, (ISSN: 0356-9403), Edited by S. Sarja. [published in or by??]

PRO 22:   *International RILEM Symposium on Joints in Timber Structures, (ISBN: 2-912143-28-4*), Edited by S. Aicher and H.-W. Reinhardt. [published in or by??]

Index